AUTONOMOUS URBANISM

The Framework

AUTONOMOUS URBANISM

Towards a New Transitopia

Evan Shieh

Introduction

Introduction

Autonomous vehicles (AVs) are beginning to appear on our roads. It is only a matter of time until the remaining technological challenges they face will be solved, and the transportation planning and urban form impacts they spark will become front and center. Today, the media coverage and public debate around AVs centers on short-term technological and safety issues. However, their long-term effects on our built environments and mobility systems remain largely underestimated.

AVs have the potential to become major catalysts for future urban transformation of our cities in both negative and positive ways. This future could evolve into many extremes. If left unregulated, AVs have the potential to exacerbate the worst tendencies of automobile-enabled city-building. These well-documented tendencies—urban sprawl, traffic congestion, and the proliferation of mono-functional mobility infrastructures—have led to a shrunken public realm, the degradation of our built and natural environments, and transportation inequities in our cities.[1] However, if implemented correctly, AVs have the latent potential to unlock human-centric urban models that prioritize vibrant, livable, and sustainable design principles. In addition to saving lives (more than 40,000 U.S. roadway deaths happen each year) and reducing traffic fatalities (94% of serious crashes are due to human error), they have the potential to reduce environmental emissions, traffic congestion, and travel time.[2] Furthermore, AVs could significantly reduce urban demand for street and lot parking, freeing up real-estate for densification and mixed land uses. The design of auto-oriented building, block, street, and highway typologies that proliferate in our cities could be transformed, while new AV-enabled urban forms could emerge. AVs will impact the ways we move around our cities, and the mobility modes, methods, and service models that we use to do so. They could improve transportation access for seniors, people with disabilities, and those who live in transit deserts.[3]

This book illustrates a vision of this potential future city and outlines the foundational steps in guiding us to achieve it. Embedded in this vision is an explicit set of core values and principles. The book argues for a future city that is not dictated by technology but rather one where technology is strategically deployed as a tool to tackle the urban issues that cities universally face today. It argues that autonomous vehicles, like automobiles and locomotives before them, need to be considered not just as technological innovations but as important infrastructural change-agents that shape urban growth patterns. It argues that how we choose to move around a city has vast implications on how that city is designed and will evolve at local and regional scales. It argues for automation to complement and enhance, rather than replace or compete with, existing transit and transportation systems. It argues for AVs to fill mobility gaps and for our mobility to be consumed as a service, rather than through private ownership models. Using the city of Los Angeles as a testbed, this book speculates on how an existing transportation model can incorporate shared, multi-modal AVs as a blueprint for a transportation paradigm shift. In doing so, the book endeavors to push our cities away from a future in which the worst tendencies of automobile-dependency become exacerbated, and towards

1 A mono-functional infrastructure is one that prioritizes a single (or mono) use (or function) above all others. Highways are a textbook example of such an infrastructure.

2 *Traffic Safety Facts*, National Highway Traffic Safety Administration, 2018.

3 A transit desert describes an area with limited access to transportation options. These areas tend to occur largely in socioeconomically disadvantaged areas of cities.

a future in which our built environments are more human-centric, multi-modal, and environmentally sustainable.

These are lofty goals that require multi-scalar design thinking, interdisciplinary methods of operation, and both top-down and bottom-up approaches. It is essential that top-down metropolitan transit planning models incorporate AVs in their networks. To fail to do so risks a proliferation of market-driven, private AV ownership, amplifying the negative externalities of an automobile dependent city.[4] At the same time, bottom-up approaches must also be used to persuade automobile-dependent motorists that there are alternatives to car culture. In particular, these social and behavioral car ownership norms, and the individualistic freedom that they represent, stand as a complex cultural challenge for these goals. If both approaches can be simultaneously undertaken—as this book attempts to do—then possible mobility paradigm shifts for the future city will emerge.

4 A negative externality is the imposition of an indirect negative cost on a population group as a byproduct of actions of another group. Driving a car produces several negative externalities, or costs, that drivers impose on others, including things such as traffic congestion, air pollution, and much more.

To take up these charges, this book is structured through a two-volume set that is meant to be read together, but can also be read separately by different audiences. Book 2 deploys the unique format of a graphic novel that depicts the *experience* of a future city. As seen bottom-up through the eyes of the everyday public, it visualizes a city that has incorporated automation into its transportation services and built environments. Book 1 (the book you are reading right now) lays the *framework* for that speculative future to potentially occur. It grounds this future city through discussions on our recent urban transportation history and its effect on the evolution of our built environments. It outlines concrete, actionable transportation policies and models to guide current trajectories of mobility technology innovation. This mobility paradigm shift creates radical *implications* on the spatial and built environment of our cities, explored through urban and architectural typologies illustrated in both books.

Armed with design visions, experiential narratives, policy, and planning models, both books serve as guidebooks for the various stakeholders that can impact a driverless revolution. These range from AV companies and industry professionals, all the way to the general public and individual mobility consumer. In particular, it is vital that the design and planning disciplines play a key, influential role in this conversation in order to ensure that the public realm of our cities is protected and well-served. Altogether, the book hopes to shift the contemporary conversation around AVs towards one in which technological advances are fore-fronted by the public good—one that transitions cities dependent on the motorized, private vehicles of today to cities designed for the shared, multi-modal, and autonomous vehicles of tomorrow. If implemented in that way, AVs offer a major opportunity to rethink the design and evolution of our city's built environments, with profound implications on urban life since automobiles replaced horse-powered travel and changed the design of cities in the prior century. In envisioning this potential urban future and delineating the steps towards achieving it, this book contends that cities can transition from the Autopias of today, to the Transitopias of tomorrow.

This is a big shift. Are cities and their inhabitants ready?

1 The Promises of Driverless Vehicles

Defining Driverless Technology

What is an autonomous vehicle (AV)? Before speculating on the promises of an urban future enabled by this technology, let's set the contextual stage for where it stands today. Generally, an AV is defined as a passenger-carrying vehicle that is capable of sensing its environment and operating itself without direct human control. The technology today relies on a variety of sensors (cameras, radar, lidar), actuators, complex algorithms, machine learning systems, and powerful processors to analyze and input live information from the physical environment around it into its software and then execute vehicular driving actions in response.

As defined by the Society of Automotive Engineers and adopted by the U.S. Department of Transportation, there are 6 levels of driving automation ranging from Level 0 (fully manual) to Level 5 (fully autonomous). Levels 0 to 2 can be generally grouped to describe the critical need for a human to monitor and be in control of the driving environment at all times while operating a vehicle. Levels 3 to 5 can be grouped to describe the ability for a vehicle's automated system to monitor and be in control of the driving environment, with varying levels of human override in place.

The majority of vehicles on our roads today are defined as Level 0 or 1. Level 0 (No Driving Automation) requires full manual control of the vehicle by a human, while Level 1 (Driver Assistance) includes forms of automated driver assistance like cruise control and lane centering systems amongst others. Level 2 (Partial Driving Automation) vehicles have several assisted driving technologies that work together to control the vehicle at a more advanced capability (controlling acceleration and deceleration, for example), but the human must actively engage these systems to monitor and intercept at a moment's notice. Importantly, vehicles defined with Level 2 are only capable of self-driving under very specific conditions (like driving on highways). Tesla's Autopilot technology is a popular example of Level 2 automation.

The next, Level 3 (Conditional Driving Automation), describes a subtle but important technological shift from the human perspective. Level 3 vehicles have environmental detection capabilities, meaning that they have the ability to make certain driving decisions *without* human judgement when certain driving conditions are met. For example, Audi AI's traffic jam pilot can self-drive the vehicle during a traffic jam, and once the traffic is cleared turns control of

The Human Monitors the Driving Environment			The Automated System Monitors the Driving Environment		
Level 0	Level 1	Level 2	Level 3	Level 4	Level 5
Manual control, human performs all driving tasks	Manual control, vehicle features a single automated system (i.e. cruise control)	Vehicle can perform limited steering and acceleration, human must take control at moment's notice	Environmental detection capabilities, vehicle can perform most driving tasks, human override required	Vehicle performs all driving tasks under specific conditions, human override and geo-fencing required	Vehicle performs all driving tasks under all circumstances, no human attention required

the vehicle back to the driver. The vehicle monitors the environment under certain conditions while all other conditions fall on the driver.

Level 4 (High Driving Automation) vehicles have full automated driving functionality in *most* conditions. In select conditions, human oversight and potential intervention is required. An important difference between Level 3 and 4 is that Level 4 vehicles have the ability to intervene and react to road conditions without requiring a human to take over driving. Level 4 vehicles can operate in self-driving mode but often (due to legislation) can only do so within a limited geo-fenced area.[1] For example, Waymo (formerly known as the Google Self-Driving Car Project) provides Level 4 self-driving services in select geo-fenced metropolitan areas. In 2020, Waymo became the first robo-taxi service to offer service to the public without safety drivers in the vehicle.

1 Geo-fencing is a virtual perimeter for a real-world geographic area. Within the context of mobility technologies, geo-fencing uses GPS and GSM technology to control, limit, or inhibit the operation of vehicles or transportation devices within a virtual perimeter based on select city neighborhoods, streets, areas, or zones.

Level 5 (Full Driving Automation) vehicles have full automation technology. They require no involvement of humans and have the ability to operate under *all* driving conditions. The jump from Level 4 to Level 5 is technologically significant—but so too are the opportunities that it enables. Since the function and monitoring of the human is completely eliminated at this stage, the design of the vehicle does not need traditional controls like a steering wheel or front-facing seats. Level 5 vehicles will be free from geo-fencing, able to drive anywhere that an equivalent human driver can.

Another note to make when defining AVs are a pair of terms that are often used interchangeably: "autonomous" versus "automated" and "self-driving" versus "driverless". "Autonomous" as a technical term has implications beyond the electro-mechanical. A fully autonomous vehicle would be theoretically self-aware, capable of its own autonomy in making its own choices. An automated vehicle, on the other hand, would follow human instructions and then drive itself based on inputted destinations. However, the distinction between "autonomous" and "automated" within the applied technology begins to blur in practice as they are simultaneously applied to a broad range of similar functions. With regards to the range of autonomy levels from 0 to 5, semi-autonomous may actually be the best technical definition as these driving systems, particularly at Levels 4 and 5, are given autonomy to make critical driving decisions *without* human permission. Self-driving technology today also deploys artificial intelligence in predictive behavior modeling in order to make real-time decisions based on the live data that is gathered from a vehicle's sensors. This further muddies the definition of who or what retains autonomy in driving decision-making. Today, both terms are used interchangeably in the industry and have come to mean the same thing culturally as well.

A "self-driving" vehicle is a term that is used to describe the general idea of a vehicle doing some or all of the work of moving itself from point-to-point. A self-driving car could be defined as having Level 3 driving automation or above. "Driverless" vehicles, on the other hand, are those defined has having Level 5 capability with the ability to remove the driver from the decision-making process of operating a vehicle, resulting in a truly driverless experience. All driverless cars are self-driving, but not all self-driving cars are driverless.

Within the blurriness of these technical and cultural definitions and their use, both "autonomous" and "automated" vehicle terms are therefore used interchangeably in the context of this book. "Driverless" is also used amply in this text, for it is particularly at this level of automation (Level 5) in which the promises of transformative change in our urban environments emerge. While differences in terminology seem minor and are used interchangeably within the industry today, I bring them up to highlight the importance of language in affecting the social and behavioral decisions around the use of technology. Tesla's Autopilot technology is defined as Level 2, which distinctly requires a human driver to monitor every aspect of the driving environment while in operation. However, the naming and marketing of the system implies a level of automated driving far above Level 2 that frees drivers from monitoring the vehicle at all times. Senior U.S. transportation officials have asserted that this misleading naming is implicated in a variety of documented traffic accidents caused by misuse of Autopilot, in many cases as a direct result of a driver's failure to monitor.[2] As this example illustrates, the language, marketing, and media promotion of technology can have important implications on the use of, behavior around, and inevitably the cultural absorption of these technologies into our everyday mobility patterns. This relationship is something that we will continue to discuss throughout this book.

2 Krisher and Powell, "Tesla shouldn't call driving system Autopilot," *Associated Press*, 2023.

Amara's Law and the Gartner Hype Cycle

There are several challenges facing full Level 5 automation today, ranging from the technical to the ethical and the legal. For example, AVs still face challenges operating without any issues in extreme weather conditions, like heavy precipitation or snow, which can obscure critical road signals like lane markings. The design of AV software to react and respond to intuitive human driving or pedestrian behavior, particularly in complex urban environments, still remains a challenge.[3] Traditionally, the automobile industry has been heavily regulated at the federal level, but current AV-policy regulation has been led at the state level, resulting in a range of diverse standards.[4] State and federal laws largely fail to specify who will be held liable for accidents that AVs are involved in, for example. Additionally, ethical quandaries arise in the design of AV software and its decision-making, such as programming software to prioritize protecting occupants versus external pedestrians in traffic accident scenarios.

3 Anthony, "Self-driving Cars still can't Mimic Human Behavior," *Quartz*, 2017.

4 Fanelli and Savrin, "State Policy guides U.S. Autonomous Driving," *Automotive World*, 2022.

However, the purpose of this book is not to expound on these current challenges. Rather, it posits that while these challenges are being solved, we must also understand the long-term spatial possibilities that will emerge—particularly in the transformation of our city's built environments. The book acknowledges today's challenges (the question of *How?*), while critically framing them within the larger opportunities that they are already beginning to enable (the equally, if not more so, important question of *Why?*).

This framing can be encapsulated by a phrase coined by Roy Amara, a researcher, scientist, and futurist at the Stanford Research Institute, who stated that:

> We tend to *overestimate* the effect of a technology in the short run, and *underestimate* it's effect in the long run.

His adage, known as Amara's Law, is tied closely to the Gartner Hype Cycle, a conceptual graphic developed and employed by the technology information firm Gartner to illustrate the maturity, adoption, and social impact of modern-day technologies on our contemporary society. The discourse and development of AVs subscribes to Amara's Law and seems to be closely following the Gartner Hype Cycle graph.

The Gartner Hype Cycle focuses on five key phases of a new technology's maturity and life cycle. During Phase 1, the Technology Trigger, a potential technology breakthrough occurs. Subsequent proof-of-concept stories and media interest trigger significant publicity, even though the technology's commercial viability is unproven. During Phase 2, the Peak of Inflated Expectations, the technology is fueled by this publicity and initial successes, becoming backed by a number of venture capitalists which sparks further startup companies beyond the early adopters. In Phase 3, the Trough of Disillusionment, interest in this technology begins to wane as it encounters robust development and implementation challenges. This phase is often sparked by the first negative press that the technology can trigger. During Phase 4, the Slope of Enlightenment, instances of how this new technology can be of long-term benefit to society begin to crystallize. The technology becomes more widely and deeply

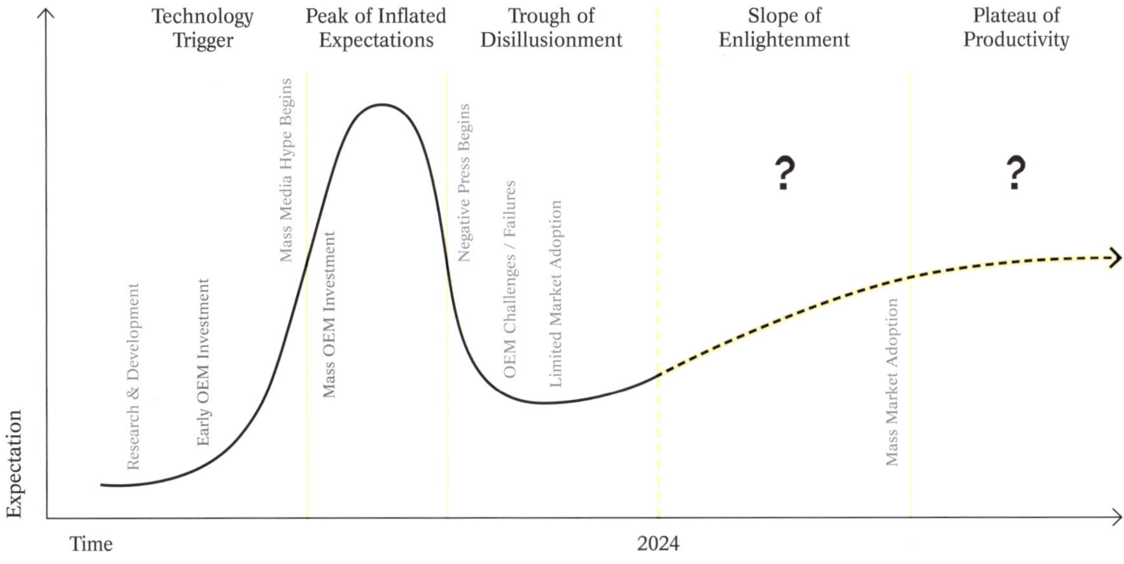

understood and developed, as second and third generation products emerge. In Phase 5, the Plateau of Productivity, the technology's transformative applicability and relevance become more fully comprehended and accepted. Mainstream society begins to adopt this technology in a ubiquitous manner.

Framed within the context of AVs, this technology appears currently, as of this writing, to be within Phase 3 of the Gartner Hype Cycle, the Trough of Disillusionment. While automating aspects of driving technology began to be dreamed up as early as the mid-1900s, the reality of this technology advanced sharply in the public's consciousness with the DARPA (Defense Advanced Research Projects Agency) Grand Challenge in the early 2000s, which kickstarted the technology's Gartner Hype Cycle Phase 1. The DARPA Grand Challenges in 2004, 2005, and 2007 were the first long distance competitions for self-driving cars in the world, with expansive research funding, prize money, and media coverage surrounding the events.[5] With the 2007 DARPA Urban Challenge, teams were tasked to test their self-driving vehicles for the first time in an urban course with rules that included obeying traffic regulations while negotiating with other vehicles and urban obstacles. Six teams successfully finished the entire course, and the challenge served as a foundation for several team members that, in the coming years, spearheaded the development of Google's Self-Driving Car Project company which now operates as Waymo. Several years later in 2018, Waymo launched their first commercial self-driving car service called Waymo One, to much anticipation in the greater metropolitan Phoenix area.

By this time, hype for AV technology had become widespread, firmly situating it near the peak of Phase 2 of the Gartner Hype Cycle. Media articles speculating on the speed with which this technology would land and transform our urban mobility abounded. By then, almost every major automobile manufacturer had invested into the technology. Transportation network companies (TNCs) like Uber and Lyft had announced self-driving robo-taxi services in various locations, and a variety of self-driving companies had come to market. It was also around this time that the first fatal accident was reported.

5 Buehler, Iagnemma, and Singh, *The 2005 DARPA Grand Challenge*, 2007.

1957 Magazine print advertisement for all-electric living with electric super-highways facilitating self-driving cars.

POWER COMPANIES BUILD FOR YOUR NEW ELECTRIC LIVING

Your air conditioner, television and other appliances are just the beginning of a new electric age.

Your food will cook in seconds instead of hours. Electricity will close your windows at the first drop of rain. Lamps will cut on and off automatically to fit the lighting needs in your rooms. Television "screens" will hang on the walls. An electric heat pump will use outside air to cool your house in summer, heat it in winter.

You will need and have much more electricity than you have today. Right now America's more than 400 independent electric light and power companies are planning and building to have twice as much electricity for you by 1965. These companies can have this power ready when you need it because they don't have to wait for an act of Congress—or for a cent of tax money—to build the plants.

The same experience, imagination and enterprise that electrified the nation in a single lifetime are at work shaping your electric future. That's why in the years to come, as in the past, you will benefit *most* when you are served by independent companies like the ones bringing you this message—America's Electric Light and Power Companies*.

2015 Waymo self-driving vehicle prototype first deployed at that time.

In March 2018, a pedestrian was struck by an Uber self-driving test vehicle in which the human safety backup driver was not monitoring the vehicle.[6] The pedestrian later died of her injuries, sparking a wave of negative press and indicating a shift into Phase 3 of the Gartner Hype Cycle, the Trough of Disillusionment. In the years since then, the promises of driverless vehicles proliferating completely into our everyday lives has come and gone with the aforementioned challenges still being resolved. Meanwhile, the media has largely moved on to debating newer technological triggers like virtual reality and artificial intelligence. In the background, however, self-driving companies are continuing to advance the technology slowly but surely, gradually shifting the technology long-term towards Phase 4 of the Gartner Hype Cycle, a phase in which I expect we will reside in for some time. In 2020, Waymo One was operating self-driving vehicles without a safety backup driver, the first service worldwide to do so. By 2023, it had expanded its operations from Phoenix (where it originally began testing) to cities including San Francisco, Los Angeles, Austin, and Bellevue, moving us towards a future in which driverless mobility operates in all cities around the globe.

6 Wakabayashi, "Self-Driving Uber Car Kills Pedestrian," *The New York Times*, 2018.

Expanding the Frontier of Knowledge

It is within this context that this book inserts itself into a cultural and media environment that has largely ignored the second half of the Gartner Hype Cycle graph. Media discourse has been primarily focused on the technology itself—its infrastructure, engineering, safety, ethics, and other related issues. Meanwhile, the long-term implications of AV technology on the urban form and mobility systems of our cities continue to be roundly under-considered and under-researched. Even when the hype around AV technology reached its peak, there was little conversation around the radical effect it could spark in the built environments of our cities. In particular, the design disciplines have largely been absent in these discussions. Any visioning that has been presented, has been done so without outlining the important planning and design steps it would take to reach those urban futures. In parallel, corporations in the mobility technology and automobile industry are commercially rolling out automated technology, *without* demonstrating to the public an understanding of the transformative impact of their inventions on the built environment. Although techno-utopists present these technologies as a great benefit to society at large, these companies are driven by highly monetized commercial and profit-generating mandates, capitalizing on AV technology to essentially sell more individual cars. Meanwhile, those in positions to regulate and impact the long-term effects of AV technology on cities—public transportation departments and agencies at federal and state levels—have largely remained slow-moving and reactionary.

In these discussions, the issue this book raises is not the development of AV technology itself but rather the lack of critical discourse surrounding its implementation and resulting implications. In a vacuum, technology itself can never be defined as either "good" or "bad". Technological advances in our history have simultaneously enriched the efficiency and experience of our urban environments, as they have in other ways devastated them socially and environmentally. Rather, questions around *how* these technologies are used and deployed, by whom and for whom, and most importantly, to *what ends*, must drive these conversations and the decisions that result from them.

By positioning itself in the latter half of Amara's Law and the Gartner Hype Cycle, this book aims to expand this frontier of knowledge through several critical lenses. Firstly, the implications of short-term technological advances on the long-term future of our cities and our urban life must be explored. The way driverless vehicles are deployed will impact not only our vehicular modal choices, mobility behaviors, and transportation networks, but also the design of our urban environments, street infrastructures, and urban form. Secondly, discussions must be framed with the public good and the ecological health of our cities at its core. What are the values that we associate with more positive urban environments (multi-modality, mixed land uses, environmental sustainability), and how can AV technology be deployed to push us towards those goals? Thirdly, any visioning must delineate the blueprint it would take to reach that future. Any seductive visions must be accompanied by important short-term design and planning steps that we must collectively take—as city regulators, technology innovators, designers of the built environment, and everyday consumers of mobility—to move us towards that long-term future. Lastly, these discussions and the decisions that emerge from them must be conducted in advance of AV technology proliferating widely on our city streets. Once certain technologies become widely used and accepted, it is far more difficult to regulate their uses and alter their impacts than it is to do so pre-proliferation. For example, this is something that happened with technological disrupters like the ride-hailing mobility apps Uber and Lyft (more on this later).

Summarized in four simple points, this book expands the narratives framing driverless technology to:

- Envision the impact and implications of driverless technology on the design of our future cities
- Establish the implicit values and principles that constitute that potential future
- Outline the actionable and concrete steps that can enable that future to transpire
- Act on that future before alternative (potentially worse) futures become a reality

Driverless-Enabled Spatial Opportunities

So, what exactly are the spatial opportunities that this AV technological leap enables? How does removing a driver from a vehicle actually change the way our cities can be designed? The technological shift to a driverless vehicle might seem small, but it comes with several critical implications. When extrapolated, these implications result in powerful new ways to think about the future of our cities. It changes the means and methods for which we move around, as well as the design of our city's built environments and physical infrastructures that facilitate that mobility.

Firstly, removing a driver implies an increase in efficiency to both travel time as well as towards the physical space utilization of our city streets and highways. In a driverless world, passengers are freed from the act of driving to use their time productively—working, resting, or socializing—reducing the psychological time associated with commuting. Driverless vehicles also have the ability to park themselves, saving additional time in daily commutes. Faster travel times could also be theoretically achieved through autonomous driving efficiencies like vehicle platooning, which describes the ability for a group of vehicles to travel very closely together, safely, and at high speeds.[7] Vehicle platooning is enabled by autonomous cars that can communicate directly with each other, thereby moving together at far closer distances than if individually driven by humans. Compounded together, these time savings may incentivize geographically farther commutes as travel times become more efficient. Collectively, this has vast implications on the structure of our cities urban cores and their relationship to their suburbs and exurbs. One of the factors influencing the spatial relationship of urban-suburban sprawl is travel time, and driverless vehicles fundamentally alter that equation. Furthermore, the physical space gained from AV driving efficiencies would also become contested spaces for alternative uses in our urban mobility infrastructures and right-of-ways.

7 Vehicle platooning describes the linking of two or more vehicles into a convoy using wireless communications and sensor technology. This capability allows vehicles to accelerate or brake simultaneously, decreasing the distance between vehicles and allowing for closer headways.

Secondly, the method and means in which automated technology is utilized in the near future implies a shift towards mobility-as-a-service (MaaS). MaaS is a transportation concept that describes travelers interacting with mobility providers through a service model, rather than consuming mobility through privately-owned modes of transportation (like a personal car). MaaS offers travelers a diversity of mobility options from carpooling, ridesharing, bike-sharing, and transit to fit their varied travel needs. MaaS has been available ever since public transit offered its mobility services as a means to get from point A to B. However, it has, in its contemporary definition, been fueled in popularity and use by a myriad of new private transportation service providers. Ride-hailing transportation network companies (TNCs) like Uber and Lyft, bicycle and scooter-sharing programs like CitiBike or Bird, and car-sharing services like Zipcar, all provide alternative transportation services that allow consumers to move about without having to *own* a personal vehicle. AVs are expected to accelerate this trend, as they are widely estimated to make MaaS services more affordable when the cost of human driving labor is eliminated. Common to all MaaS services, including public transit, is the deployment of large fleets of vehicles, often owned and operated by a central transportation agency or company. Imagine a potential future in which we forgo owning individual vehicles because MaaS has become a far more convenient and affordable option to move around in. This has vast implications on the space that vehicles take up in our cities. It is estimated that there are up to eight parking spaces allocated for every car owned in the United States. In the U.S., 24% of households own three or more vehicles.[8] These parking spots are distributed and dispersed widely across the urban landscape. They are often located in highly valuable land due to the need for individuals to conveniently access their vacant, parked car at a moment's notice from their home or workplace. With a shift towards MaaS, these fleets of vehicles are no longer privately stored on distributed individual lots but collectively stored in centralized depots, freeing up valuable distributed real estate for other potential urban uses.

8 *Personal Transportation Factsheet*, University of Michigan Center for Sustainable Systems, 2023.

Thirdly, automated technology sparks an opportunity to revisit the design of the vehicle itself. This has important implications on how many passengers can occupy and be transported per each ride. AVs release the necessity for front-facing drivers in vehicles, which allows the external and internal design of the vehicle to position passengers in any direction. This also allows AVs to freely move in either direction without the typical front- or back-facing vehicle orientation. Vehicle designers could then be freed to rethink, from the ground-up, how passenger transport vehicles should be designed. The critical part of this conversation with regards to its larger spatial implications, is *how many* passengers each vehicle is designed to transport, and therefore what various vehicle *sizes* might accommodate a range of transportation needs. With a transition towards MaaS, one can imagine a potential future in which a range of diverse vehicle sizes are deployed on our streets depending on the urban densities they are serving.

Autonomous shuttles (carrying three to five passengers) or local autonomous buses carrying passengers in the tens or more could operate in denser neighborhoods on fixed routes. On high-capacity corridors, autonomous articulated bus rapid transit (BRT) could carry passengers upwards of 50 to even the 100s (if vehicles are platooned), rivaling the capacity of light rail.[9] Other less dense areas may see more personalized, smaller, on-demand, mini AVs carrying only one to two passengers. Implementing a scale of diverse vehicle sizes has important ramifications on the spatial dimensions of our urban infrastructures that support them. Today, a personal vehicle typically transports a single individual, yet is often designed to carry up to four or five at all times of operation. The majority of our urban infrastructures, including streets and highways, are designed to accommodate the particular dimensions and turning radii of these standard capacity cars. Therefore, a range of vehicle sizes—either smaller vehicles demanding less urban space, or larger vehicles consolidating more passengers into less total space—could spark

9 Articulated buses are comprised of two or more rigid sections linked by a pivoting zone (articulation) that allows a longer legal length than rigid-bodied buses. These buses are generally used in bus rapid transit or express lines, due to their higher passenger carrying capacity.

opportunities to reclaim street right-of-ways for other uses. This transition from majority single-occupancy towards multi-occupancy vehicles, coupled with a shift towards all-electric fleets, is also a critical opportunity in which our urban mobility can become far more efficiently deployed. This shift could have vast ramifications on the environmental sustainability of our cities.

Fourthly, removing the driver from the car has cost-of-labor ramifications that spark opportunities to expand the frequency and service quality of our urban mobility. In the context of MaaS, driverless vehicles have the potential to significantly reduce operational costs by removing a large portion of driving labor costs. While some of these cost savings will inevitably be offset by capital investments from purchasing these vehicles, a recent estimate found that automating a single bus could provide over $3 million in savings over that bus's 12-year lifespan.[10] Labor costs are typically one of the largest components of mobility service operating costs, something that also holds true of both public transit and private transportation MaaS. A report by the Rocky Mountain Institute estimated that over 60% of the average cost per passenger-miles traveled for private ride-hailing companies is attributed to driver costs.[11] Such substantial cost savings in both private and public transportation MaaS could be used to make significant improvements in service quality and frequency by increasing route coverage. This is important because it is often service frequency that is a major contributor towards consumers choosing to drive a personal vehicle over taking public transit, particularly in the American context. Taking public transit typically results in travel times being twice as long on average than driving individually; much of this time is attributed to long wait times, low travel speeds, and multiple transfers.[12] Automating mobility therefore has the opportunity to make MaaS much more competitive to private automobile ownership both economically and temporally, thereby shifting people and vehicles away from the significant real estate dedicated to their individual uses.

Lastly, of all the aforementioned opportunities discussed, the most spatially transformative potential offered by AVs is the reduction in the demands and requirements for parking in our cities. In addition to the estimate that for every car owned there are up to eight distributed parking spots dedicated to storing them, there are nearly a quarter of U.S. households that own three or more vehicles, demanding 24 or more distributed parking spaces per household in cities. The surface area dedicated to parking takes up on average 20% of American city centers.[13] In some of the most extreme cities like in Arlington, Texas and Las Vegas, Nevada, this number reaches a staggering 42% and 32%, respectively. Huge swaths of our cities, through surface parking lots, vertical parking garages, and street parking, are used for the storage of individual private automobiles. In almost all cases, this land is swathed in black asphalt, significantly contributing to urban heat islands and forming uninterrupted impermeable surfaces that don't allow rainfall to pass through or vegetation to grow.[14] Altogether, there are an estimated two billion parking spots in the U.S. alone, each one taking an average of 320 square feet of space. Meanwhile, cities struggle to procure enough land to build affordable housing to keep up with the demands of 21st century urbanization. For comparisons sake, an average studio

10 Crute et al., *Planning for Autonomous Mobility*, American Planning Association, 2018.

11 Johnson and Walker, *Peak Car Ownership*, Rocky Mountain Institute, 2016.

12 Liao et al., "Disparities in Travel Times between Car and Transit," *Scientific Reports*, 2020.

13 Scharnhorst, *Quantified Parking*, Research Institute for Housing America, 2018.

14 An urban heat island is an effect in which an urban area witnesses significantly warmer temperatures than its surrounding rural areas. This is due to the activities generated by humans, as well as infrastructural surfaces such as buildings, roads, and parking lots that absorb and re-emit heat more than natural landscapes such as forests and water bodies.

A working parking lot at Disney World in Orlando, Florida.

apartment in the U.S. is estimated to be 514 square feet. Imagine the opportunities provided by this land if even a portion of it were freed from the constraints and demands of an unoccupied vehicle—for pedestrians, natural systems, and housing just to name a few. Automobiles today spend around 95% of their lifetime parked and vacant and only 5% of their lifetime in use.[15] Driverless electric vehicles, on the other hand, could theoretically operate at far greater use efficiencies, transporting passengers for the majority of their daily operation and only needing to park when recharging. Compounded with further efficiencies provided by shared and service-driven mobility models, a significant demand for parking and their associated land requirements could be eliminated. Facilitating these changes would of course require the right policies (the removal of arcane parking minimum mandates, for example) which are further discussed in Chapter 5.[16] These policies would certainly have an impact on public revenue streams, as a large percentage of city revenue is collected through public parking fees and fines. But with the right regulatory frameworks, these compounded spatial opportunities could begin to definitively address the urban, environmental, and density issues facing cities today.

15 Scharnhorst, *Quantified Parking*, Research Institute for Housing America, 2018.

16 Margolies, "Cities Rethink their Parking Needs," *The New York Times*, 2023.

The spatial opportunities enabled by land released from parking requirements unlocks multiple scales of rethinking our cities' urban form. At the scale of architecture and building typologies, a plethora of exciting opportunities arise. How can the myriad of automobile-oriented typologies—the fast-food drive-in, the parking garage structure, the motel, the strip mall—spatially adapt to a driverless future? From a land use and zoning perspective, how do altered parking requirements affect the development of our block configurations and the urban forms they engender? Furthermore, when our mobility infrastructures have been allocated and designed to respond to the dimensions and operating maneuvers of cars, automating these vehicles presents an opportunity to revisit the structure, configuration, and right-of-ways that facilitate their movement. Lane reconfiguration, dynamic lane management, street reclamation, and highway conversions are all spatial opportunities ripe for examination.[17] At the scale of our movement patterns, how should metropolitan-scale public and private transportation networks absorb and integrate driverless mobility? Cumulatively, how can these collective changes form a blueprint for the urban growth, densification, and development of our cities and their metropolitan structure?

17 Dynamic lane management describes the infrastructural ability for street lanes to be dynamically adjusted at different times of day. Possibilities of lane management strategies include reversing traffic directions, changing lane designations, or even closing lanes for alternative uses during under-capacity times of day.

Altogether, these are the exciting urban opportunities that this book spatially explores, through typological drawings and the visualizations that bring them to life. While these opportunities are presented as hinging on the technological advances of automation, many of the ideas in this book can be realized even without the immediate catalyst of the driverless vehicle. The myriad of urban spaces that have been devoted to the demands of the single-occupancy vehicle, and the policies and cultural frameworks that enable their proliferation, can already begin to be reclaimed for other uses. The height of the Covid-19 pandemic, for example, saw decreased car use on street right-of-ways which led certain cities and businesses to reclaim street parking space to support alternative outdoor-dining uses. While AVs trigger the rethinking of parking and multi-occupancy mobility, these critical shifts can also be simultaneously initiated by progressive transportation policies and the designs that spatialize them. Driverless technology, and the media attention and industry funding that fuels it, can only spur these urban changes to greater and more transformative heights. AVs will become integrated into our everyday mobility within our lifetime, and likely much sooner, but the work to push us towards those urban futures begins now.

While the opportunities described in this chapter are tantalizing, it is also important to remember that driverless vehicles are *not* a silver bullet to our urban mobility woes.[18] In fact, they have the powerful potential to exacerbate them, especially if we do not intervene now to guide them in more human-centric directions. While much of what is described in this chapter are so-called "heaven" scenarios, in which the dreams and promises of a driverless-enabled urban future have come to fruition, we need not look further than our recent transportation history to understand how potential "hell" scenarios might very well transpire. This is precisely what we will examine closely in the next chapter.

18 A silver bullet is a metaphorical term to describe a simple and seemingly magical solution to a complicated problem.

2 A Historically Informed Transportation Trajectory

The Age of the Automobile: Negative Urban Externalities

Without understanding its context and history, any projective act of visioning risks the potential for utopic naivety or even worse, the chance that mistakes from the past are repeated. In order to design a transformative future enabled by mobility technologies, it is critical to understand the history that conditioned their inventions. As such, let's take a look back at the last transportation revolution that widely transformed our urban environments, brought on by yet another technological impetus: the combustion-engine automobile.

It could be argued that no invention affected society in the 20th century more than the automobile itself. While driverless technology has advanced quickly in our recent history, it pales in comparison to the technological and societal changes brought on by the automobile in the 20th century. In the early 1900s, horse-drawn carriages, trolleys, and locomotive travel were the primary means of transportation. The few automobiles that did exist then were steam-powered, since gasoline was still considered a waste product of the kerosene industry. Then, in 1916, Henry Ford revolutionized the mass production of the automobile and the economies of scale that came with it. From that point on, the automobile industry exploded in growth. The automobile transformed from an almost non-existent product to one that was accessible, affordable, and ubiquitous to the majority of consumers. By the end of the 1920s, there were an estimated 23 million automobiles on the road, adopted by, developing with, and integrating into our society at a ferocious pace.

1920 Ford General Motors Model T automobile assembly line.

Since the history of the automobile is extensively well-documented in considerable literature elsewhere, let's focus on the spatial impacts of its invention on our built environments so thatparallels may be drawn to driverless vehicles today. The automobile transformed not only the means and methods of moving from one point to the next but it also transformed the experience of, behaviors around, and cultural identities associated with this new form of mobility. Many

1977 *Time* Magazine advertisment for Buick's LeSabre custom sedan.

of these transformations benefited society greatly, expanding travel efficiencies and distances through comfort and convenience previously unheard of. The automobile gave people more personal freedom, access to jobs and services, and contributed greatly to the rise of leisure activities. Fueled in part by persuasive marketing, the automobile became inextricably linked with American cultural notions of freedom, flexibility, and self-determination while also aligning itself intimately to social class, privilege, and notions of home ownership. It promised and delivered to the post-war burgeoning middle class the ability to work, shop, and play almost anywhere at any time. The automobile also dramatically changed the economy of our society, sparking an industry that has become one of the largest in the world. It has contributed to the rise of the industries that fuel it, including gas and oil, and largely advanced the standardization of production assembly lines, manufacturing, and the construction of asphalted roads that it travels on. The automobile also inherently transformed the very fabric of our urban and suburban spaces, the infrastructure that facilitated its movement, and the urban form that evolved around it.

But what also came hand-in-hand with the economy and convenience of the personal automobile, were the root causes of the many cultural, physical, economic, and environmental issues we face in our cities today. The personal automobile directly imposed extensive negative externalities on the built environments it was introduced into. A negative externality is defined as the imposition of an indirect cost on a group or population as a derivative byproduct of actions of another group in a negative way. Driving and owning a personal automobile produces a wide expanse of unintended but devastating externalities that cities grapple with today on a daily basis.

Negative externalities from the age of the automobile:

Economic (time-spent in traffic congestion)

Environmental (vehicular emissions)

Cultural (auto-fueled sub-urban sprawl)

Spatial (street right-of-way and parking allocations)

One of the most glaring negative externalities that the invention of automobiles produced are the overwhelming number of deaths and injuries resulting from vehicle accidents. Across the globe, 1.3 million people are killed in car accidents every year, and another 20–50 million are estimated to be injured.[1] Astoundingly, road traffic crashes from human driving are a leading cause of death in the U.S. for those under the age of 54. It is no wonder that one of the first statistics championed by driverless vehicle proponents is the ability for automated driving to drastically lower and even theoretically eliminate these millions of deaths and injuries.

1 *Traffic Safety Facts*, National Highway Traffic Safety Administration, 2018.

The environmental impact that is caused by automobile externalities is also astounding. Gasoline-powered, these vehicles emit a wide range of emissions into the air and atmosphere. Collectively, automobile pollution is one of the major contributors towards the global warming of our planet. The amount of embodied carbon resulting from heavy industrial processes in the steel and metal used to construct these vehicles in order to protect passengers in case of accidents is substantial. Automobile emissions endanger human health and can contribute to asthma for those exposed frequently enough to these vehicles (populations living adjacent to highways or major thoroughfares, for example).[2] Automobiles create noise pollution—vehicular traffic contributes to about 55% of total urban noise—and the exhaust heat radiating from automobile engine blocks and pipes are documented as a key contributor to urban heat island effects.[3] Unsurprisingly, AV proponents also champion the opportunity for the *electrification* of driverless vehicles to address these environmental impacts, reducing their contribution to the climate crisis and improving urban sustainability in our cities.

2 Hauptman et al., "Proximity to Major Roadways and Asthma Symptoms," *Journal of Allergy and Clinical Immunology*, 2020.

3 Vijay et al., "Assessment of Honking Impact on Traffic Noise," *Journal of Environmental Health Science & Engineering*, 2015.

Another major externality of the automobile is its effect on urban traffic and congestion in cities. It is estimated that on average, U.S. drivers spend the equivalent of 2.5 work weeks sitting in traffic a year.[4] This number dramatically increases in major metropolitan cities. Los Angeles commuters, for example, have to endure one of the worst traffic congestion levels of any U.S. city, spending upwards an equivalent of four work weeks in traffic on average, or 44 minutes of *extra* travel time in traffic every day.[5] This has economic (time-wasted) and social effects on the daily lives of drivers. Driving an automobile not only means that one is often stuck in traffic, it also means that one is contributing to the cause of that traffic. As one advertising campaign once coined, "you are not *stuck* in traffic, you *are* traffic." This circular feedback loop is exhibited in the phenomenon called "induced demand" that occurs in modern day traffic engineering. When roadways and highways are expanded in capacity to "alleviate" congestion, inevitably traffic volume will increase to match it, thereby resulting in more traffic in more space. Road and highway expansions meant to "relieve" traffic ironically induce more traffic to occur. AV's effect on congestion is less clear, but could result in several possible outcomes. If utilized in the same way we use our personal, single-occupancy automobiles today, there are concerns that automation will increase congestion. With no need to park, AVs may drop off passengers and either return home or circle city streets, adding more miles to the road and contributing to the rise of "zombie cars."[6] These possibilities will be expanded in more depth in the next chapter, forming a critical basis

4 Bradford, "The Average American spends 2.5 Work Weeks in Traffic," *Marketplace*, 2019.

5 McCarthy, "L.A. Commuters Spend the Most Time Stuck in Traffic," *Statista*, 2020.

6 A "zombie car" is a colloquial term used by automated vehicles critics to suggest the rise of driverless vehicles that roam our city streets en masse carrying no passengers, contributing to worsening traffic gridlock in our cities.

2016 Railway advertising billboard.

for the book's charge to move cities towards more shared forms of automated transit.

Finally, the most transformative *spatial* externality produced by the automobile is its effect on the urban form, development patterns, and infrastructure of our cities. In particular, these widespread spatial externalities transformed our city's mobility infrastructures at two scales: that of the street's cross-sectional use allocation and design, and the construction of the interstate highway system. Both distinctly modified the metropolitan structure and development paradigms of our cities and their hinterlands. In order to extrapolate AV's distinct impact on these spatial dimensions, let's examine these scales and their recent history in more detail next.

A Contested Public Realm: The Fight over Our Streets

The continual expansion of streets, boulevards, and highway systems in our cities is one of the most striking physical impacts of automobile use in the 20th century. These widespread *spatial* externalities are one of the most transformative—and yet have become normalized today despite their fraught histories. The primary use of the street by automobiles is generally accepted by modern society as part of the DNA of our city's urban fabric. However, this was not always the case. As the automobile proliferated, famed historian, urban planner, and architectural critic Lewis Mumford decried dramatically in his 1961 book *The City in History* that "the assumed right of the private motor car to go any place in the city and park anywhere is nothing less than a license to destroy the city."[7]

7 Mumford, *The City in History*, 1961.

While our cities' recent memory might suggest that the street's primary function is to facilitate the movement of an automobile from point-to-point, it is easy to forget that our use of the cross-sectional composition of streets is far more diverse. Before the arrival of the automobile, streets were shared by several overlapping systems and uses. Pedestrians, streetcars, and horse-drawn vehicles used them for local mobility. Street-facing businesses and pushcart vendors spilled onto the street, using them for places of formal and informal commerce. Pedestrians and children used the street as places for play and recreation. City services like telephone lines, water and electrical

1914 Hester Street, NYC, pre-proliferation of the automobile.

supply depended on the infrastructure of the street to facilitate their connectivity and use. This balance between uses was delicate and sometimes unstable but always in negotiation.

In ways, these diverse uses of the street continue today in some parts of our cities. But the dramatic change that occurred with the arrival of the automobile was the disruption to this balance, and the fight to establish *who* and *what* had the primary legal and psychological right to streets for their use. Of these fights, the most relentless and deadliest was the feud between pedestrians and motorists. Pedestrian

1924 *The New York Times* cover article depicting automobiles as killing machines.

NATION ROUSED AGAINST MOTOR KILLINGS

Secretary Hoover's Conference Will Suggest Many Ways to Check The Alarming Increase of Automobile Fatalities.—Studying Huge Problem

deaths skyrocketed following the introduction of automobiles. The main cause of this was that at the time, the rules of the street were vastly different than they are today. The street was utilized more akin to a pedestrian mall, or even a city plaza, a public space in which one could move in any direction at any point, freely without fear of personal safety. Automobiles introduced high-speed, moving vehicles into the equation of a street governed by informal and far more low-speed uses. To activists at the time, the way to make the streets safe for humans was to restrict the newcomer. Measures were enacted to limit automobiles to pre-motor age speeds. Police-imposed speed limits on automobiles were always low (even by 1919, still in the median range of 8–10 miles per hour maximum).[8] Judges and juries commonly defended pedestrian's right to the road in accident cases involving a motorist and a pedestrian. At the time, the motorist was the usurper, and therefore was to be restricted in the delicate balance of the street and its use.

8 Norton, *Fighting Traffic*, 2008.

Of course, many of these restrictions negated the very advantage of the automobile—speed and efficiency of movement—as considered by motorists and manufacturers. Any rules put in place in favor of the automobile, for example, crosswalks that required pedestrians to cross at designated areas of the street, were struck down in the name of supporting pedestrians' priority right to the street. It was at this time that large interest groups and organizations created a national coalition under the American Automobile Association self-named Motordom. Motordom was formed to actively promote the benefits and use of the automobile in order to fuel its commercialization and acceptance into society. Invoking the American ideals of political and economic freedom, and deploying the tools of advertising to define language, they promoted the car as the way of the future. Motordom deployed several influential tools to assist their cause. They brought engineers to solve the safety dilemmas plaguing the fight over streets, engineers who proclaimed that they could rebuild the city for the car. They largely took over school safety initiatives and partnered with groups like the Boy Scouts of America to teach children how to behave around speeding cars in

Government jaywalking safety posters in the 1920s and '30s.

streets. Peter Norton in *Fighting Traffic: the Dawn of the Motor Age in the American City* notes that by 1924, "Motordom realized that the war over the streets depended on redefining the 'misuse' of the street by the car as in fact a primary and legitimate 'use'."[9] Using its substantial resources, Motordom went about persuading pedestrians that streets should be used for vehicles to run upon, putting the onus of safety responsibility back onto the pedestrian. In response to pedestrians' use of terms like "joyriding" or "speed maniacs"—terms connoting irresponsible and reckless motorists—Motordom countered by deploying the term "jaywalker."[10] This term was negatively used to describe a "jay," a country boor out of place in the city who did not know how to walk correctly in modern automobile-filled city streets. By the late 1930s, anti-jaywalking laws became the norm in many cities, ushering in and securing the street's legal use-right for the automobile, fining pedestrians for crossing streets illegally. These promotional approaches were a key lever with which Motordom influenced transportation policy and public opinion around the street and who had the right to use it. Their influence was widespread, contributing to the general agreement by the 1930s onwards that the street's primary thoroughfare was for the car.

I bring this historical discussion up for several reasons in order to contextualize the impending arrival of the driverless vehicle in our cities today. Firstly, to highlight the critical importance of media, language, and methods of communication in shaping mobility behavior and societal norms around the users of such technology. Throughout the 20th century, Motordom used these communication tools to recast the individual automobile as not only having a right to the street-use, but also as one of social privilege, class, and desire. They did so in order to fuel the sale of more and more cars, linking ownership of one as tied to upward economic mobility and the American cultural ideals of freedom, individualism, and home ownership. Furthermore, Norton elucidates that "users are 'agents of technological change'… chang(ing) (and are changed by) the artifacts they use… Misuse shapes artifacts as much as use, and that struggle between rival social groups to fix

9 Norton, *Fighting Traffic*, 2008.

10 Norton, "Street Rivals," *Journal of Technology and Culture*, 2007.

1925 Midtown Manhattan, increasing automobile traffic claims street use and right-of-way.

the meaning of an artifact in ways they prefer, often take the form of struggles to define use and misuse."[11] Just as we saw that struggle play out in the early 20th century over automobile technology, today we see this struggle being played out yet again, spurred by the technological catalyst of the driverless vehicle.

11 Ibid.

In 2017, the Ford Motor Company released a controversial statement coining the term the "petextrian" in order to protect cars from distracted pedestrians. In response to rising traffic fatalities since the new century, Ford claimed these injuries were equated to the global influx of petextrians, defined as pedestrians who simultaneously walk, text, and use their phone. According to Ford, this creates a massive new safety problem for drivers and motorists. In 2017, Honolulu passed the nation's first distracted walking law, approving fines for pedestrians who look at a phone or other digital devices while using a crosswalk. However, there is not much evidence to support that rising traffic deaths are the direct result of distracted walking. According to studies conducted by the National Transportation Safety Board, motor-vehicle speed is the factor most heavily correlated with death and injury on the road, not pedestrian behavior.[12] But yet again, we see language and subsequent policy being deployed to cast blame for contested use of the street onto the pedestrian.

12 *Safety Study*, National Transportation Safety Board, 2017.

The motivating force behind this new round of pedestrian-shaming tactics is the advent of the driverless car. Ford introduced the term "petextrian" in their advertising efforts to promote "Pre-Collision Assist" technology in their cars.[13] This technology scans the roadway and identify objects blocking it in order to "defend" the driver against the petextrian. Accurately detecting and predicting the movements of pedestrians and cyclists remain one of the final hurdles facing autonomous mobility technology. AV technologists champion the idea that the driverless vehicle is rational, driven by the logics of a software designed to follow the rules of the road. According to them, it is the pedestrian that is unpredictable, prone to unplanned and haphazard behaviors on city streets that the predictable vehicle cannot make sense of. Certain technologists even argue controversially

13 Fraade, "Who is Afraid of the Petextrian," *The Baffler*, 2018.

14 Kahn, "Robot Driving, Some Want to Reprogram Pedestrians," *Bloomberg News*, 2018.

15 Reid, "Biden's Infrastructure Bill Hastens Beacons," *Forbes*, 2021.

16 Conger, "Driver Charged in Uber's Fatal 2018 Autonomous Car Crash," *The New York Times*, 2020.

that pedestrians must be reprogrammed to behave more predictably around new technologies like AVs.[14] Another techno-utopian solution involves discussion and even recent federal funding allocated towards beaconization.[15] This describes a strategy to equip cyclists and pedestrians with transponder beacons that can communicate automatically with sensor-equipped cars including AVs in order to "solve" our traffic safety woes. Will pedestrians (for cost or privacy reasons), who choose not to outfit themselves with a beacon, be blamed for a traffic collision caused by a driverless vehicle? Again, must the onus of safety responsibility be cast on the human rather than the technology itself? These are issues of equity, privacy, and ethical priorities to the use-right of the street that its history has already bore witness to.

The phantom menace of the distracted petextrian is the contemporary version of the same tactic deployed at the advent of the automobile. Its goal is to condition and alter human behavior around new technological advances in order to facilitate the commercialization and acceptance of that technology into mainstream societal use. Yet again, language and media are being deployed as tools to condition and redefine these socio-technical systems. In the previously mentioned case of the fatal autonomous Uber strike in 2018, the human tasked with monitoring the self-driving car was found guilty of endangerment, instead of the vehicle's software.[16] Meanwhile, the Uber corporation itself avoided any criminal charges for liability. A more cynical reading of this outcome might be that the corporations developing self-driving technology profit off of their creations while washing their hands of the responsibility of those inventions if they were to actually endanger or hurt a human. The unspoken agendas behind new distracted pedestrian laws and the language that surround their implementation, are actually about maintaining the status quo privileges of automobiles' right to the road. These are rights that have been established ever since the arrival of Motordom. When cities are safer for pedestrians however—even distracted pedestrians—they are safer for everyone. The cities with the lowest rate of traffic deaths correlates to cities where pedestrians are more common and their right to the

Pedestrian-shaming artist illustration of the distracted petextrian.

street is less restricted. Moreover, the *design* of our streets and the urban environments they condition—the sidewalk, building and business frontages, their cross-sectional composition—exert a powerful influence in determining the priority and use-right of those streets.

This leads me to the next critical opportunity that driverless vehicles bring to the design of our streets. AV technology has the potential to act as a catalyst to reverse the hierarchies of the street in its modern age, giving more of its space to pedestrians, cyclists, and public transit and the associated urban benefits they bring. Streets today are overlooked as a major design opportunity for transformation. They take up an extraordinary proportion of cities, occupying on average 18% of the total land area of American cities, while averaging 55 feet in width.[17] In some city districts, this figure reaches upwards of 30% of urban land area. If one were to consider streets as an essential form of public space—which they are—then they would make up more than an astounding 80% on average of a city's overall public space.[18] With the arrival of AVs, the space usage of our streets are once again being contested. The cross-sectional composition of the street itself and the land uses that rely on it changed dramatically with the proliferation of automobiles into society. AVs present a new catalyst that will disrupt the current use of our streets, giving us an opportunity to reprioritize what we value about this fundamental urban space.

17 Millard-Ball, "The Width and Value of Residential Streets," *Journal of the American Planning Association*, 2021.

18 NACTO, *Urban Street Design Guide*, 2013.

Imagine if a portion of our streets could be permanently reclaimed or dynamically allocated for other uses: recreational, commercial, or ecological. This could be made possible by infrastructure like dynamic lane management, which works hand-in-hand with driverless vehicle technologies to adjust in real-time the use of the street based on its demand. Our streets today are physically fixed infrastructures, static in their sectional use allocation and in their physical design at all times of day. Dynamic lane management, on the other hand, describes the flexible infrastructural ability for street lanes to be dynamically adjusted at different times of day. Possibilities of lane management strategies include reversing traffic directions to improve traffic flow or closing lanes for alternative uses during under-capacity times of day. Imagine if during lunchtime, on a local retail street, the majority of its lanes were to dynamically change, signaling to AVs that those lanes were closed to vehicular traffic. One lane might be marked for an automated bus to use, and the rest marked as an extension of the sidewalk for pedestrians. Restaurant dining would be able to expand outdoors to accommodate increased lunch hour needs while other parts of the street serve as public recreational spaces for a noontime break for office workers. At other times of day, when commuting needs are high, lanes would accommodate to their use instead. Multiple lanes could facilitate the movement of automated buses at both rapid and local speeds while other lanes could be dedicated to cyclists commuting home.

The diverse scales of our street typologies could incorporate the benefits sparked by driverless vehicles in other ways as well. These possibilities affect the varied hierarchies of our existing street network. Urban boulevards, for example, could consolidate lanes from parking and private AVs into lanes for automated transit or high-occupancy AVs. This could free up the street right-of-way for cycling lanes or expanded sidewalks. At the scale of local

Rendering of a potential reclaimed multi-modal street with autonomous mobility.

residential streets and intersections, the space of the street could be reclaimed even more dramatically. Right-turn only lanes for low-speed driverless mini-shuttles could be implemented in underutilized intersections to reclaim the center of the street for safe pedestrian crossings. Streets freed of street parking demands could be converted to pedestrian-only spaces as well. Outside of just servicing mobility, the many commercial, social, and environmental uses of the street could be expanded upon. These diverse uses could range from street vending and outdoor dining, to expanded public spaces and street furniture, to tree plantings, rain gardens, bioswales, and other urban landscaping.[19] These exciting possibilities illustrates how a space—that has today, primarily dedicated the majority of its use for the private automobile at all times of day regardless of need—could be reclaimed for urban uses that are far more democratic, sustainable, and beneficial to the health and vibrancy of our cities.

19 Bioswales are linear, vegetated channels designed to concentrate and convey stormwater runoff from adjacent streets or impermeable surfaces, while also filtering and removing debris. An accumulation of bioswales can help reduce stormwater flooding that many cities face during extreme hydrological events.

These spatial changes are enabled by a transition towards shared forms of mobility, a transition to mobility-as-a-service to drive down street parking demands, and progressive transportation policies to incentive these behavioral shifts. This is both a design and engineering opportunity, as it is a planning and policy one. Much of it falls on the political will of regulators and the constituents they represent. Some of these exciting design opportunities are explored further illustratively in Chapter 8 of this book, while the policies that support their implementation are expanded on in Chapter 5.

While new mobility technologies could spark these transformations, the concepts behind them are nowhere new. The fight to reclaim our streets has already begun in many cities around our world. Recognizing the deleterious effects of automobiles on our streets, contemporary movements like the Complete Streets Coalition or PlayStreets Programs endeavor to reclaim street right-of-ways for community amenities (including larger sidewalks, bike and bus lanes, and landscaped recreational areas). These movements recognize that not only are these changes brought about by planning policy and political will, but that the design of streets is critical in indicating who and for what purposes they serve. Urban street design can be far more influential on safety and driving behaviors than, for example, any speed limit postings.

One of the more well-documented struggles to reclaim the space of the street is described in Janette Sadik-Khan's *Streetfight: Handbook for an Urban Revolution*. In her book, she describes the hard-fought efforts of New York City's Department of Transportation, under her leadership as transportation commissioner during the Bloomberg mayorship, to convert some of the city's most congested streets for pedestrian, cyclist, and public spaces.[20] Some of the most contested spaces in the city, including Broadway at Times Square and Herald Square, were pedestrianized through tactical interventions that eventually became permanent due to their resounding success. Alongside an investment into bike-sharing infrastructure, the expansion of rapid bus routes, and forward-thinking transportation policies, those efforts transformed the way New Yorkers use those vibrant public spaces today.

20 Sadik-Khan, *Streetfight*, 2016.

Another trigger that unexpectedly assisted these efforts were the changes brought about by the Covid-19 pandemic. When the demand for vehicular use disappeared almost overnight, due to world-wide lockdowns, the demand for that space for other public uses skyrocketed, especially in cities where immediate access to outdoor public space remained low. In New York City, for example, the emergency Open Restaurants Program was established to allow restaurants to use business-fronting roadway spaces that were previously dedicated to parking spots, for dedicated outdoor dining instead. What was born out of necessity during the pandemic, became a boon for businesses across the city. The gains brought by decreased automobile use during lockdowns in NYC are now in the process of becoming a more permanent fixture of the city.[21] The motorists that decried their loss of parking spaces have largely adapted to these changes. Even though only about 8,550 of the city's 3 million parking spots have been transformed to outdoor dining areas (or 0.3%), the effect of those changes for economic street uses has been far greater.[22] Opportunities to reclaim the street by triggers like Covid-19, or visionaries like Sadik-Khan and others around the world, have been incremental but hugely transformative when they have been achieved. The other important lesson that they both offer is the value that temporary, easily reversible "experiments" can have on convincing the public to make these implementations permanent. In the case of New York City, these various projects were given a chance despite their initial opposition, because they were presented as temporary and reversible. Once the experiment was conducted, the outcomes spoke for themselves, converted many naysayers, won ample support, and eventually became a permanent fixture of the urban fabric. Their results support that when streets serve more democratic uses, more sustainable, vibrant urban environments can emerge. Streets are, as Sadik-Khan describes, places "where life and history happen, and that places transportation at the cultural, social, and political center of cities."[23]

21 Novini, "What the Future Holds for NYC Outdoor Dining Sheds," *NBC New York*, 2023.

22 "NYC's Temporary Open Restaurants Program," NYC Department of Transportation, 2023.

23 Sadik-Khan, *Streetfight*, 2016.

The arrival of AVs will again disrupt the balance of the street, acting as another important trigger in our near future. As the history of mobility technological triggers like the automobile has shown, they can have deleterious effects on urban environments.

Street transformations in NYC under the leadership of DoT Commissioner Janette Sadik-Khan.

2009 Broadway Avenue transformation at Times Square in Manhattan.

2009 Herald Square transformation in Midtown, Manhattan.

2007 Pearl Street Plaza transformation in Dumbo, Brooklyn.

Reclamation of Dyckman Street, NYC, for outdoor dining during the Covid-19 pandemic.

Reclamation of Main Street, Huntington Beach, CA for dining and commercial uses during the Covid-19 pandemic.

If implemented with the right design values and regulatory frameworks though, they can embolden and enable more democratic street uses. How can we embrace the benefits that driverless technology will bring, while acknowledging our fraught history with socio-technical systems in order to minimize their negative externalities? How can we collectively wrestle back the vision of our future cities from corporations deploying advertising and media to advance private commercial interests? How can we take control of the language and narratives surrounding these issues to make our cities and their streets better places for commerce, work, recreation, and movement? New mobility technologies must be used as tools to support and advance these gains, rather than as tools to reinforce the opposite. As Sadik-Khan so succinctly and ambitiously put it, "if you can change the street, you can change the world."[24]

24 Ibid.

Mono-functional Highways and Suburban Sprawl

Our city streets bore witness to the hugely transformative nature of mobility technologies on their design and use. The other major transformative spatial impact that automobiles had was through the expansion of highways: both in their physical infrastructural design and in their impact on the larger metropolitan structure and development patterns of cities. Again, the arrival of AVs presents a critical trigger to impact the urban growth structure of our city's future evolution. Once again, let's reflect on our cities' recent histories in order to frame how new technologies can impact the spatial design of their possible futures.

The automobile's spatial effect on our land went even beyond the city core and its immediate surroundings, it played a key part in the transformation of our cities' hinterlands—the suburbs. To be clear, the process of suburbanization began well before automobiles arrived. Electric streetcars during the pre-automobile era played a prominent role in turning cities' hinterlands into speculative real-estate. This form of mobility brought a large amount of land into commuting limits far beyond what horse-powered travel could. To be sure, cultural, economic, environmental, and racial factors also pushed populations out of urban cores and into a mix of areas around cities. The cultural promise of the "American dream" of land ownership, affordable home prices, lower property taxes, open space free from city pollution, and racially fueled stereotypes of cities being "crime ridden" and "lower class," all contributed to the effects of white flight into the suburbs.[25] The effects of this phenomenon are still felt today, as they determined much of the spatial structure and make-up of modern metropolitan urbanization.

25 White flight describes the phenomenon of white populations moving out of urban areas, particularly those that were becoming more racially and culturally diverse, into suburban areas.

However, it was the automobile that was the enabling technology that permitted an even greater dispersion of the population into wider metropolitan areas. With a voracious appetite for land, automobiles were a force in feeding the impulse of populations and urban form to sprawl—so much so, that it eventually become

1957 Aerial view of newly built suburban tract homes in Los Angeles.

1939 Futurama exhibit at the World's Fair.

synonymous with the term in our contemporary cultural milieu. In this history, Motordom was yet again crucial in influencing and bringing about this vision for this future that eventually came to pass. In 1939, at the World's Fair in New York under the theme "The World of Tomorrow", the General Motors Corporations displayed an exhibit called Futurama which presented a motor-age city 20 years into the future. Through large-scale dioramas and images, Futurama prophesized a techno-utopian city free of congestion or accidents, enabled by individual car owners with semi-automated vehicles. A key component to this vision was the implementation of large-scale highway infrastructures. Models of these new infrastructures advertised their capability in isolating automobiles from the congestion of the street, promoting high-speed usage while sequestering low-speed pedestrians into constrained "units" of the city. Widely influential, this promotional exhibit garnered great enthusiasm, inspiring public support for the transformative construction of our cities' highway and street networks in the ensuing eras. The power of visioning in media and promotional methods of communication was yet again on full display.

Many key aspects of Futurama's vision for the future came to pass. One of the largest accelerators of this vision was the implementation of the Interstate Freeway Act of 1956. Post-World War II, some scholars have speculated that President Eisenhower's observation of the efficiency and effectiveness of the German Autobahn (federal motorway) for wartime uses contributed to his vision of an America crisscrossed by a similar infrastructural network. Whether that is true or not, Eisenhower's vision for the highway system was realized with this federal act, when from 1956 to 1990, construction of nearly 50,000 miles of standardized, limited-access highways radically reshaped the metropolitan landscape of the continental U.S. Considered by many at the time as a feat of public works engineering, the act consolidated many of the highways built previously, while dramatically expanding their reach under common federal standards

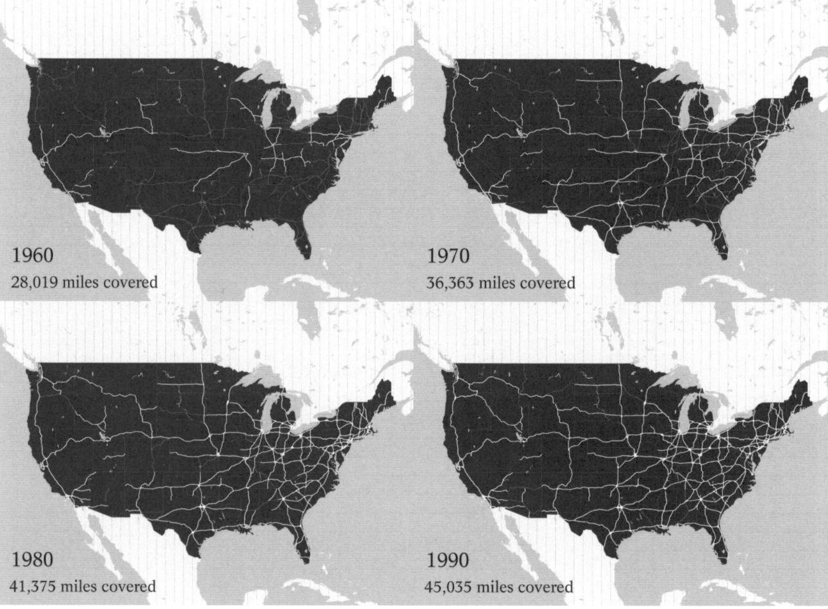

Expansion of the national interstate highway system in the U.S.

and funding. The interstate highway system could be considered a culmination of the coming of age of automobile culture, reaching all corners of the nation. It accelerated the decentralization of cities into wider metropolitan structures by connecting emerging low-density suburbs to high-density urban cores. The highway system also undermined the U.S.' federally-sponsored mass transit programs at the time, a critical component of this transportation story which we will discuss further in the next chapter.

The era of highways and their impact on the urban structure of cities in the U.S. can be broken down into multiple stages: from suburban bedroom communities, to the development of office park campuses that followed them, and to the evolution of edge cities.[26] This history is illuminated in other expansive literature, but for our purposes, let's focus on their infrastructural spatial effects in order to shed light on the potential impact of driverless mobility technologies.[27]

The physical, spatial, and environmental effects of the interstate highway system on our built environments were profound. Like streets, highways became an essential physical and experiential link between people, their homes, their jobs, and their destinations. Unlike

26 An edge city, or a suburban business district, describes the development of a concentrated center outside of a traditional city downtown or central business district, in what had previously been a suburban area. Often, these emerge at the confluence of major freeway intersections, fueled by the connectivity they provide, consisting of mid-rise office towers surrounded by surface parking and residential suburbs that support this density.

27 Hanson, *The Geography of Urban Transportation*, 1995.

1962 Construction of the Santa Monica and Harbor Freeway interchange in Los Angeles.

streets, however, highways restrict their sole use to motor vehicles moving at high speeds. Highways are often grade-separated from the rest of the city in either elevated, bermed, or sunken conditions. These sectional level changes highly restrict direct abutter access to highway edges by buildings and their associated land uses, not to mention introducing difficulties crossing perpendicularly. Highways are a textbook form of mono-functional infrastructures, defined as infrastructure designed for a single (or mono) use (or function). With few intersections and access points, they were expressly designed with the sole function of transporting the commuting automobile or long-distance freight truck, in order to alleviate urban congestion on standard city streets. In many cities, even public bus transit is often not permitted to use highways. However, not only did this infrastructure not "solve congestion," it exacerbated it (recall our previous discussion around induced demand).[28] The increasing use and number of automobiles in the ever-expanding metropolitan landscape inevitably created congestion bottlenecks that intensified traffic gridlock. Like many forms of transportation, they also turned the urban landscape into real estate. Previously vacant land connected

28 Induced demand is a phenomenon that occurs when road and highway expansions meant to relieve traffic instead induce more of it to occur. When these infrastructures are expanded in capacity to alleviate congestion, inevitably traffic volume will increase to match it, thereby resulting in more traffic in more space.

by streets and roads became commodified by landowners, developers, and corporations. Highways contributed to the urbanization and land-speculation of the areas they connected.

Highways also led to the devastation of many of the urban communities and ecological systems they were planned through, with significant racial and environmental ramifications. Planners reinforcing racial segregation routinely sited highways through the heart of minority socioeconomically disadvantaged populations without the economic and political capital to fight back. These infrastructures generally benefited upper-class white neighborhoods they connected to and that could afford to own an automobile. Highways were also often planned along boundaries between communities of different socioeconomic classes, reinforcing spatial inequality. They were utilized to bolster widespread racist planning practices such as redlining.[29] For those whose homes and neighborhoods were not demolished to make way for permanent swaths of highway infrastructure, their long-established social connections were cleaved in two, often leading to the complete disintegration of these communities over time. Those that did survive had to contend with increased physical, economic, and psychological divides—not to mention increased pollution and decreased land values that came with living right adjacent to a limited access highway.

29 In the U.S., redlining is a discriminatory planning practice in which services are withheld from potential customers who reside in neighborhoods classified as "hazardous" to investment; these neighborhoods are often where a majority of lower-income racial and ethnic minorities reside.

Highways also had significant environmental ramifications on our built and natural landscapes. At the scale of the intercity and

1959 Aerial view of Hastings Street in the predominantly black neighborhood of Black Bottom, Detroit.

1961 Aerial view (same approximate location) of the I-375 freeway replacing Hastings Street in Black Bottom, Detroit.

interstate, highways disrupted plant life ecosystems and contributed to habitat loss. For wildlife, highways create a disproportionately larger amount of forest edge, which can be detrimental to forest interior species by segmenting animal populations, not to mention the number of animals that are killed by vehicle collisions. The construction of highways often involves dramatic regrading of natural terrain, resulting in soil erosion, sediment deposition, and water runoff issues. Suburban sprawl fueled by highway expansion also has broader environmental ramifications. Suburban low-density neighborhoods consume more energy per capita than their high-density counterparts in the city's core. Modern suburban dwellings are typically larger than their counterparts in cities and they also share little to no proximity with neighboring units or homes. This results in more passive energy loss to the atmosphere, requiring more energy to heat them in the winter and cool them in the summer. Suburban developments, being of low-density, require a more extensive proportion of asphalt roads, streetlights, and other associated infrastructure to sustain them while also serving far fewer populations of people. Each of these environmental externalities collectively contribute to global climate change while serving fewer and more privileged populations taking up more land space. This is on top of the aforementioned greenhouse gas emissions produced by the millions of combustion engine automobiles facilitated by highways. In recognition of these impacts, Lewis Mumford in *The City in History* dramatically decried that the building of a highway "has about the same result upon vegetation and human structure as the passage of a tornado or the blast of an atom bomb."[30]

30 Mumford, *The City in History*, 1961.

This discussion serves to illuminate an important point framing the arrival of the driverless vehicle. Firstly, that highways are a prime example of engineers and techno-utopists designing major infrastructure around the sole limitations and extents of a single technology. The built and natural environments are then forced to adapt to the constraints of that technology. Highways do one thing arguably well, getting a vehicle from point A to point B at far distances and at fast speeds. Everything else—humans, habits, landscapes, cities, society—must contend with the ramifications of prioritizing that singular technological efficiency. These documented effects dramatically worsened our built and natural environments. Certainly, highways and automobiles brought key benefits and societal advancements—one which this book does not deny—but at what costs?

Mono-functional infrastructures designed around singular technologies provide a warning flag to AV techno-utopists and technology developers today. These figures proclaim that we should adapt our cities, their infrastructures, and the humans that occupy them to the technological advantages and limitations of the AV. Andrew Ng, a technology futurist who was chief scientist at Baidu at the time, proclaimed in a controversial 2016 op-ed that in order to facilitate Baidu's plan to put commercial self-driving cars on the roads, we should make "changes to our infrastructure... and teach the public new ways to interact with them. These changes can accommodate autonomous cars' weaknesses..." to compensate for the limitations of the technology as it existed then.[31] In critique of these remarks (and the corporations and technology innovators that might share these views), this book suggests that the debatable

31 Ng, "Self-Driving Cars Won't Work until we Change our Roads—and Attitudes," *Wired*, 2016.

question should *not* be, "do we design our cities and their infrastructures around technology?", or even, "should we adapt technology to the design of cities and their infrastructures?" Technological progress is relentless, no matter whether you like it or not. Instead, we should be asking ourselves, "how can technology be deployed to make our cities and their built environments better places to live, work, and move about?" What defines these "better" places—equitable use, environmental sustainability, spatial vibrancy, and beautiful design—are expanded further in Chapter 4. But if this question were to frame all AV-related decisions and their impacts, then we can avoid repeating the story that played out in our cities' histories with highways and the technology they were designed to expedite.

The second point this historically framed discussion raises, is the opportunity for AVs to spatially impact the future transformation and use of our highways in our urban cores. Highways are very much a present fixture of our cities and societies today. Despite their considerable negative externalities, they are already built. Thankfully, most cities have largely stopped building new highways in recognition of their fraught history. Just like contemporary efforts to reclaim our streets from the automobile, there have been recent incremental efforts to reverse the harms of highways on our built environments. How can driverless vehicles be deployed as technologies to assist and accelerate those efforts?

When highways were in the process of being constructed, many neighborhoods and activists recognized their detrimental effects and revolted against their implementation and the urban renewal they brought with them. Some of these well-documented highway revolts resulted in the successful halt of expansive freeway construction. These include the San Francisco Freeway Revolt of 1966, the D.C. Freeway Revolts of 1965, and the famed battles between Robert Moses and Jane Jacob-led activists over the construction of the Cross-Manhattan Expressways in 1960s New York, amongst several others. While we can thank these activists for their successful battles in preserving those areas of the city that remain vibrant today, highway proponents by and large won the war, implementing highways across the country, our urban cores, and the metropolitan landscape.

In recent decades, efforts to reverse or remove highways continues. By and large, these highway removal efforts today fall under three main categories. The first category, "A Failure to Maintain", describes highways that through age and maintenance issues were demolished out of necessity and never rebuilt due to public efforts. An example of this is the San Francisco Embarcadero Freeway removal in the early 1990s that resulted from a significant earthquake. The 1989 Loma Prieta earthquake damaged the freeway and eventually led to its permanent removal and replacement with a surface-grade boulevard due to community lobbying. The second category, "Burying the Problem", describes highways that were buried or rerouted underground. The famed rerouting of the Central Artery of I-93 in Downtown Boston of the 1990s, nicknamed the "Big Dig", is such an example. This substantial project replaced an outdated, elevated expressway with an underground, sunken expressway while the space at-grade became surface-level streets and public space for the city. The last category, "Low Hanging Fruit", describes highways that were transformed into surface-grade

1965 D.C. Freeway Revolt protesters demanding the construction of rapid rail transit instead of highways.

1959 Artist illustration of the proposed Midtown Manhattan expressway, never implemented.

A Failure to Maintain: 1990–1991 Pre- and post-removal of the Embarcadero Freeway in San Francisco.

Burying the Problem: 1982–2007 Pre- and post-submersion of the Central Artery in Boston.

Low Hanging Fruit: 2014–2017 Pre- and post-removal of portions of Rochester's Inner Loop highway.

boulevards due to low traffic volumes and undercapacity lane usage. As an example, the removal of a portion of Rochester's Inner Loop highway I-490 was replaced with surface-grade streets, bike lanes, and mixed-use development. This was made possible in part by the decrease in population of the city in the late 20th and early 21st century, resulting in far lower traffic volumes than the original highway was built for.

Each of these highway removal types make incremental steps towards a spatial future of our cities that is less disrupted by monofunctional highways. The documented political and coalition-building efforts to negotiate, fund, and bring about their eventual changes was incredibly immense. *However*, most of the few highway removals that have be achieved don't address the root of the issue: our dependency on single-occupancy automobile usage as the source of spatial contestation in our urban environments. In many of these examples, and in many highway removal proposals put forth today, automobile capacity is replaced with surface-grade, multi-lane boulevards that still prioritize automobile movement. Sometimes, these capacities facilitate the same amount or even *more* automobile traffic than pre-highway removal. Antiquated traffic engineers bemoan the need for reallocation of traffic capacity while political figures acquiesce to motorists and other automobile interest groups. Few projects are so forward-thinking as to dramatically reclaim the majority of the right-of-way for non-automobile uses. One celebrated example, however, is the Cheonggyecheon Stream Restoration Project in Seoul, South Korea. This progressive project, completed in the early 2000s, saw the Cheonggye freeway removed in favor of daylighting the area's historic stream.[32] The majority of the right-of-way of this prior highway was reconstructed as an urban park and ecological waterway, stitching together an urban fabric that had been previously severed. To address traffic reallocation apprehensions, the project went hand-in-hand with city investments into multi-modal transportation policies. There was a strong focus on expanding Seoul's bus

32 Within the context of landscape design, daylighting refers to the act of restoring above-ground water flow and access to a stream or waterway that had been previously buried or diverted below-ground. Typically, daylighting a waterway can restore its riparian environment for habitats and other ecological or recreational uses.

2003–2005 Cheonggyecheon Stream Restoration Project in Seoul.

transit systems to incentivize residents away from private vehicular use and highway demand. The results were transformative, ushering in dramatic mobility, economic, environmental, and cultural changes to not only the area but to Seoul at large, which went on to demolish nearly 15 other expressways in the city following the overwhelming success of Cheonggyecheon.

Can AVs introduce another opportunity to accelerate progressive efforts to reclaim the space of our highways for more beneficial uses, following in the footsteps of Cheonggyecheon? Can we consider another category of highway conversion previously unexplored, one that begins to address the root cause of highway dependencies? One potential spatial opportunity is through that of infrastructural adaptive conversion or reuse. The few highway transformation projects that exist today consider only a full-scale removal of the physical infrastructure of the highway in its entirety. Most neglect the fact that these built structures hold a huge amount of embodied carbon. Embodied carbon describes the carbon emission metrics released during the holistic lifecycle of a particular construction material, infrastructure, or building including metrics related to the manufacturing, transportation, construction, and disposal of that material's carbon emission footprint. In particular, highways constructed primarily of concrete and steel have materials that hold far higher embodied carbon footprints in comparison to their bio-material counterparts like timber and wood. Demolishing highways entirely neglects not only their embodied carbon (as any removal and new construction would expend further emissions) but also removes the uninterrupted long-distance connectivities for moving vehicles that they do still provide.

How can driverless vehicles help reduce the negative externalities associated with highways while taking advantage of certain benefits? Imagine if instead of complete removal, adaptive reuse strategies were employed to reclaim the structure of highways for more democratic uses. Several lanes could be converted to automated bus rapid transit for longer-distance commutes, with new dedicated bus stops integrated into the vertical cross-section of highway structures. Bus rapid transit (BRT) often does not have dedicated lanes on city streets and must contend with the slower-speeds of local traffic, limiting their effectiveness and use. Automated BRT running on highways could be platooned, reaching carrying capacities that could rival light rail or subway. New multi-modal transit hubs could be introduced and designed into the infrastructure of the highway, transferring passengers from automated BRTs that leverage higher-speed highways, to local automated vehicles and buses that run on streets and fill the first/last mile gap.[33]

33 A first or last mile gap refers to the area not serviced by transit, in the mile from a traveler's starting location to a transit stop or from a transit stop to a final destination.

Shifting people out of private cars into automated transit could also free up the major use-allocation of multiple lanes of the highway for alternative uses. Some could be demolished and the space reclaimed at surface-grade. Others could be kept and converted into eco-infrastructures for productive landscapes like urban farming or greenways that expand the city's public realm. These design opportunities could take advantage of certain highway's elevated spatial condition for views while providing uninterrupted linear connectivity for pedestrians or other habitats. Other highways that are trenched could be decked over with new public spaces or converted to waterways with detention basins that utilize their sunken topographies as ecological and hydrological infrastructure. With vehicle electrification, pollution and noise from passing traffic could be reduced, allowing for the space underneath elevated highways to be redeveloped for more diverse land uses. Neighborhoods previously devastated by highways, turning their backs to their physical impacts, could now benefit directly from their conversion into an urban amenity. Their expansive offramps and curved interchanges, which are calibrated to the dimensions and radii of human-driven reaction speeds, could also be rethought, given the space and operation efficiencies that AVs bring.

Bringing about these transformations requires more than just design and engineering planning—driven just as much by economic, policy, and political factors—but the possibilities suggest that they are worth pursuing. If implemented correctly, AV technology could spark a widespread modal-shift towards shared forms of automated mobility and mobility-as-a-service. Supported by progressive transportation policies and the compounded space efficiencies of driverless technology, the massive cross-section of highways in urban cores could be redesigned. Automating our mobility can trigger us to utilize the space for their movement far more efficiently, while simultaneously taking advantage of certain connectivity benefits that existing highway infrastructure provide. These transformations could prove especially critical in the urban cores of our cities, where space is highly contested and immensely valuable, and where the negative externalities of highways are most distinctly felt. This is certainly a significant cultural and political shift (particularly in the context of the U.S.), as it is a design challenge (explored in Chapter 8), but one in which the gains would be incredibly valuable as Cheonggyecheon and other progressive highway conversions around the world have proved.

Urban Form and Transportation: A Synergistic Co-dependency

Through these varied historical discussions, a critical relationship emerges when considering the impact of a driverless urban future. As the lessons from our street and highway histories reveal, our city development patterns and the urban form that evolve in those structures are closely tied to the means and modes of the transportation that engenders them. Changing the way you move about a city fundamentally changes the type of city you get. This relationship plays out in the multiple scales of the city, from its metropolitan growth structure all the way to the blocks and buildings that constitute it.

Prior to the industrial revolution, cities were scaled primarily for walking and the intermittent horse and carriage "technology" at the time. Areas of cities that evolved to facilitate these movements were generally far more compact and irregular in their urban pattern and form. With the invention of the locomotive, and later the streetcar, city geographies expanded outward, spurred by land speculation and real-estate development that densified alongside major rail corridors. When automobile technology arrived, the structural evolution of urban geographies responded accordingly. Areas of cities that were either built or re-built for automobiles exhibit urban characteristics to facilitate that transportation method. The untethered freedom of the automobile hierarchically "flattened" grid networks, expanding the reaches of our cities even further outwards, linking dormitory suburbs that organized land and urban form to the constraints of the car. The history of city planning and transportation is far more expansive than this paragraph or book can give justice to. A city's urban structure is influenced by not only transportation but also economic, political, cultural, and a wide range of other important factors. However, these generalized summaries express that the evolution of urban fabric has varied over time and that their evolutions were greatly impacted by each era's respective transportation technology.

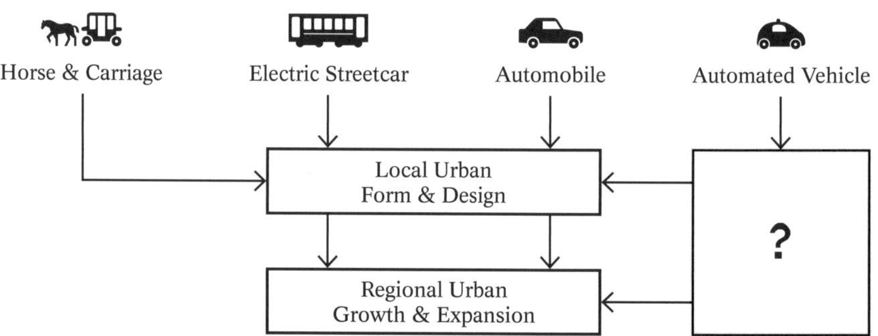

In many cases, the growth of the city expanded from center to periphery, with each successive transportation technology trigger. Various transportation historians and theorists have noted that successive transport technologies in their respective eras typically leave more significant effects in expanding urban areas than in preexisting urban fabric.[34] The more radical the changes in transport technology, the more dramatic the alterations in urban form. As a result, cities with longer histories are often composed of a variety of urban fabrics between their core and their outward peripheries, as

34 Muller, "Transportation and Urban Form," in *The Geography of Urban Transportation*, 1995.
Rodrigue et al., *The Geography of Transport System*, 2020.
Adams, "Residential Structure of Midwestern Cities," *Annals of the Association of American Geographers*, 1970.

each successive wave of urbanization was impacted and planned. Additionally, as each transportation technology is introduced, it often enables further expansion of the structural limits of cities.

Within the context of this spatial relationship, let's consider the impact driverless vehicle may have. As has been previously discussed, a large extent of these transformations in our preexisting urban fabric will come through adaptations and transformations in our mobility infrastructures, streets, and highways. However, areas of cities undergoing new urban expansion in the decades following the introduction of AVs may develop an urban structure more particular to the spatial dimensionalities and patterns of this new technology. AVs may also fuel the expansion of existing city geographies and limits into new exurbs. AVs that can travel at higher speeds more safely through vehicle platooning and vehicle-to-vehicle communication may allow commuters to live farther and farther away from their respective jobs. A 2014 study by MIT researchers show that commuting times stay constant even as distance changes.[35] According to this study, commuters assess the time it takes to complete a trip, generally independent of the distance they have to travel. This suggests that faster or more efficient transportation commutes allow for geographically farther commuting distances as long as commute times remain relatively constant. Furthermore, freed of driving responsibility and requirements to monitor their driving environments, commuters could use their time for other activities like resting, working, or socializing, therefore making any additional commuting times more tolerable.

35 Ratti et al., "Exploring Universal Patterns in Human Home-Work Commuting," *PLoS ONE*, 2014.

Compounded together, farther geographic expansion of our metropolitan limits along with new or altered structural patterns could emerge. This AV-driven development could become even more diffused than the suburbs of today, flattened by the non-hierarchical nature of a personal autonomous car. Exurban geo-fenced AV enclaves of higher-density islands could emerge, akin to the edge cities of today but based on different AV transportation and infrastructural hierarchies. AVs could also be deployed in new transit-oriented development (TOD) areas, in which the calibration between the land uses, densities, and block structures around high-density transit stops would invariably be disrupted. AV-based urban developments would also be conditioned not only by transportation but also market forces, mobility policies, and other factors including the local context of the cities that deploy them. While it may be nearly impossible to predict the exact form these AV-based urban growth patterns will take, it is clear that driverless transportation has an important role to play in determining the structure of our cities future metropolitan form.

The other scale at which our transportation choices engender urban form is that of the block and building level. Nowhere did this co-dependent relationship play itself out more distinctly than in the mid-20th century. During the age of the automobile, architects and planners began designing urban form around the spatial limits and constraints of this transportation technology. This relationship was made especially distinct during the era of Modernism, which has wide spatial influences and impacts in our cities even into the 21st century. In many ways, the ideals of the private motor-vehicle and Motordom that championed it, aligned succinctly with the tenants of modernist architects and planners of the time. Modernists saw the city, its urban form, and the societies it engendered as disorganized and chaotic. According to them, the progressive city for the new age should be organized as orderly and efficiently as possible in order to provide a better quality of life. The automobile, which demanded straight, ordered lines of movement free of the chaos of the uses of the street that preceded it, was the perfect technological tool for shaping that urban form. At the 1933 CIAM IV (International Congress of Modern Architecture), Le Corbusier—a Swiss-French architect who is regarded as one of the pioneers of modernism—proclaimed the ideal city as one that "incorporates modern traffic techniques and is directly proportionate to its purposes and usage."[36]

36 Le Corbusier, *Charte d'Athenes*, 1973.

The modernist school, along with many members of CIAM, rejected the urban solutions for future cities proposed by the movements that preceded them. Influential movements like the garden city movement of the late 1800s and early 1900s sought to combine the benefits of both the "country" and the "town" together. The garden city was an idealized city structure comprised of clusters of development radially removed from a city's core and concentrically connected by the technological means of inter-municipal railways. In contrast, to address the urban blight facing cities of their era, modernists sought to renew the inner city instead, rebuilding them with the latest technologies of their time which included steel, glass, the elevator (enabling greater vertical densities of urban form), and the automobile to achieve these ends.

Their auto-enabled approach to organize the city and its urban form played out through several spatial ideas. They principally sought to separate the functions of the street into delaminated vertical layers in the name of efficiency. Ludwig Hilberseimer, an influential modernist architect and planner, developed ideas about transportation hierarchy and street classifications in his book *The New City: Principles of Planning* which explored the relationship of traffic speeds to city fabric.[37] These ideas were previously explored in his unbuilt *Vertical City* proposal, which illustrated a city structure in which urban form elevated the circulatory plane of the pedestrian several stories above the ground. The ground was then released as a continuous plane for unencumbered use by the automobile. In his various urban visions, including his 1925 *Plan Voisin* and 1924 *Ville Radieuse*, Le Corbusier also took various approaches to grade-separating pedestrian movement independent from the automobile. He either elevated the automobile vertically, in which motor traffic could operate on two levels by higher or lower speeds, or raised the pedestrian on elevated podiums. Modernists valued speed as a symbol of the advanced society and insisted that the automobile must be separated vertically from other traffic and humans that might impede it.

37 Hilberseimer, *The New City*, 1944.

These proposals to disentangle urban form from the many layers of the street was an attempt to resolve the conflict between the ever-increasing speed and large-scale geometries of the automobile and the much finer grain and slower-speed of the traditional street for humans. Issuing emphatic proclamations like "death to the street!", Le Corbusier and other modernists attempted to resolve the contestations over the street that were being fought since the introduction of the automobile, by removing the human from the

1924 Illustration of Hilberseimer's Vertical City proposal, unbuilt.

1924 Illustration of Le Corbusier's Ville Radieuse proposal, unbuilt.

equation. In his 1947 book, *The Four Routes*, Le Corbusier proclaimed, "the current situation of confusion between pedestrians and the automobile must disappear forever. The automobile and the pedestrian must be separated. The current network of streets is old-fashioned, created in part before the appearance of cars in the city."[38] The invention and eventual implementation of highways in cities was a manifestation of these very ideals.

38 Le Corbusier, *The Four Routes*, 1947.

Freed from the constraints of dialoguing with the motorway, modernist urban form was able to turn its back on the street. In traditional urban fabric, architectural form can be described of as in dialogue with the street that forms it. Architecture, land use, and zoning work in balance to activate (and *be* activated by) the street and its pedestrians, resulting in urban form that more often than not creates continuous street frontages. In modernist urban fabric, architectural form became independent of the street pattern. Modernist urban form, conditioned from the high speed of the automobile, retreated from engaging with the grid of the local street. In many of Le Corbusier's visions for the future ideal city, high-density skyscrapers are scattered in landscaped parks free from this

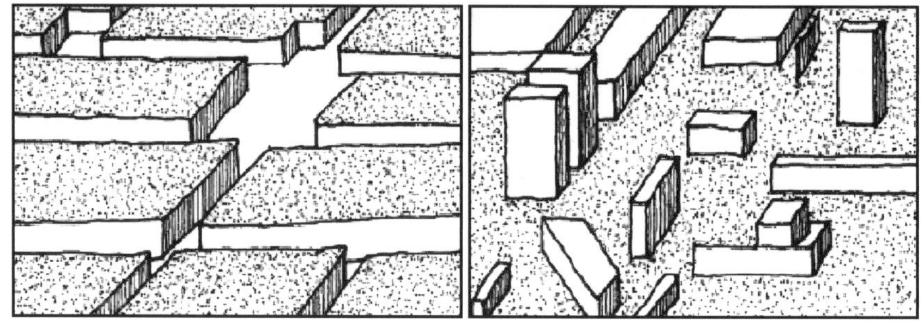

Diagram comparing traditional city form versus modernist city form.

engagement. These areas, spanning multiple traditional blocks, were outlined by major arterials for the purpose of fast-speed automobile movement, forming the structure of the modernist superblock. These spatial ideas worked in concert with parallel ideals to zone the city into separate uses, composed of isolated superblock clusters separating residential from commercial or industrial land uses, reconnected only by the super-network of the automobile.

These modernist visions were widely influential in the planning and design of cities and urban form around the world. They inspired city planners, like Robert Moses of mid-century New York, to imagine and implement an exquisitely rationalized metropolis through the logic of the automobile. They influenced the design of public housing superblock projects around North America and Europe in the 20th century, nicknamed "towers in the park." Modernists themselves were commissioned to design and plan new cities around the world like Chandigarh, India and Brasilia, the capital city of Brazil. Modernist ideas that delaminated the layers of mobility play out even today, with pedestrian over- or under-passes proposed as traffic-engineering solutions to solve the contestation between cars and the pedestrian.

Unfortunately, the implemented results of these spatial ideas at the urban scale have proven more often than not to be failures. They have resulted in cities and neighborhoods with less vibrant public realms, prioritizing automobile traffic at the expense of the pedestrian and all else. The morphology of the modernist "towers in the park" is widely criticized today for contributing to the social fragmentation of the populations they housed. This was not only due to a failure of maintenance and disinvestment by public housing agencies who managed them but also due to the failure of its urban form to engage with and contribute to the vibrancy of the ground, the street, and the social interactions that that engenders. Cities like Brasilia, which were built from the ground-up on modernist planning principles, are criticized for the antiseptic quality of its superblocks. They were also designed with aesthetically monumental buildings, urban forms, and wide boulevards that were scaled to and accessible only by the passing vehicle, rather than to the walking human. These cities exhibit an overreliance on highways and large-scale boulevards that prioritized cars at the expense of vibrant pedestrian-activated public spaces. Pedestrian infrastructures like overpasses, underpasses, sidewalk bollards, and barriers are also often deployed in these auto-dependent cities. These "protective measures" are designed to further eliminate the contestation between pedestrians and cars by prioritizing the automobile's claim to the road right-of-way.[39]

39 Cortright, "The Myth of Pedestrian Infrastructure in a World of Cars," *Strong Towns*, 2020.

1955 Pruitt-Igoe apartments, an infamous modernist urban-renewal public housing development in St. Louis. Its eventual decline and demolition in 1976 came to symbolize the failure of modernist urban form.

Aerial view of Brasilia, exhibiting its wide boulevards and modernist superblocks.

Today, many of these spatial concepts and modernist principles live on in new forms. New utopian cities are being planned and developed based on the technological trigger of the driverless vehicle. Masdar City, a ground-up city designed by Foster + Partners that is funded and constructed by the Abu Dhabi government in the United Arab Emirates, is one such example. It is designed to elevate the city's pedestrian ground plane and street network, on a podium 23 feet vertically to allow and facilitate a network of individual AVs to operate uninterrupted underneath. Passenger elevators are deployed to reconnect the pedestrian level to the automated mobility network. Another example, announced in 2020, dubbed the Woven City, envisions a "smart city" from the ground up located near Japan's Fujiyama. The development of this city is funded by the Toyota Motor Corporation, who are not coincidentally using the city to test their self-driving vehicle technologies in real-time. Toyota CEO Akio Toyoda announced the city as a "unique opportunity to create an entire community... from the ground up and allow us to build an infrastructure of the future that... freely tests technology such as autonomy, mobility-as-a-service, personal mobility... artificial intelligence and more in a real-world environment."[40] Bjarke Ingels Group, the architects of this utopic—or dystopic city, depending on your positionality—describe how the three components of a typical road are peeled apart into a new urban fabric, resulting in a street that is optimized for AVs while others are optimized for pedestrians.

40 Ravenscroft, "BIG and Toyota Reveal City of the Future," *Dezeen*, 2020.

These spatial relationships sound eerily familiar to those who recognize the fraught historical relationships between automobiles, pedestrians, and urban form, as we have discussed at length in this chapter. History, yet again, serves as a warning sign to techno-utopists who fail to recognize the negative externalities of built environments that separate the pedestrian and their mobility networks from each other, prioritizing new technological efficiencies above all else. This historical context and its modern-day impacts are critical to understand in light of new transportation technologies in order to avoid the documented mistakes of the past.

The evolution of our urban form, and what forms of transportation we decide to prioritize in shaping them, go hand-in-hand. One cannot be designed without understanding its implications and impacts on the other. It is this particular observation that underlies the illustrated studies in the latter half of this book. The automobile has, in its present form, engendered a wide range of urban form impacts. Automating these vehicles presents an opportunity to disrupt and question their spatial impacts. How will our current and future urban form be impacted by driverless mobility? What forms produce better environments for the vibrancy, vitality, and health of our cities? How can AVs catalyze spatial relationships that reclaim our urban spaces and right-of-ways for a diversity of multifunctional uses?

At the scale of the block, this relationship has led to block structures that are calibrated to support the proportional spatial requirements of automobile parking. For example, residential blocks balance density limitations and parking requirement ratios between the number of residents who live in those buildings, and the number of cars owned by residents that need to be parked. Other block structures, like the office business park or retail commercial blocks, have carefully calibrated ratios between their respective land uses and the parking spots and lots that support those particular user densities. If our reliance on automobiles was reduced through automated transit and shifted towards mobility-as-a-service, how can these spatial relationships be rethought?

At the scale of the architectural building form, what adaptive conversion strategies could emerge for buildings that have been designed around the dimensional qualities of the private automobile? Drive-through fast-food restaurants, for example, are an infamous building type that are sited and oriented both internally and externally around the spatial queuing constraints of the automobile. Another example, single-family homes, are designed with front-facing garage structures and driveways dedicated to housing the automobile and facilitating its access. Other building types like motels, gas stations, and parking garages also exhibit calibrated spatial relationships between the car and the architecture that houses it. Freed from parking requirements, how can these architectural types evolve? Could new inventive building types emerge from the spatial opportunities that AV technology bring? How can all these transformations prioritize more diverse land uses or public realm opportunities that make our built environments better places to experience? Taken together, and visually explored further in the second half of this book, the scale of these cumulative urban form impacts is wide-ranging and could be transformative for cities.

As this chapter's various historical contexts have surfaced, the impact of the staggering outward thrust of automobile technology in our cities is monumental. From the suburbs to our nation's expansive freeway network, the primary mobility technology of the prior century has shaped our built environments in ways previously unimaginable. This symbiosis between the individual car and the collective city has become so seamless that, in many cases, people are unaware of the powerful historic forces and contestations that blended the two. Normalized today by media forces and corporate interests, it is easy to forget how motorized vehicles have shaped our cities to often extremely negative ends. Automobiles, whether conventional or autonomous, can certainly be one of many useful tools in a diverse variety of mobility options. The threat, however, is automobile *dependency*, wherein no other choices besides the private car exist, and where urban form reshapes itself around that choice so completely that it subsumes all other priorities. These documented negative externalities span across environmental, economic, cultural, and spatial factors to shape the very tone and texture of our urban fabric.

Technological innovation will inevitably continue to disrupt and impact these symbiotic relationships. It is clear that automating our mobility will have implications on the way we plan, design, and shape the spatial dimensions of our built environments. How can we capture the benefits that automated mobility technologies will bring, while minimizing their negative externalities? Can driverless vehicles be deployed to address and even reverse the negative externalities caused by its predecessor? Who is delineating the vision for that future city, and how can we ensure it is one that is better to live in? By understanding the arrival to our present, the decisions we make today can determine if our future cities are more environmentally sustainable, equitable, and spatially delightful.

3 An Automated Transit-Enabled Future

30
NO VEHICLES
ON TRACKS
Vehículos
no permitidos
sobre vías

Competing Futures of Yet Many Unknowns

Confronted with this history, it is easy for technology skeptics to be fearful of new technological triggers. It is also easy for technology dreamers to ignore that history, running the risk of believing that new technologies can in-and-of-themselves solve the urban ills and issues of the present day. On one side of the contemporary conversation, AV-skeptics fear the proliferation of "zombie cars" onto our streets, proclaiming that the technology will be the demise of cities, even more so than the automobile of its past eras. On the other extreme, AV utopists effuse that driverless technology will do the opposite—solving our transportation woes by enabling pedestrians and vehicles to seamlessly share the road while allowing new cities and societies previously unimagined to be designed and built. Rather than react on either extreme, it is important to untangle why and how these technologies can affect our cities in both positive and negative ways. This will help us make more informed decisions on when and where to deploy these technologies for better outcomes. What must be highlighted, however, is the amount of uncertainty that is yet unknown around the current development, implementation, and deployment of AV tech.

Firstly, will AVs still be gasoline-dependent, powered in part by the combustion-engine, or will the promise of fully electric-powered vehicles be fulfilled? While the majority of AV tech developers have invested in the electrification of the technology, the level to which that would be achievable is unclear. A recent study by the University of Michigan Center for Sustainable Systems found that today's battery technology may lack the storage capability that limits the range of AVs due to the extensive energy requirements of their sensors, computers, and transmitters.[1] Some companies, like the Ford Motor Corporation, have invested into hybrid vehicles as the basis for their AV strategy, while others, like Waymo, are in the process of transitioning from hybrid vehicles to all-electric fleets in their geo-fenced service locations.[2]

Secondly, will AVs still require a measure of human oversight, requiring a human to be present and vigilant at the steering wheel, ready to take control of an AV in the case of an emergency? Generally, this is where the technology stands today, though some companies like Waymo have achieved Level 4 automation capabilities, freeing drivers from the wheel within geo-fenced areas. The elusive dream of automation Level 5 still remains right out of reach.

Thirdly, how will we own and use these vehicles? Will they be majority privately owned and operated, sold as individual vehicles for individual consumption, maintenance, and use? Or will driverless vehicles be deployed primarily through a service model, in which fleets of vehicles are owned by private transportation companies or public entities that maintain and operate these vehicles for public consumption? Many automobile manufacturers, like Tesla, who are developing AV technology are pushing for the former scenario in a bid to sell more individual vehicles to each household. Meanwhile, transportation network companies, including Waymo and Uber, are pushing for the latter scenario in order to expand their reach as mobility-as-a-service providers. It is certainly no coincidence that these ownership visions align with the respective business models of

1 Gawron et al., "Life Cycle Assessment of Automated Vehicles," *Journal of Environmental Health Science & Engineering*, 2018.

2 Doll, "Waymo to Retire Chrysler Hybrids in Transition to All-Electric," *Electrek*, 2023.

each company, and the consumer market shares and growth they are respectively targeting.

Fourthly, will AVs remain majority single-occupancy, carrying a single person for the majority of their use just like the non-automated vehicle of today? Or will they transition to majority multi-occupancy, incentivizing a shift towards fitting more passengers per vehicle, with vehicle sizes correlated to passenger occupancy? It is estimated that on average, the occupancy of an automobile sits at 1.5 persons per trip in the U.S.[3] Ride-hailing services, despite their claims otherwise, also exhibit similar metrics, with average vehicle occupancies at 1.4 passengers per ride.[4] This number drops even lower to 0.8 passengers when accounting for deadheading, which incorporates the empty trips made by a ride-hailing vehicle when there are no passengers in the vehicle being serviced.

3 *Personal Transportation Factsheet*, University of Michigan Center for Sustainable Systems, 2023.

4 Schaller, *The New Automobility*, Schaller Consulting, 2018.

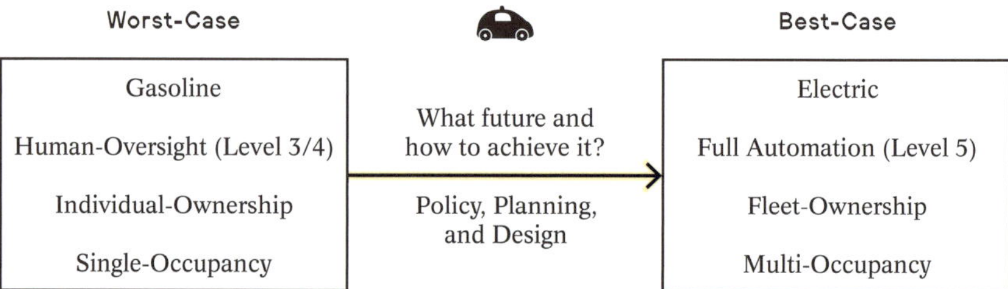

Collectively listed in the diagram shown above, we realize that even with the introduction of driverless vehicles, a future that evolves into the "worst-case" path delineated on the left essentially represents the same way we use automobiles in our cities today: gasoline-dependent, requiring human oversight, majority privately-owned, and single-occupancy. This very possible worst-case future could spark a second motorization wave in our cities, a future fueling further automobile dependency or automated "cars on steroids." The gasoline car, the electric car, and the automated car impart the same spatial externalities on urban space, *if* they end up being used in the very same way.

Given these unknowns and what we know now about the spatial externalities of automobiles, the question that presents itself is, what future outcomes do we aspire towards? And if we do aspire towards the "best-case" future—electric, driverless, majority multi-occupancy mobility-as-a-service—what is the framework that will allow us to reach it? It is easy to proclaim a best-case scenario resulting in an AV-enabled city without delineating the critical steps to reach that future. Similarly, it is easy to retreat to a worst-case vision, rejecting the possibilities of other outcomes to occur.

It is also important to highlight that the future may very well manifest through a range of permutations between the two extremes listed here. Each of these competing futures have vastly different effects on our cities' urban development and form. As a thought experiment, let's play some of these possibilities out to extrapolate the futures they would result in. Let's say that the remaining technology barriers and speedbumps facing driverless vehicles are solved, and Level 5 automation and full electrification is achieved. Scenario A titled "AV-Exurbs", describes if electric driverless vehicles were to remain privately owned and majority single-occupancy. Scenario B titled "The Death of

Space needed to transport 60 passengers by bicycles, cars, or bus (from left to right).

Space needed to transport 60 passengers by cars, electric cars, or automated cars.

Transit", describes if electric driverless vehicles were to shift to fleet ownership and be consumed as a service, but still remain majority single-occupancy. Lastly, Scenario C titled "Towards a New Transitopia", describes a best-case outcome for driverless technology, making a case for why the technology should support our transit systems as well as expand to fill in transportation mobility gaps.

These competing potential scenarios highlight the multiple futures that could emerge, given the uncertainty around the implementation and use of driverless technology. Therefore, how we get from the worst-case future to a best-case one requires careful consideration and critical steps that engage policy, planning, and spatial design frameworks. It is precisely these steps which concern the latter half of this book, with the hope that the deployment of AVs can be guided towards this path before graver outcomes manifest. Let's discuss these three scenarios in more detail next in the following sections of this chapter.

Worst-Case		Best-Case
Gasoline Human-Oversight (Level 3/4) Individual-Ownership Single-Occupancy	Scenario A: AV-Exurbs	Electric Full Automation (Level 5) Fleet-Ownership Multi-Occupancy
Gasoline Human-Oversight (Level 3/4) Individual-Ownership Single-Occupancy	Scenario B: The Death of Transit	Electric Full Automation (Level 5) Fleet-Ownership Multi-Occupancy
Gasoline Human-Oversight (Level 3/4) Individual-Ownership Single-Occupancy	Scenario C: Towards a New Transitopia	Electric Full Automation (Level 5) Fleet-Ownership Multi-Occupancy

Given the cultural and behavioral norms embedded in the way automobiles are used today, it is not difficult to imagine a future in which the way we own and use vehicles for transportation continues in the very same way. This is Scenario A titled "AV-Exurbs", which is essentially a future version of the current status quo. This scenario is particularly acute in auto-dependent cities and countries where transportation modal choices are few and where households depend solely on the automobile for access to their livelihoods because few alternative mobility options exist. It is also the likely scenario to emerge if automobile manufacturing corporations were to dominate the market commercialization of AV technology. In this potential future, electric driverless technology has materialized but it remains mostly privately owned and used. Automobile manufacturers, like Tesla and Ford, are able to commercially roll out, market, and sell driverless vehicles at a mass scale. Individual households replace their individually-owned automobile for an automated upgrade and continue to engage with mobility through similar behavioral patterns. A household's AV shuttles an individual from their home to a destination. Freed from the requirement to park, the AV either circles the block waiting for its owner, or is sent all the way back home to park and charge until it is recalled back to the original destination to pick up its owner for their next trip. In both scenarios, the vehicle remains empty while traveling city streets and travels for even more miles than its non-automated predecessor. Traffic on our city streets skyrockets due to the dramatic increase in vehicle miles traveled on roads—many of these trips conducted by empty AVs not carrying a single passenger. The average occupancy of these private AVs plummets. In a more dystopian version of this future, Motordom has cordoned off city sidewalks separating pedestrians from AVs with traffic barriers in an effort to reduce congestion by increasing the maximum speed limit of AV travel. The claim to the right-of-way use of the street by the private vehicle becomes even further solidified.

The effects of Scenario A on the growth and planning of our cities has been described and alluded to several times previously in this book. To expand on that logic, the spatial impacts of this method of AV-use fuel further outward explosion of our city's metropolitan reach into the exurbs and hinterlands. This AV-enabled land speculation, like the automobile and streetcar before it, is made possible by the construction of AV-superhighways shuttling driverless vehicles at maximum speeds. This expansion is further fueled by a household's willingness to live farther and farther away, as passenger commuting times can be used for other productive uses like working, resting, or recreation. With increased commuting distances, these AV-exurbs built around the privately-owned AV continue to fuel the development patterns of suburban mono-enclaves. Households become further isolated from each other and from their destinations while *dependency* on that individual driverless vehicle and AV superhighways to reconnect them increases. As previously discussed, these low-density exurban areas continue to consume more energy and emit more carbon than their denser counterparts, further contributing to disproportionately

Artist illustration of a self-driving car and superhighway, produced for automotive industry advertisements.

increasing energy consumption and global warming. While this scenario is certainly dependent on several assumptions to occur, the logic behind it remains plausible, given that they build on the culture, ownership, and spatial externalities of our current automobile use today. One can not only easily imagine this future occurring but can even argue that it may be a likely one if automobile manufacturers proliferate AV tech in an unregulated free market. And if it were to occur, the worst trends of 20th century automobile-driven urbanism might very well be exacerbated, fueling ever-worsening congestion, urban sprawl, and mono-functional uses of our streets and cities.

Let's entertain Scenario B next, wherein electric driverless vehicles have arrived but with them a shift towards fleet ownership. In this potential scenario, driverless technology is rolled out primarily as a service model where transportation network companies own and operate a fleet of AVs that deliver on-demand mobility for use by consumers. In this future, many households have sold their privately-owned automobile in favor of the ability to call for an AV based on their travel needs and uses. As driving labor costs have been virtually eliminated, mobility-as-a-service becomes widely economically viable, offering on-demand mobility that is cheaper and more cost-effective than owning a personal car and the insurance, maintenance, and parking costs that entails. As mentioned previously, driver labor costs account for over 60% of the average cost per passenger-miles traveled by MaaS companies.[5] *However*, in this future Scenario B, the carrying capacity of most AVs remain the same as the automobiles of today, and subsequently they end up shuttling on average one to two passengers per trip despite being designed to carry more.

This potential future—electric, driverless, fleet-owned, majority single-occupancy—is dramatically titled "The Death of Transit" when their implications are projected forwards. In many ways, this future scenario is based on the effects of private ride-hailing models of transportation that cities have witnessed today. This is also a

5 Johnson and Walker, *Peak Car Ownership*, Rocky Mountain Institute, 2016.

likely future if private transportation network companies (TNCs), like Uber and Lyft, realize their visions for automated mobility ride-hailing. Let's explore further why this extrapolated potential might not be such a boon for the future of our cities, as one might have originally expected.

The transportation method of ride-hailing is not a new phenomenon. Reaching as far back as the early 1830s, the for-hire horse-drawn carriage was eventually replaced by the modern taxicab, which proliferated around the world in the early 20th century with the invention of the automobile. However, what TNCs like Uber and Lyft did to ride-hailing in the early 2010s was make this modal choice far more accessible and convenient (through user-friendly digital apps), as well as more economically affordable (subsidized by TNC venture-capitalist backers) than the ride-hailing that came before it. TNCs also capitalized on a younger market demographic that was far more receptive to new technologies. It sparked the wider urban population to participate in the sharing economy, and to forgo individual car ownership for alternative mobility choices.[6] According to studies cited by Lyft, in 2017 alone almost a quarter of a million of their ride-hailing passengers had sold their personal car or abandoned replacing their current car due to the availability of ride-hailing services.[7] The accessibility, convenience, and affordability of TNC services are factors that will be accelerated by driverless vehicles, which is why, to no surprise, these companies have heavily invested in advancing and integrating this technology into their service models.

6 The sharing economy describes an economic model that allows users to share resources, goods, services, or skills, often enabled by shared online platform technologies. Uber is one of the best-known examples, in which a resource (an automobile and the mobility it provides) becomes a shared resource and service available to more than just the car-owner, via a smart-phone enabled app.

7 Etherington, "Lyft says nearly 250k of its Passengers ditched a Personal Car," *TechCrunch*, 2017.

However, the widespread proliferation of TNC services into our cities has resulted in several unintended and only recently recognized externalities. The first externality, is a dramatic increase in city traffic congestion. Prior to the proliferation of TNCs, ride-hailing services were typically closely regulated by public agencies. In large cities like New York where taxicab use was pervasive, a taxi medallion licensing system was implemented requiring taxi drivers to purchase a permit in order to operate a taxicab for service. The medallion system was meant to intentionally constrain the supply of taxicabs in operation, limiting oversaturation of the market for drivers and taxi companies. They also had the effect of limiting the total number of taxi vehicles that could operate in a city area. When TNCs arrived in cities in the early 2010s, they did so without similar regulations and protections, saturating the market with tens of thousands of vehicles while adding billions of miles of driving per year by the time their use became widespread. For example, in New York City, TNC usage peaked at nearly 800,000 trips per day between 2018 and 2020, compared to traditional taxicabs that were estimated at around 200,000 trips per day at the same time.[8] While the lockdowns and health concerns brought by the Covid-19 pandemic significantly reduced ride-hailing use for a time, TNC ridership numbers have almost climbed back to pre-pandemic levels as of the writing of this book.

8 Schneider, "Taxi and Ridehailing Usage in New York City," Monthly Data Report, 2023.

These effects were documented in a 2017 report by the UC Davis Institute of Transportation Studies, which determined that TNCs caused an overall 180% increase in driving on city streets.[9] This is due to the second important externality caused by TNC use: TNCs compete primarily with public transit, walking, and biking, drawing consumers away from these non-auto modes of

9 Clewlow and Mishra, *Disruptive Transportation*, UC Davis Institute of Transportation Studies, 2017.

Would have used if TNC was not available for ride:

Personal Car	Taxicab	Transit or Carpool	Walk or Bike	Not Made Trip
21%	1%	33%	23%	22%
22%		78%		

transportation *rather than* the personal automobile. According to these findings, about 78% of TNC users in large cities would have taken public transit, carpooled, walked, biked, or would not have made the trip if the TNC had not been available for this trip. In the range of cities studied, 33% of TNC users would have taken an alternative public transportation service or carpooled. TNCs replace automobiles on the road only when parking is expensive or difficult to find. In other words, TNCs are not taking a majority of miles away from automobile use. Instead, they are adding more miles to the road by drawing users away from public transit and other more sustainable modes of mobility. The same report found that ride-hailing services, up to that point, had resulted in a 9% reduction in Americans' usage of bus and light rail transit services. Furthermore, as ride-hailing vehicles take up to two to ten times more space per person than their transit counterparts, these modes of transportation are a far less efficient use of the contested urban space of our streets and cities.

These metrics are further supported by other studies on transit use decline in our cities. The charts shown below illustrate that, despite relative constant economic growth in each of these four major U.S. cities in which public transit remains a modal choice, ridership numbers take a nosedive around the years 2014–2015. Around that time, in each of these four cities, Uber and Lyft became ubiquitous and widespread, suggesting that as soon as consumers are able to reach an income threshold in which they are able to

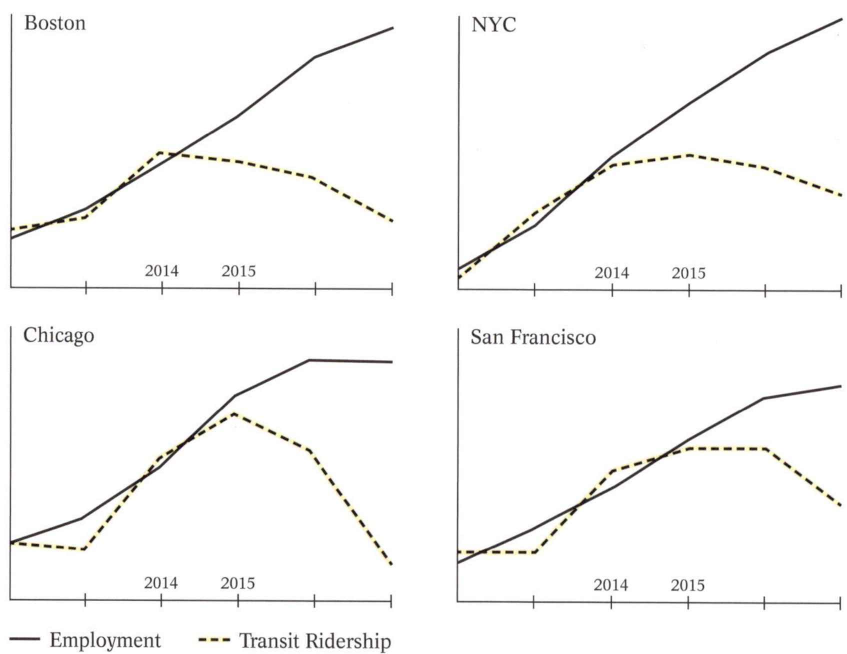

afford private ride-hailing services, they opt to do so over public transportation alternatives.

Despite these documented statistics, TNC's marketing campaigns aim to convince the public that ride-hailing services do the opposite. Lyft's Co-Founder and President, John Zimmer, stated in 2018, "we believe that Lyft as a platform can virtually eliminate [congestion and pollution] long term."[10] Prior to the Covid-19 pandemic, Uber and Lyft's marketing promoted their respective services' ride-*sharing* options. Ride-hailed ride-*sharing* was a service that grouped passengers traveling in similar directions at similar times in the same vehicle for a discounted cost. Matched through TNC software algorithms, these services were essentially offering modern-day carpooling. A Lyft company spokesperson stated, "Lyft is focused on getting more people to share rides," while Uber CEO Dara Khosrowshahi claimed in 2018, "shared ride and express pool rides [are] something we're very much behind."[11,12] At the time, implementing UberPOOL (now UberX Share) and Lyft Shared options was an effort to optimize these networks by moving people more efficiently. They were also a way, not coincidentally, to improve TNC's profit margins on each ride. However, a 2018 report conducted by transportation policy expert Bruce Schaller (former deputy commissioner at NYC's Department of Transportation) documented that these shared ride services, while touted as reducing traffic, in fact add mileage to city streets and do *not* offset the traffic-clogging impacts of private TNC use. While private TNC trips produced an overall 180% increase in vehicle mileage, if the number of TNC rides were 20% shared, this still resulted in a 160% increase in vehicle miles traveled as compared to a privately owned and operated vehicle. Even if Lyft's stated goal of 50% shared rides was met, the study showed that this would still result in an increase of 120% in driving on our city streets.[13]

10 Hawkins, "Lyft Thinks we can end Traffic Congestion," *The Verge*, 2018.

11 Ibid.

12 Griswold, "Getting People to Share Rides is a 'Battle' against 'Societal Norms'," *Quartz*, 2018.

13 Schaller, *The New Automobility*, Schaller Consulting, 2018.

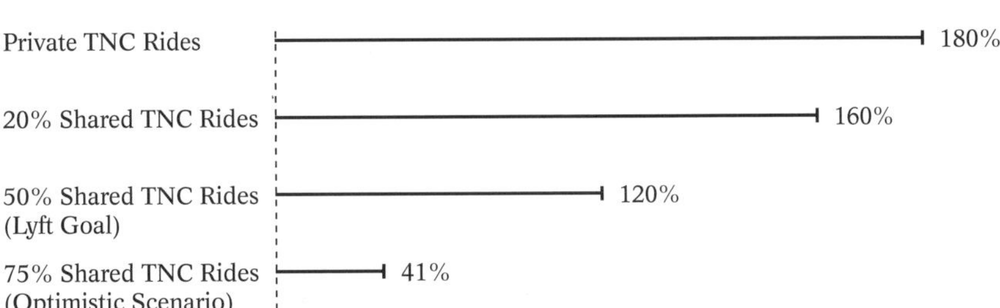

Any potential "reduction" in the increase in vehicle miles traveled by passengers electing for *shared* rides with strangers, were further eradicated by the social and health norms brought by Covid-19 pandemic. The reason for these metrics is that in shared rides, consumers are still being drawn from non-auto modes while portions of shared rides still involve just one passenger (between the first and subsequent pickups). Additionally, in both shared and private TNC rides, significant mileage is also added to roads in the time and distance between trips, as drivers drive to pick up their next dispatch from the app.

Altogether, the results are undeniable. Transportation writers have proclaimed that "ridesharing appears to be delivering a body blow to public transportation. Ridership on public mass transit is down in nearly every major U.S. city... [and] ridership decline can be attributed to ride-hailing."[14] TNCs are able to do this because a large percentage of their fares, up to 50% by certain estimates, are actually subsidized by their venture capitalist backers. Uber's business model is to subsidize fares in order to flood cities with their services so that they can corner the market and eventually gain market pricing power. That business strategy has, along the way, not only driven down traditional taxicab use (the value of taxi medallions has plummeted since the introduction of TNCs) but also public transportation and other modes of mobility.

14 Hill, "Ridesharing versus Public Transit," *The American Prospect*, 2018.

In our hypothetical Scenario B, all these trends and effects instigated by mass TNC use in our urban centers are further exacerbated. In this potential future, TNCs have replaced their human-driven vehicles with a fleet of driverless ones, further decreasing their operational costs and fares. Traditional fixed-route services like the local bus network have been replaced with driverless, on-demand, ride-hailing services that still remain majority single-occupancy at the carrying capacity of the traditional automobile. Traffic congestion in our urban cores continues to increase, inflicting with it the aforementioned negative spatial externalities on our streets and built environments. It is clear then that Scenario B may not be the boon to cities that AV-proponents and companies might believe. As Bruce Schaller stated, "without intervention, the likelihood is that the autonomous future mirrors today's reality, more automobility, more traffic, less transit, less equity."[15]

15 Schaller, *The New Automobility*, Schaller Consulting, 2018.

Extrapolating Scenarios A and B to their logical future outcomes leads us finally to Scenario C, "Towards a New Transitopia", which this book endeavors to push cities towards. In this best-case scenario, electric driverless vehicles are fleet-owned and operated by public transportation agencies, or public-private partnerships models, that have adopted the technology to automate and expand their transit operations at multiple scales. Any private TNCs that operate in the network do so in partnership with and under the oversight of public agencies who work together to provide holistic mobility services for the broad population geographies of the city. This shift in owner-operation has not only allowed but has incentivized AVs to operate closer to their full-occupancy capacities. Transportation services have expanded the diversity of their vehicle sizes and services, ranging from subway/light rail and automated express rapid bus to automated local buses, smaller mini-shuttles, and bike/scooter-sharing services. Altogether, transportation services work together to serve a wide range of trip types and geographic areas. The benefits of driverless technology are used to not only support and expand public transit services but to also address geographic areas where transit previously did not service, filling in mobility gaps or underserved zones.

This potential future enables the wide range of spatial outcomes illustrated in this publication's graphic novel to occur, including the adaptive transformation of our mobility infrastructures and urban form at the block and building scales. They also suggest important mobility behavioral shifts around the holistic use of our transportation network overall. This in turn has vast implications on the future urban growth structure of cities.

The potential outcomes described by this scenario are based on several important assumptions that need to be discussed further. Why are cities predicated on a range of transportation choices, including public transit, more desirable places to live in? How can passenger behaviors be incentivized to use a wider range of modal choices over a private automobile or private AV? This is the last missing critical portion of the transportation story previously undiscussed: the important impacts that mass transit plays in determining our cities spatial form and use. Let's begin with a simple example and revisit our recent history yet again.

To contextualize our mobility behaviors, while transportation technology has evolved significantly in our recent history, our available transportation modal choices have in many ways remained constant. Take for example, New York City in the 1870s. At this time, the fastest way to travel long distances for the majority of urban dwellers was to either (1) flag down an empty, for-hire, horse-drawn carriage, called a hansom cab, or (2) take a shared-seat, horse-drawn tram that could be boarded at any number of designated carriage stands throughout the city. The trams generally followed predefined routes and charged much lower fixed-rates according to distance traveled. Hansom cabs, on the other hand, charged passengers rates based on distance at typically higher fees but provided more convenient direct service. In subsequent eras, the individual automobile replaced the private hansom cab, while the electric streetcar replaced

1870's Horse-drawn hansom cab (left) versus horse-drawn tram (right).

1900's Motorized taxicab (left) versus electric streetcar (right).

Modern-day taxicab (left) versus transit bus (right).

the horse-drawn tram. The hansom cabs of the 19th century are basically analogous to the ride-hailing services of today, taxicabs and TNCs like Uber and Lyft. Horse-drawn trams and electric streetcars are essentially analogous to current-day public bus or rail systems. This example highlights that individual modal choices have not evolved that much, despite our recent technological shifts. Driverless vehicles, while certainly representing the next technological trigger, are just another development in the push and pull our modal choices have on our mobility behaviors.

These choices have distinct spatial impacts on the urban form and structure of our cities. Should we structure urban planning and design around the mobility services of the privatized automobile or on public transit? As we've discussed, automobile dependency has wrought vast spatial, environmental, economic, and social consequences in cities dependent on them. This is a history that has played itself out in the majority of North American cities. What about cities developed primarily on transit-based mobility?

Today, especially in the context of American cities, we think of our urban environments synonymously with the automobile. It is easy to forget that public transit has actually been an important part of the urban landscape in North American history, ever since the horse-drawn tram was popularized in the mid-1800s. Furthermore, automobile technology is actually considered the *second* transportation revolution in the nation's history. The first was the electric streetcar, also referred to as the street railway or the trolley. The first electric trolley line opened in 1888, and by the early 1890s, was the dominant mode of urban transit, replacing horse-drawn trams as the new form of transportation in cities. This technology extended the range and practical use of the transit that came before it, which in turn brought a large amount of new land into city commuting limits. Inevitably, the streetcar also went hand-in-hand with land speculation and urban development. Real estate developers built streetcar lines in order to capitalize on the new housing suburbs that they enabled. The streetcar was one of the first industrial technologies that also gave workers the ability to move away from the factories that they worked at. Transportation historian Sam Bass Warner in his book, *Streetcar Suburbs,* describes how streetcars formed the genesis of development-oriented transit, creating a two-part city: a city of work separated from a city of homes.[16] Trolley lines were built

16 Warner, *Streetcar Suburbs*, 1978.

1901 Streetcar suburb of Sawtelle in Los Angeles.

radially outward from city cores, sparking the first wave of outward expansion in many cities of that time. Streetcar-driven development was so prevalent in the U.S. that prior to 1916, the nation was documented to be the world's leader in transit rail miles, streetcar ridership, and many other transit metrics.

This position did not endure long though. Once mass-produced automobile technology arrived, most cities in the U.S. began disinvesting in public transit infrastructure during the Great Depression and World War II. With the affordability of the private automobile suited more for the cultural aspirations of the country's citizens (i.e. freedom of movement and the expression of that freedom), real estate developers realized that they no longer needed to predicate their land developments on the expansion of trolley lines. Without this financial underpinning, along with the dominant rise of Fordism and the expansion of the nation's Interstate Highway System, the streetcar quickly disappeared, leaving but few physical traces of its prior existence in cities.[17] For some cities, public management of city transit systems switched principally towards bus networks, which along with the proliferation of the automobile, gave priority to roads designed and built for the auto-vehicle. In the U.S., national disinvestment in public transit as a whole left many cities and areas vastly underserved by affordable transportation options.

17 Fordism describes the industrial and manufacturing system that served as the basis of modern social and labor economic systems supporting mass production and consumption of goods; it was pioneered in the early 20th century by the Ford Motor Company through the proliferation of the automobile.

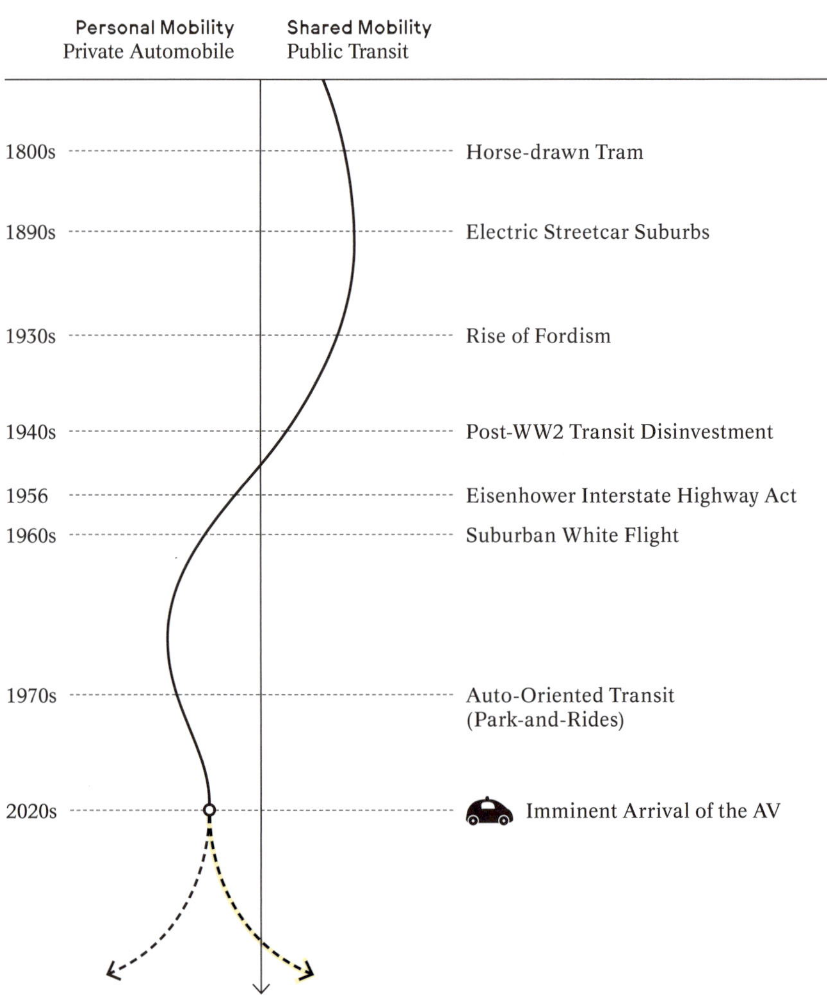

However, the major transit disinvestment that happened in the U.S. during the automobile's rise didn't occur correspondingly all around the nation nor all around the world. While all cities were invariably affected by the use and spread of automobiles, many cities also invested in public transit following the general disappearance of the streetcar. Once the electrification of rapid transportation systems occurred, some cities began tunneling their railway systems, forming the basis of many modern-day light rail and subway systems. Cities also began recognizing the deleterious effects that automobile-dependency was incurring on its urban environments. Today, subway expansions in cities like Tokyo or Moscow, refurbishment of outdated systems in cities like Paris, newly built subway systems in Mexico City or Hong Kong, and even American cities like New York City have modern transit systems that serve as a critical backbone for their urban mobility and metropolitan structures. Other cities, like Curitiba, invested heavily into bus rapid transit systems (BRT) which was the first in the world at its time. The eventual success of BRT in Curitiba influenced the implementation of express bus systems in many Latin America cities and others around the globe.

What kind of spatial externalities do transit-rich cities benefit from, in comparison to their automobile-dependent counterparts? Cities that have urban cores considered more vibrant, walkable, and densely packed with a diversity of land uses and activities, are not coincidentally, cities that have *not* catered their urban fabrics solely to the automobile. Built environments are produced from a cyclical feedback loop with the density of their land uses and their mobility mode shares. Areas serviced by higher public transit and walking/biking modal shares exhibit denser retail patterns and land uses (more shops per square mile) to service them. Denser retail patterns lead to higher population densities overall (more residents and jobs per square mile), which results in more compact and mixed-use urban form that house those densities. Higher densities contribute to shorter average trips taken, as more land-uses and destinations are in closer proximity to one's home or work. Short trip distances incentivize higher walking and transit mode shares, which completes this circular feedback loop.[18]

18 Sevtsuk, *Street Commerce*, 2020.

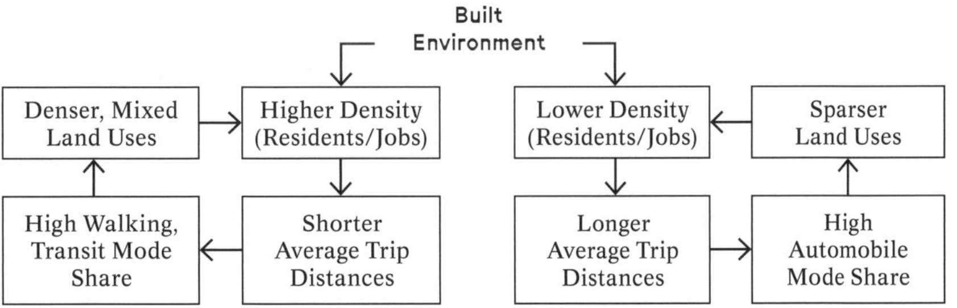

In contrast, cities with high automobile dependency exhibit the inverse circular relationship. Cities that exhibit lower densities and less compact land uses require longer average trip distances to reach, resulting in higher automobile mode shares. In turn, the urban form of these cities caters to these spatial tendencies. In Los Angeles, for example, its urban form commonly retreats from the street grid and the sidewalk, as a result of its dependency on high automobile use and the easily accessible and vast parking lots required to support parking them in less dense areas of the city. New York, on the other hand, exhibits the opposite spatial trait, in which more mixed-use, densely-packed urban forms front the street grid and sidewalks, activated by—and in turn supporting—higher walking and transit mode proximities. Meanwhile, businesses and other land uses don't require expansive lots for parking to support their economic success. Public transit also contributes to, and is typically supported by, walking or cycling modes in the first and last legs of a trip. Taken together, these effects result in cities that invariably condition more vibrant, activated, and diverse public realms. The recognition of these cyclical relationships and the benefits that multi-modal transportation brings to urban environments has led to contemporary movements today championing the walkable city. Contemporary urban planning movements like the 15-Minute City and New Urbanism are based on these policy and design principles of pedestrian-friendly, transit-oriented environments with the goal to produce denser, mixed-use, and accessible neighborhoods.[19]

19 The 15-Minute City is an urban planning concept which describes an approach where the majority of one's daily necessities and services can be easily reached by a 15-minute walk, bike ride, or public transit ride from any point in the city.
New Urbanism is a contemporary planning approach based on the principles of how cities and towns had been historically built—with walkable blocks and streets, housing and shopping in close proximity, accessible public spaces, and human-scaled urban design.

Transit-based cities also exhibit far more environmentally sustainable metrics as compared to their automobile-dependent counterparts. Transit, when well-utilized, produces far less greenhouse gas emissions and lower-energy requirements on a per passenger basis. Because emissions from internal-combustion engines are proportional to its gasoline fuel consumption, a full bus, for example, will produce less pollution per person-trip than an automobile. The average carbon emissions by transport type—measured by passenger kilometers traveled and incorporating their embodied manufacturing, roadway, maintenance, and direct/indirect operation—is far more costly by gasoline-powered cars (at 210 grams) than it is for buses (33–38 grams), light rail (11–25 grams), or bikes (8 grams). Operationally, some modes of transit that are entirely electric can dramatically reduce carbon emissions generated per vehicle, per trip, making their initial carbon investment in production and manufacturing pay off long-term. Transit's spatial

Carbon Emissions by Mode
Average in Grams per Passenger-Kilometer

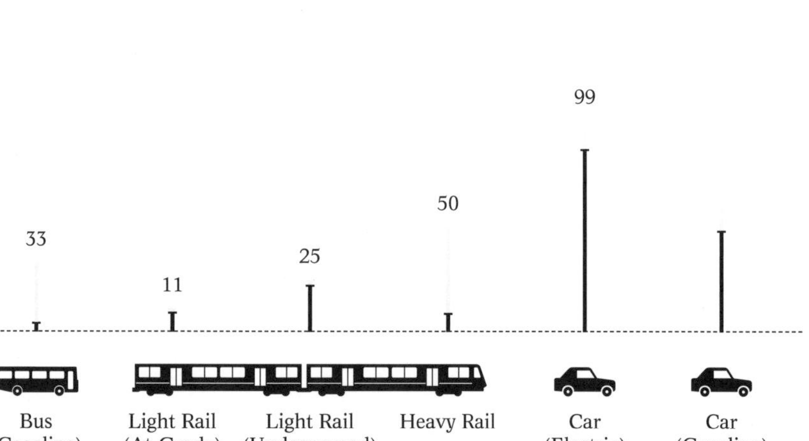

Vehicle Manufacture & Disposal, Maintenance, Roadway   Operation (Direct/Indirect)

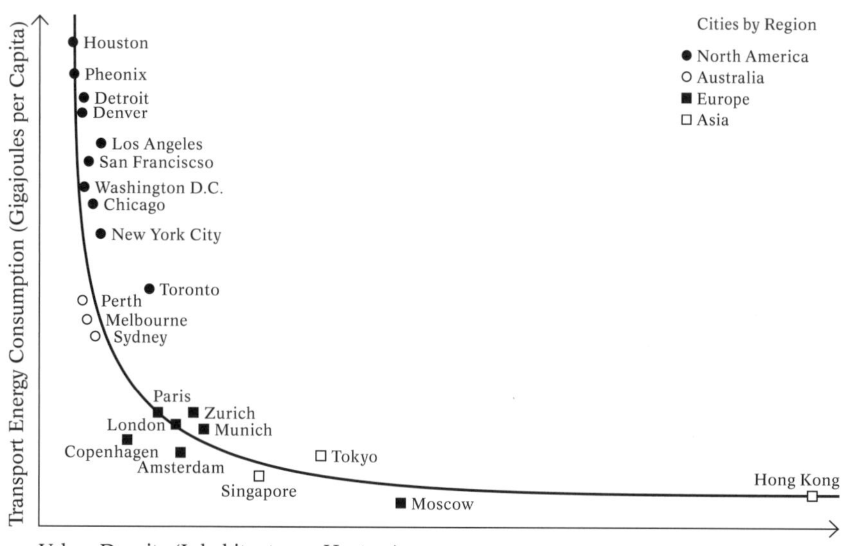

efficiencies, along with its contribution to reducing parking demands and denser land uses in urban cores, means that less land consumption is required to support the same population density as compared to auto-dependent transportation. Altogether, these factors contribute to the fact that transit-rich cities exhibit far less transport energy consumption than their automobile-dependent counterparts overall. As shown in the graph above, cities with the policies, economies, and cultural behaviors that support mass transit use also support far denser inhabitants per hectare, which is inversely correlated with their transportation energy consumption per capita.[20] While there are certainly many other factors that contribute to a city's energy consumption and its environmental sustainability, it is clear that mass transit has a critical role in contributing to those efforts.

20 Kenworthy and Newman, *Cities and Automobile Dependence*, 1989.

These cumulative benefits can be traced to the fact that transit services more passengers in far greater capacities in far less space. When operating near peak capacities, buses, BRT, light rail, and subways can facilitate the passage of anywhere from 9,000 to 40,000+ passengers per hour, per direction, per single 10-foot width-lane of road space. Meanwhile, single-occupancy cars carry only around 2,000 passengers in the same amount of time and space. Let's flip this equation for an even more dramatic spatial representation of this difference. In order to carry the same number of passengers, a single width-lane of a subway or light rail system, or around 4 to 5 width-lanes of equivalent capacity buses, would require upwards of 20 lanes of cars. As a result, the spatial formation of much of our urban environments has adapted to all that extra space demanded by the car, both during the trip, as demonstrated in the diagrams on the next page, but also before and after dedicating vast spaces for parking.

The discussion in this section therefore illustrates the underlying stance with which this book positions itself, that urban growth predicated on hierarchical multi-modal transportation—from rail to walking—produces better urban environments than those that are predicated on the flattened transportation choices of the private automobile. If prioritized, multi-modal transportation made up of a range of public transit options make cities more sustainable, democratic, and vibrant places to move around in.

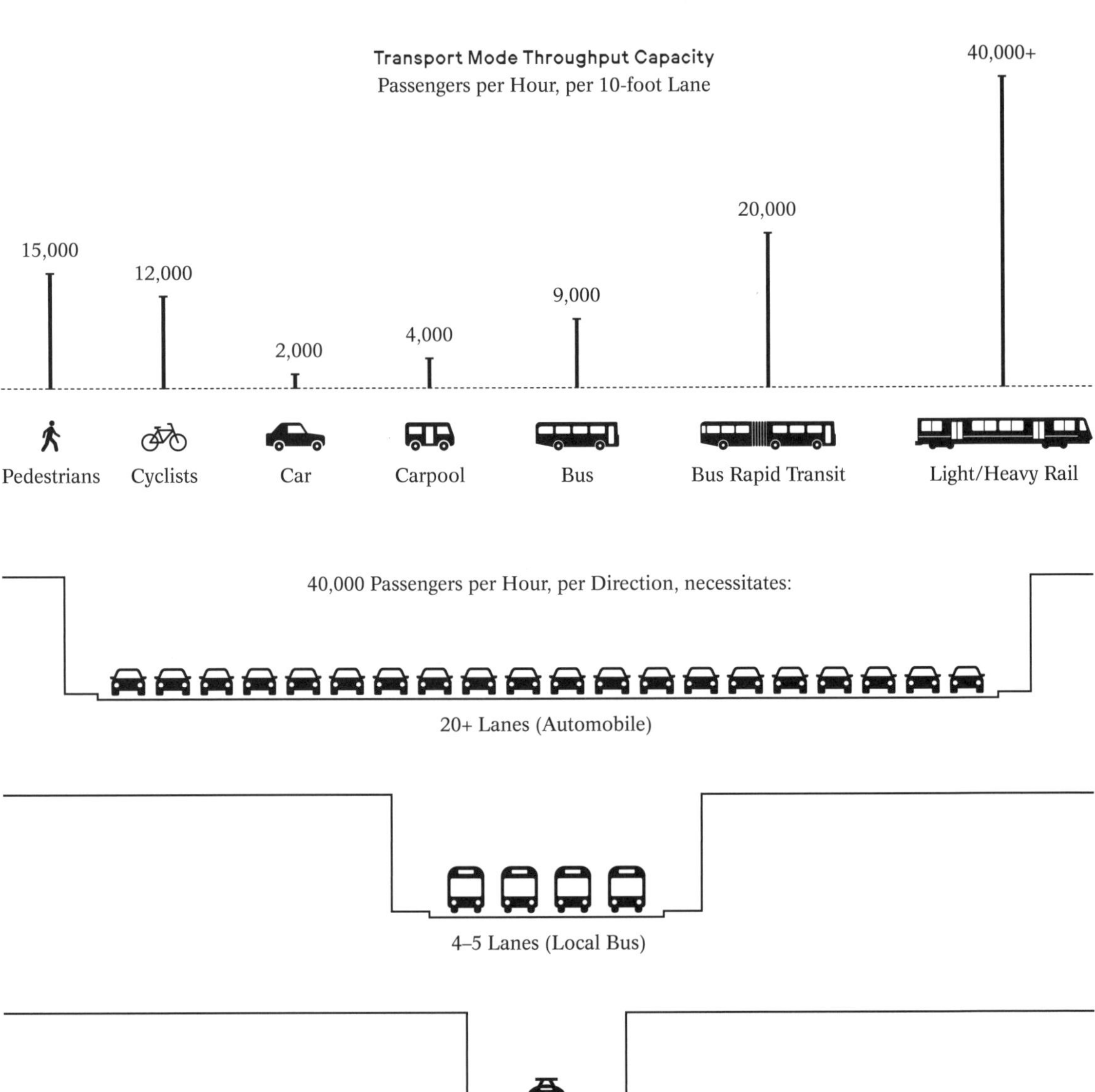
Transport Mode Throughput Capacity
Passengers per Hour, per 10-foot Lane
15,000
12,000
2,000
4,000
9,000
20,000
40,000+
Pedestrians
Cyclists
Car
Carpool
Bus
Bus Rapid Transit
Light/Heavy Rail
40,000 Passengers per Hour, per Direction, necessitates:
20+ Lanes (Automobile)
4–5 Lanes (Local Bus)
1 Lane (Light Rail)

Despite the documented benefits that transit-enabled urbanism exhibits, there are particular challenges that cities often face in implementing it. In many cities, it is very difficult for transit networks to service the entire metropolitan reach of the city, especially in areas that are of lower population density. Geographic, economic, and financial factors contribute to these challenges. Travel time is often one of the cited barriers inhibiting public transit use. Operationally, fixed transit schedules, fixed routes, frequency of service, and transfer times contribute to the fact that overall, transit use today takes on average 1.4 to 2.6 times longer than driving a car, even despite the documented traffic congestion and time wasted metrics faced by car commuters.[21] Buses also encounter traffic congestion, as many cities do not dedicate any of their street right-of-ways to prioritize their efficient movement. Cultural and behavioral attitudes towards transit are also significant barriers. Some cities and societies have historically prioritized the automobile not only through public policies, funding priorities, and urban design, but also through media and advertisements. This has led to cultural associations of public transit with those of less means. In the U.S., the automobile is a status symbol of wealth, individualism, and freedom—something that has historically permeated into the cultural mindset. This association with class and status leads households to purchase a car as soon as they are economically able to. All of these factors contribute to the fact that commutes today are still are dominated by car-use despite the documented spatial and environmental benefits of transit. Roughly 73% of Americans use an individual car for their commute while only 13% elect to regularly use public transportation.[22]

21 Liao et al., "Disparities in Travel Times between Car and Transit," *Scientific Reports*, 2020.

22 Richter, "Cars Still Dominate the American Commute," *Statista*, 2023.

Driverless mobility has the latent potential to address many of these challenges. One of the key differences between the previously discussed Scenarios A, B, and C in this chapter are that driverless vehicles in our ideal Scenario C are deployed to *support* and *work with* transit systems, rather than compete with them. In Scenarios A and B, either privately owned driverless cars or privately-consumed automated TNCs have proliferated at a scale of use that is irrespective of the transportation hierarchies of transit. Shown on the diagram series in the following pages, Scenarios A and B exhibit how the personal automated vehicle is used for every single trip irrespective of trip distance. Solely automating a personal automobile will not replace the long-distance efficiencies (per passenger, per square foot) of high-capacity mass transit transportation modes. As a result, the individual personal vehicle imposes the same spatial externalities on the built environment for every length of trip, from longer-distance commutes to short-distance errands.

Alternatively, shared mobility services move passengers at greater urban space-efficiencies, pairing trip length with vehicle capacity by grouping passengers traveling to similar geographic destinations together. Often, it is in this first and final trip leg, however, in which transit can fail to provide transportation services. One of the principal barriers to widespread transit-use are passenger's unwillingness and/or inability to walk this first and final leg past a half to one mile radius. This leg of a trip is called the "First/

Mobility Scalar Gaps

0-1 Mile
0-3 Miles
3+ Miles
5+ Miles
Shared Mobility
Personal Mobility
Private Automobile

Automation Fears

0-1 Mile
0-3 Miles
3+ Miles
5+ Miles
Shared Mobility
Personal Mobility
Private Automated Vehicle

Filling Mobility Gaps

0-1 Mile
0-3 Miles
3+ Miles
5+ Miles
Shared Mobility
Personal Mobility
Automated Mini-Shuttle

Reconceptualizing Bus Transit

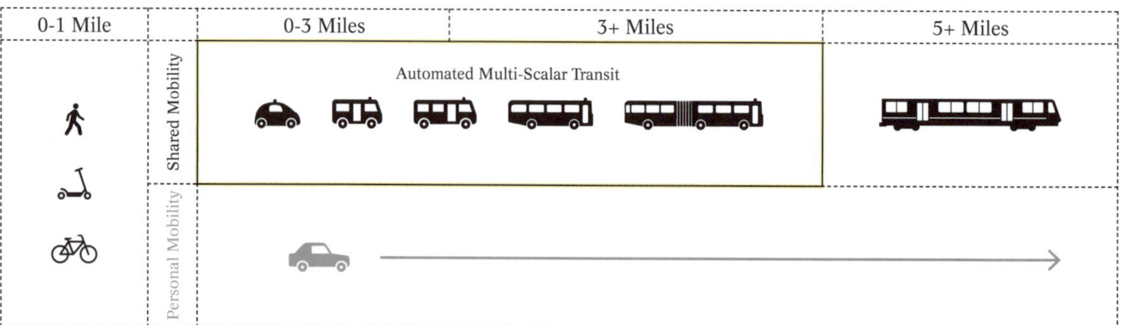

Last Mile Gap" in transportation terminology, which describes the first or final leg occurring both from one's home to a transit stop and from a transit stop to a final destination. The distance of this gap is further compounded by other factors, including the land use and urban design facilitating that walk, or cultural behaviors and social habits. When the first or final leg of a trip is filled with vibrant storefronts full of commercial uses, beautifully designed and shaded public spaces, or comfortably walkable sidewalks that service the pedestrian, the perceived length of this walk decreases. When that walking experience is filled with parking lots, driveways, and urban form that services the car, the opposite occurs. Today, transit services face operational and geographic challenges in servicing this gap, resulting in areas of cities that are not within walking distance of public transportation options. This often results in increased reliance on personal vehicles by trip-goers. Sometimes in cities, the first/last mile gap is serviced by micro-mobility options like personal bicycling or bike- and scooter-sharing programs. However, streets must be designed to facilitate their safe movement and these modes are also often unable to service those with mobility disabilities or the elderly.

A whole new range of driverless shared-mobility vehicles could be introduced to support the expansion of public transit and fill in its service gaps. To address the first/last mile, autonomous mini-shuttles at the scale of one to two passengers akin to today's "SmartCar", officially called the Smart Fortwo, could be implemented. These mini-shuttles could provide flexible and on-demand service in areas of the city that traditional transit service fails to reach. Mini-shuttle services could be adopted by public agencies or even TNCs that operate in close coordination and under the oversight of larger public transit services. AV mini-shuttles could expand the reach and range of this gap, in order to shuttle passengers to transportation hubs or higher-capacity transit stops that service high-capacity rail and subway lines.

This expanded network reach has implications on two development strategies that cities have used to increase transit ridership: transit-oriented development (TOD) and park-and-ride (PnR) station typologies. Transit-oriented development describes compact, walkable, and dense neighborhoods that are developed around transit stops in which the area of development is typically predicated on up to a mile walking radius, or 15-minute walk. Park-and-rides, on the other hand, describe the allocation of a high density of parking lot and parking garage facilities adjacent to transit stations, which are developed to serve a much wider area of typically suburban households. These households will drive their car and park at a PnR facility for a short commute that is just out of range of the first/last mile, and then switch to rail transit for a longer distance commute to their final destination. Both strategies are used to increase transit ridership in geographic areas of the city less typically served by transit, and both could be impacted by the arrival of the driverless vehicle. In PnRs, the vast swaths of parking lots and parking garages surrounding typical park-and-ride typologies could be rezoned and densified once automated transit shuttles replace the need for individually parked cars. This will have spatial ramifications on urban development and the geographic reach of PnRs, as well as reducing

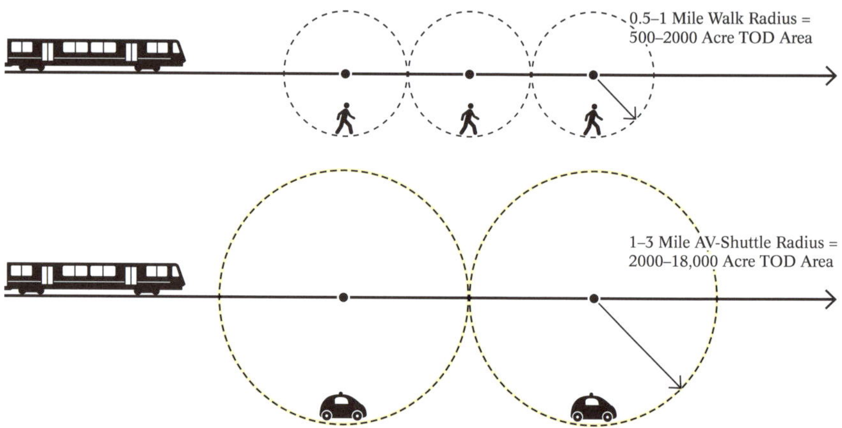

individual car use and the suburban development sprawl that PnRs are often linked with. In TODs, the radius of urban development serviced by new automated transit shuttles deployed around these stations could expand and densify the area of development past the initial first mile. As a result, the distance between future newly implemented TOD stations could increase, with implications on expanding the overall metropolitan reach of our transit network in cities. This type of service use and deployment in the first/last mile gap is already something that Waymo has tested in coordination with Phoenix, Arizona's Valley Metro transit system in recent years.[23] It demonstrates an important example of multiple transportation systems and service providers trying to complement, rather than disrupt or compete with each other and the infrastructures that support them.

23 Marshall, "Can Waymo Self-Driving Cars Help Fix Phoenix's Public Transit?" *Wired*, 2018.

Automation has implications to support transit beyond just the first/last mile gap as well, at the scale of local and rapid bus transit. As driving labor accounts for a significant portion of transit service costs, automating transit vehicles could significantly reduce these overheads. The resultant cost-savings could be put towards increasing transit service frequency (more vehicles per route, per hour) and geographic coverage (expanding routes). Bus transit could also increase its potential carrying capacity. AV platooning could chain multiple buses together safely and at high speeds for certain legs of trips that service overlapping routes while individually separating in others. This factor, in particular, has important implications for bus rapid transit service, which deploy articulated vehicles that comprise two or more rigid sections linked together to provide higher passenger capacities. Platooning multiple articulated buses together could begin to approach the carrying capacities of light right or subway, while still retaining the benefits of bus service. Bus rapid transit (BRT) is far cheaper to implement compared with the infrastructure and cost of building out light rail or subway. BRT would also retain its ability to change service lines, distances, and routes dynamically based on demand, while rail infrastructure once built is more or less fixed. Automated BRT could serve as important precursor lines to rail transit expansion, in which high-capacity BRT lines could test route expansions before cities invest in the fixed infrastructure of rail.

This type of BRT network has proven successful in cities like Curitiba, which were implemented as a system that gave buses many of the functional advantages of urban train systems as possible, while

Downtown Curitiba, bus rapid transit station.

dramatically reducing cost. In Curitiba, it was estimated that implementing its BRT system cost 50 times less than implementing rail transit. Integrated, dedicated bus lanes were designed along the city's main street arteries with stations placed on route medians in order to allow buses to run at speeds comparable to light rail. The implementation of BRT in Curitiba is also noteworthy as an example of a successful public-private partnership working together to provide city-wide transportation services in a coordinated effort. Private bus operators provide vehicles and operate them, while the city provides the infrastructure and regulatory oversight to facilitate their efficient use. Curitiba is also an example of a city in which transit-oriented development responds to its BRT arterial corridors, a testament to the carrying capacity and use of its network. Automating this type of service could accelerate its benefits, as well as acting as a catalyst for cities considering expanding their mass transit services.

Automation at multiple transit scales can also dynamically adjust live to meet changing consumer preferences on quicker timelines. A wider range of vehicle sizes, the ability for vehicles to be platooned, and an increase in service frequency and coverage could result in an overall system that is more agile and responsive than the public transit services of today. For example, nighttime service could be replaced by a higher proportion of on-demand vehicles, while peak commute hours could see a higher proportion of platooned vehicles servicing high-demand routes. Automated mass transit can become a viable alternative travel choice if its service is of high value to its consumers. Taken together, these changes could revolutionize public transit, creating more competitive value for the mobility consumer to rival individual car ownership.

The foundation of this ideal future Scenario C, is that all forms and scales of mobility work together in tandem. On-demand ride-hailing mobility plays a convenient role in servicing point-to-point travel when it is crucial, while shared transit plays a critical role in mobility space efficiency. These changes must go hand-in-hand with key transportation policies to incentivize these passenger modal choices. Policies like certificates of entitlements, transit subsidies, or high-occupancy vehicle lanes are a few examples, amongst others discussed further in Chapter 5, that lay the foundation for mass transit to become more widely implemented, accepted, and used for the majority of trips that we take. Other times when on-demand AV ride-hailing is needed, congestion pricing, transport zone, or curbside pick-up and drop-off fees could be levied based on the spatial cost they incur on the built environment.

In the end, the three future scenarios discussed in this chapter are not mutually exclusive but may push and pull at each other simultaneously. It may be that in the case of our city's urban cores, some combination of Scenarios B and C evolve, while Scenario A manifests in metropolitan peripheries like the suburbs. Today, public transit and TNCs dominate the market share in large, densely populated metro areas where space is contested and parking is expensive, while residents of suburban and rural areas still rely majority on private vehicle ownership. Ultimately, cities are made up of a complex web of spatial geographies that are influenced by economic, political, and cultural factors beyond just mobility. The point, however, is that these mobility choices have vast spatial impacts on the design of our built environments. Urban form shaped itself around streetcars and railway transit and was reshaped again when automobiles proliferated. If cities and their societies can collectively reset their priorities again, this time around automated transit, urban form will adapt yet again. If cities can limit the negative externalities of driverless technology, while taking advantage of their benefits to revolutionize transit, they could not only move their inhabitants around more efficiently but could also be designed in more democratic, sustainable, and vibrant ways.

The future of automated transit presented in Scenario C is exciting because of what is at stake. Supporting and expanding transit services will be increasingly important as driverless vehicles continue to be deployed on our city streets. What remains, is an urgent message for those whose decisions will impact the future transformation of our cities. In particular, public transit agencies must become far more agile, proactive, and visionary in integrating and steering autonomous mobility technologies. Otherwise, they will become the victim of growing competition pushing AVs to reinforce individual vehicle ownership and use.

AVs are not only a new technological trigger, but represent a new form of mobility infrastructure in this transportation struggle. Automobile and locomotive technologies were used as foundational

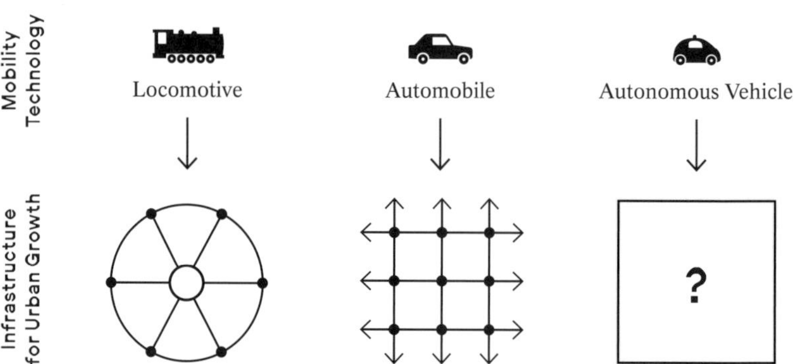

infrastructures for the planning of cities in their respective eras. They fueled new forms of urban growth, restructuring, and evolution—often with unintended impacts. AVs could become the next form of mobility infrastructure for urban growth, but only if they are deployed in spatially, socially, and environmentally beneficial ways. If done in this way though, car-dependent cities could shift their mobility cultures from the Autopias of today, towards the Transitopias of tomorrow.

4 Values, Responsibilities, and Technological Intentions

Guiding Values and Spatial Principles

Embedded in the varied discussions in this book, an underlying message highlighting the importance of establishing a guiding set of values and principles emerges. What constitutes a "better" future city? What makes cities more positive places to live and move around in? Who benefits from new technological innovations? What urban ambitions do we collectively aspire towards? This value proposition *must* underscore every decision we make, shaping how we go about making them. The more explicit they are, the more easily cities can evaluate their success and measure intervening benchmarks. Only by first determining our shared societal values, should technological tools be then deployed to push cities towards those particular futures.

Many of the values put forth in this book question the status quo principles that our cities currently operate under. For example, as a car-dependent culture we have chosen to "build around private modes of transportation, but with massive public investment in the infrastructure that allows those private uses."[1] This tradeoff reveals how we have astonishingly routinized and accepted that the primary right to use urban space and infrastructure is a right held by private car owners. Massive public tax funding towards maintenance, construction, and policing goes to advancing this use-right, while drivers themselves seldom pay for the spatial and fiscal road capacity costs they incur. Minute and indirect taxation on gasoline and car purchases cover only a fraction of these costs while also being disconnected from the value of the mobility infrastructure used. This societal value has been completely accepted, culturally proliferating in cities today. Meanwhile, the same proponents of car culture ironically condemn public transit as an extravagant waste of public funds, despite the fact that it cannot begin to rival the extent to which automobile use has been subsidized in cities. Those same proponents fight tooth-and-nail when it is suggested that we should toll congestion, reclaim street parking, or implement curbside management fees. That Motordom has, by and large, won the car culture war in our nation's history only serves to further highlight how important it is to question these status quo values.

1 Lutz and Fernandez, *Carjacked*, 2010.

Is the right to mobility in a city a public or private right? Who and what has the primary use-right to mobility infrastructures and public spaces? How should public-private mobility services and urban spaces be spatially distributed? And importantly, how should cities be designed to facilitate those priorities? These are not easy questions to answer, but ones that must be grappled with. As incoming AVs land on our shared public resources—urban space and infrastructure—we must define the rights that they and their users have in order to guide their deployment and absorption. These discussions must involve all disciplines and stakeholders, not just those currently advancing technological solutions.

Outlined in the subsequent spread, these principles range from broader mobility concepts to spatial implications for urban design and planning. They guide the typological explorations in the second half of this book, and the narrative experiences illustrated in Book 2, steering cities towards more spatially vibrant, multi-modal, human-centric, and environmentally sustainable urban futures.

Mobility Values

Move People Not Cars

Mobility infrastructure must be measured not by vehicle throughput capacity, but rather by passenger movement through space—which presents a far more complete picture of a mobility system's relationship to urban space. Additionally, mobility infrastructures must also facilitate important placemaking, social, and environmental uses in parallel to moving people and goods.

Technology is only a Tool

Driverless technologies are only a tool amongst a variety of other mobility, planning, policy, and design tools that cities must utilize. AVs are not a silver bullet to our urban mobility woes. They can be a powerful tool—and a tool that must not be ignored—but can only assist us in solving our own problems ourselves.

Consume Mobility as a Service

Mobility must transition primarily from ownership models to service models in order to free households from dependency on private vehicles and the expansive associated parking demands that come with car ownership. Automated mobility services must be more convenient and affordable then car ownership, offering multi-modal options for multi-trip needs and lengths.

Reconceptualize Mobility from End-to-End

Mobility services, and in particular transit, must facilitate passengers to their various destinations in more holistic ways. Mobility gaps, including the first/last mile trip legs, must be filled with more options. Multi-modal transport chains must integrate multiple legs of a trip together into a coordinated and unified journey—spatially, service-wise, and through digital platforms and apps.

Provide Diverse Mobility Options and in particular Transit

Use automation to expand the range of mobility options offered, including AV mini-shuttles serving one to two passengers all the way up to autonomous articulated bus rapid transit. Use automation to support and expand transit service, and in particular bus transit, at multiple scales. Coordinate trip lengths to vehicles size and occupancy: the longer the trip, the larger the vehicle used with more shared passengers in order to incur less spatial costs on valuable urban space and infrastructure.

Use Economic Incentives where Necessary

Along with expanding automated transit services and options, economic incentives and penalties can prove effective in inducing modal shift. Use policy, pricing fees, or subsidies to shift mobility behaviors in support of the spatial principles outlined on the next page. Fiscal incentives help passengers understand the spatial costs and externalities that are incurred on the urban space of cities, depending on what transportation modes they take.

Ensure Transit Equity and Affordability
Automated technologies must benefit not just those who can afford them. They must be used to expand public mobility service operations, access, and reach to transit deserts, occurring largely in socioeconomically disadvantaged areas. Incentives and subsidies must be provided to ensure that AVs serve all income ranges, ages (including seniors), abilities (including those that are mobility challenged), and demographics.

Spatial Principles

Prioritize the Pedestrian and the Public Realm
Any AV-enabled spatial transformations must prioritize the expansion and reclamation of the rights-of-ways in our cities for public uses—including shared forms of movement and pedestrian activities—rather than for expanding privatized individual vehicular use.

Catalyze Density and Land Use Changes
Mobility and urban form must be understood as in dialogue, where changes to transportation modes and methods catalyze changes to density, land uses, and spatial forms—and vice versa. Zoning and land use policies must work in tandem with the city's transportation network to allow for denser and more mixed-use urban forms.

Reclaim and Repurpose Parking as an Asset
Driverless vehicles will reduce demand for parking. Surface parking lots, street parking, and parking garages make up an inordinate percentage of the urban landscape. As this land becomes released for alternative uses, it must be treated as an important asset to improve and address contemporary urban issues facing cities, including expanding its public realm, introducing blue-green landscape systems, and densifying with affordable housing.

Explore Adaptive Reuse of Urban Form and Infrastructure
A plethora of existing auto-oriented buildings and infrastructures, from highways to drive-through restaurants, collectively hold large quantities of embodied carbon in their material and construction. Instead of being completely demolished and replaced, these urban typologies should be adaptively reconfigured for a driverless world as their uses change and adapt.

Be Vigilant with New Typological Inventions
The invention of the automobile gave rise to new urban typologies designed specifically around their dimensions, turning radii, and use. Any new spatial typologies that could emerge from driverless technology (like dynamic lane-managed streets) must not be solely designed around the technological and spatial limits of AVs but rather use technological advances to prioritize human-centric forms and spaces.

Integrate Urban and Environmental Sustainability

Electrification of AV fleets is essential to reducing carbon emissions and other pollutants produced by combustion-engine powered transport. However, the environmental externalities that come with the global production of electric batteries for electric vehicles—which include the extraction of lithium and the mining of cobalt—must also be reckoned with. Furthermore, AVs must encourage more sustainable forms of transport through shared or mass transit and less energy consumptive forms of urban development. Finally, blue-green landscape systems—including bioswales, rain gardens, and permeable surfaces—must be integrated into the design of our streets, public spaces, and buildings in order to help alleviate urban heat island effects, assist with stormwater management, and increase biodiversity in cities.

In particular, one of the mobility values listed in the prior section deserves an expanded discussion, as it has only been lightly touched on so far. Like many other technological disruptors that came before them, driverless vehicles carry with them the potential to exacerbate numerous social and equity issues in our built environments. Some of these historical injustices have been discussed in earlier chapters, which include the spatial practices of redlining, the devastation that highway construction levied on communities of color, and the unequal distribution of transportation services creating transit deserts in cities.

Current innovation driving driverless technology exists in the private sector. As such, their rollouts could mean that they become primarily available only to those who can afford them, with the dangerous potential to continue the lineage of 20th century urbanism in which private automobile transport has been prioritized over more affordable options. This results in cities that are easy to move around in if you can afford to own a car and difficult, or even dangerous, to do so for those who cannot afford or are unable to drive. Within car dependent cities, low-income households pay a disproportionate percentage of their total expenditures on transportation. In the U.S., it is estimated that the lowest earning 20% of the population spend on average almost one-third (29%) of their income on transportation costs.[2] Meanwhile, as Americans move from lower to higher income brackets, the portion of their spending on transportation decreases. This intuitively makes sense in the context of car-dominant cultures that in essence require low-income households to purchase a car for any hope of job access and economic livelihood. This is because the cost to purchase, own, maintain, insure, park, and fuel a car—on average of $9,700 annually—is a typically fixed cost and is far more expensive than the cost of a yearly transit pass.[3] Car ownership is an expensive endeavor, including the fact that automobiles are a depreciating asset with a monetary value that decreases with more use. For low-income households with few other transportation options besides purchasing a car, a huge percentage of their income must go towards doing so. Countries in which transit is a viable travel option exhibit the inverse relationship. In European countries for example, the poorest 20% of its population pay around 7% of their income on transportation. This percentage actually rises as household income increases as well.[4] This is because firstly, more affordable transport options like transit are viable travel options for lower income households, and secondly, policies that make driving a more expensive endeavor exist in more abundance—parking is more expensive and tolls roads abound. The surprising finding is that despite this fact, transportation expenditure is still on average less expensive overall in Europe (11%), than for American households (13%).[5] This can only be the case because the various fees and taxes associated with driving in European cities are an important funding source that is allocated towards more affordable and sustainable transportation options rather than just driving up the cost of mobility altogether.

Parking allocation is another example of unequal redistribution of mobility resources. Despite the fact that car owners tend to be middle to upper class, free street parking has been made available

2 "The High Cost of Transportation," Institute for Transportation & Development Policy, 2019.

3 Ibid.

4 Ibid.

5 Ibid.

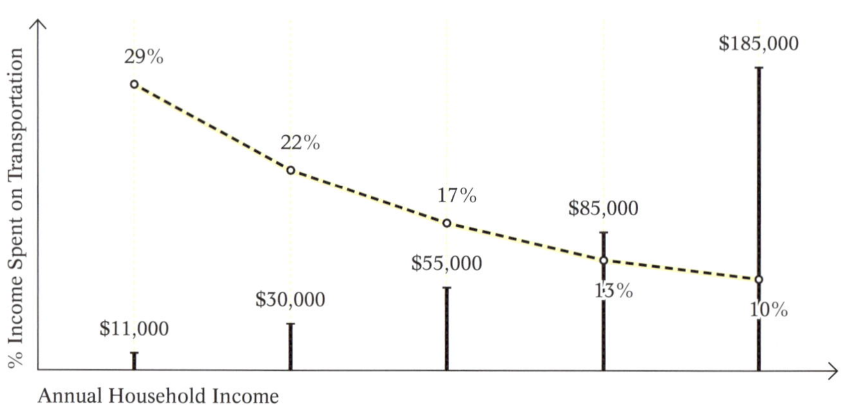

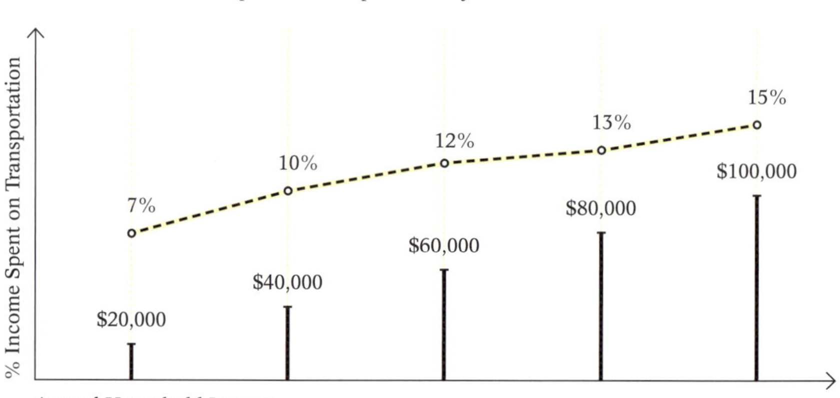

to them. A sizeable percentage of the use-right of our public streets is essentially allocated to storing private cars. The construction and maintenance of street parking is subsidized by a larger tax base, many of whom do not drive. This is especially acute in cities like New York, in which the majority of residents (54%) do not own a car but pay taxes that subsidize free street parking for the other residents who do. Pricing free street parking, and various other policy levers outlined in the next chapter, can be a way to redistribute mobility and urban space resources more democratically. However, this can only occur if the funds gained from them are reinvested in diversifying the modal options and price points available to all citizens.

Transportation equity involves not just defining which social or economic groups receive capital investments and resources, but also which geographies in cities benefit. Urban car dependency has led to spatial, zoning, and land use practices that also reinforce unequal distribution of access and resources. For example, city zoning laws began as seemingly simple rules that dictate where and what land uses can be built in cities. When automobiles proliferated, they formed a symbiotic relationship with single-family land use and zoning practices in cities, resulting in tools that have become racially weaponized. Because single-family zoning prohibits the construction of higher-density multi-family apartment complexes and dwellings, the practice contributes to urban sprawl and rising home/land ownership costs which reinforce dependency on cars and those able to afford them. It has been used as a tool to keep particular neighborhoods more homogenous and unaffordable by pricing lower-income socioeconomic groups out. These geographic inequalities also play themselves out in more urbanized areas, with more affluent areas of cities receiving expanded transportation and infrastructure investments, improvements, and maintenance over others. The racist and now illegal spatial practices of redlining, which occurred in over 200 cities in the 1930s by the Home Owners' Loan Corporation, is a mobility-enabled example of this. Meanwhile, other less socioeconomically affluent areas are left to contend with localized negative externalities like increased pollution and health hazards, to lowered property values when highways and larger arterial thoroughfares connecting more affluent suburban developments to urban centers were constructed through the heart of their neighborhoods. Additionally, these communities routinely share a history of underinvestment in street infrastructure like lighting, crosswalks, tree plantings, and sidewalks. Studies show that pedestrians and cyclists of color were more than 1.5 to 2 times more likely to be struck and killed by a vehicle as those who live in predominantly white neighborhoods.[6] This is because there are a higher

6 Susaneck, "American Road Deaths Shown an Alarming Racial Gap," *The New York Times*, 2023.

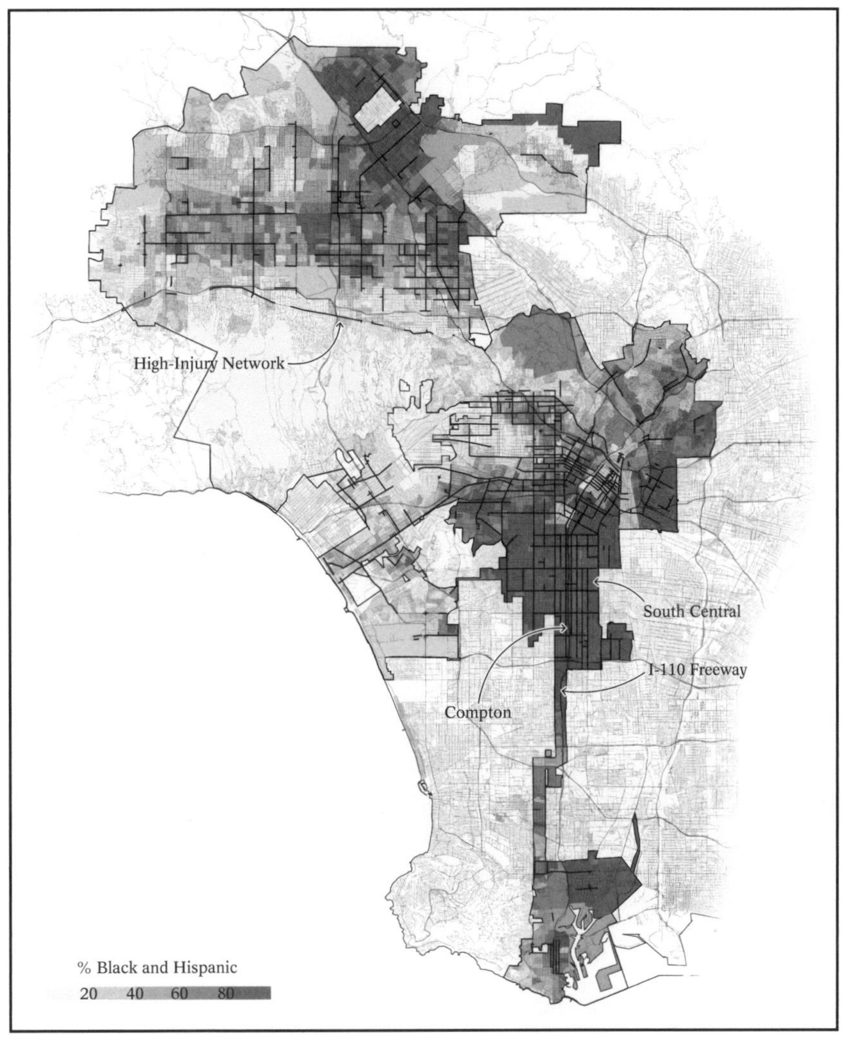

Map of Los Angeles highlighting streets with the highest traffic deaths and injuries, overlaid on demographic percentage of Black and Hispanic residents.

proportion of high-injury network streets in communities of color, especially with high-speed frontage roads that abut highways.

How will driverless vehicles impact these issues? The technological switch of automating a car in-and-of-itself, does nothing to address these spatial equity concerns. However, automation has the potential to bring a great deal of cost savings to the mobility industry, even when taking into consideration larger upfront manufacturing and technological expenses. As previously mentioned, vehicle operation labor is the largest cost associated with mobility services including both ride-hailing and public transit. What these cost savings go towards and how they are reutilized will go a long way in determining the technology's ability to address transportation inequities. If AVs replace the non-automated vehicles of today, with private service companies pocketing profits to reinforce their business models, then many of the aforementioned spatial inequalities will continue to be reinforced. One is already seeing this trend with current private MaaS providers like Uber and Lyft, which are at a price point that are typically only affordable to households of higher means. Additionally, studies have shown that ride-share users pay more when traveling to and from low-income and minority neighborhoods, due to dynamic or surge pricing machine algorithms that individualize all pricing for riders based on geography, location, traffic, and other conditions.[7] These disconcerting discoveries demonstrate how hyper-aware we must be of the ways in which our technologies can amplify societal prejudices, intentional or not.

7 Steinhardt, "Ride-share Users Pay More in Low-Income and Minority Neighborhoods," *George Washington University Today*, 2020.

However, if AVs and their associated cost savings are used to reinvest, update, and expand multi-modal options and shared transit, then they could become a useful tool for targeting populations in transit deserts—those in dire need of affordable and accessible mobility. Historically socioeconomically disadvantaged communities could also benefit from expanded fare subsidy programs that further lower the cost barriers to using public transit. Just as much as our private transportation modes have been given free use-rights in our cities, transit could be fully subsidized and fare-free all-together. Income from transit fares typically make up a minority percentage of transit operations, which could be compensated for by the cost savings brought by AVs and the various pricing strategies outlined in the next chapter. This could excitingly demonstrate a reprioritization of societal transportation values, something that entire countries, like Luxembourg and Malta, have already tested and implemented. These potential benefits would be in addition to the drastic reduction in traffic fatalities that removing human driving error will bring, as well as the potential aid they could provide in expanding para-transit service to mobility-disadvantaged populations (seniors, the visually impaired, or those with physical mobility challenges).[8] In particular, it is often these populations that are disproportionately affected by the first/last mile gap in transportation. If automated transit were to encompass a broader service range and geography, particularly targeting transit deserts and their first/last mile gaps, these populations could be better addressed.

8 Para-transit describes door-to-door transportation services provided for people who are mobility challenged and are therefore unable to use fixed-route transit or drive themselves in individual cars.

Labor is another large concern of many AV-skeptics, with driverless technology's ability to replace a human driver. This function may be of relief to the average commuter, but it poses a threat to the economic livelihood of vehicle operators in transit and MaaS.

Additionally, the many jobs that the trucking and freight industries employ will also be dramatically impacted. Automation also has a distinct potential to impact the delivery and movement of goods in inter-urban freight, especially since the technology is already ready to deploy on less complex driving environments including regional interstate highways so critical to shipping routes. Representing almost 3% of the American workforce, many of the drivers in trucking, bus transit, and ride-hailing are disproportionally overrepresented by people of color, with the risk of their jobs being affected.

While the comprehensive topical scope of labor equity can be found in other expert literature, there are a few brief points to highlight. Firstly, that for freight, the last-mile package delivery—in particular the last 50 feet—are likely to still remain in human hands due to the complexities involved in that space. This includes complex loading and unloading of goods, and other tasks that require delivery people to enter and exit buildings to ensure safe package handoff. There may be instances where new typological forms evolve, in which mail and package delivery kiosks become integrated street-side so that AV delivery vehicles can deposit

Last-mile package delivery by local freight truck in NYC.

packages directly for later pickup by owners, similar to Amazon Locker's service kiosks. This spatial transformation is explored in some of the typologies illustrated in the latter half of this book. However, by and large, existing urban form and fabric may still require humans for last-mile delivery.

The second point to mention, is that while automated transit will certainly affect bus and shuttle driving jobs, instead of being lost to automation, these jobs could transition into new service roles. In the case of transit, it is important to understand the role bus drivers play. They are required to not only drive but to also oversee vehicle and passenger safety while providing passenger services like route directions, assistance to the elderly and disabled, and others. These are taxing, multi-tasking roles which can inevitably fall to the way-side when drivers need to concentrate on navigating large vehicles in complex urban environments. In an automated future, bus drivers could be freed of their driving tasks to instead concentrate on passenger service and safety. This would be a critical component to expanding the service, convenience, and accessibility of transit overall in cities, which could also address concerns of job loss in labor. The experiential aspect of this new role, is excitingly explored in Book 2 of this two-volume publication.

It is important to note that there will also be a vast number of jobs that the expanding AV industry will create. In addition to various new technology innovation roles, vehicle operator jobs will also need to transition from direct to support operations. Even without a driver, AVs will still need humans to remotely oversee and support their operation, particularly in the event of emergencies. In an ideal automated future, expansion of transit operations and geographies will also create new jobs through service growth. While it may still be too early to determine the overall effect that AVs will have on labor, it is clear that they will have an impact. This is a larger conversation that is continually being had with all technological advances and service automations, mobility included. Cities must engage with various labor unions and transit workforces to examine where and how AV technology will impact our societies and economies, in order to prepare these workers to transition to the new roles and needs of tomorrow.

Lastly, who gets access to all of the new infrastructure investments and spatial transformations explored in this book? Cities, stakeholders, and citizens must make sure that any technological catalysts begin with those who were adversely impacted by the technologies of our prior eras. This is why the adaptive transformation of our urban highway infrastructures, explored in Chapter 8, are especially exciting. The benefits of these imagined spatial changes begin in the heart of the very same communities that were socially, economically, and physically devastated by highway construction in the first place. Communities that have turned their back on 20th century mobility infrastructures, could now benefit directly from AV-enabled urban greenways, transit corridors, and reclaimed public spaces of new 21st century highway reclamation projects.

Transportation is one of the biggest—and often one of the most commonly overlooked—tools that must be used to address equity in cities. Conventional transit planning will often consider vehicle operating costs and transit fares, but neglect to take a more holistic view of mobility overall, including the spatial, geographic, and fiscal opportunities that multi-modality and technologies could provide. As a society, we must be especially attune to the spatial inequalities that AVs could not only reinforce but even create. Instead, driverless technologies must be used to do the opposite, addressing the many existing geographic and service inequities faced by our urban environments. This starts with designing cities not for technology, nor for economics, but for people—and importantly defining who those people are as our driverless urban futures unfold.

What Disciplines Govern Design of Urban Mobility and Space?

As has been discussed in the preceding chapters, our built environments are highly contested spaces. This contestation occurs not only in defining use-rights and functionalities, but also in the varied disciplines that claim design autonomy over it. It is this very interrelation between professions and outcomes, that one can trace why our urban spaces—and in particular our mobility infrastructures—have resulted in their current configurations today. This is an important discussion to have because only by doing so, can we question and redefine the disciplinary conversations driving driverless technologies and their spatial impacts. It also highlights disciplinary barriers to potential solutions and a subsequent call for interdisciplinary blurring.

The design of urban mobility and space as a whole has always been in flux, especially in our modern history. While traditionally falling under the purview of city planning, and later transportation planning when that discipline was more formally established, other professions including architecture, urban design, landscape architecture, and engineering have at various points in time lay claim over the design of our urban spaces and streets.

The modern discipline of planning itself can be dated to the National Conference on City Planning held in 1909 in Washington D.C. Governance over planning the city has of course existed ever since the formation of cities themselves in ancient times. However, it was at the turn of the 20th century that the profession more formally defined itself and delineated its organizational role in configuring the relationships between transportation, land use, zoning, and others. In the U.S., various professional organizations like the American Planning Association or the Regional Planning Association of America were formed to define autonomy, lay claim, and outline the responsibilities of the discipline. At various points in history, influential urban planners have dramatically shaped urban mobility, ranging from Georges-Eugène Haussmann's plan for Paris in the late 1800s, Ebenezer Howard's garden city movement in the early 1900s, to Robert Moses' various infrastructural undertakings in mid-19th century New York, amongst numerous others. Wielding vast influence—often to deleterious ends when concentrated in the hands of select individuals—urban planning distinctly shapes the experience of moving through cities.

Given the highly complex and comprehensive nature of the profession, various parallel and sub-disciplines have emerged in the years since. Transportation planning—of which its origins can be found around the 1940s and '50s when the National Committee on Urban Transportation was created—arose once it was apparent that relationships between mobility and land use could be measured and therefore projective. Transportation planning in the modern era is directly responsible for designing, overseeing, and implementing the transport systems of cities, including its roads and associated infrastructures. Shortly after this profession was defined, the discipline of urban design was also birthed in a 1956 founding conference at Harvard University's Graduate School of Design. Organizer and Dean of the school, José Luis Sert, stated that the field "is that part of city planning which deals with the physical form of the city" and that its purpose was to “find the common basis for the joint work of the Architect, the Landscape Architect, and the City Planner... Urban Design [being] wider than the scope of these three professions.”[9] Authors Kreiger and Saunders capture and synthesize the origins of the profession in *Urban Design* (2009), charging the discipline to address “both the pragmatic level of calibrating demands for mobility with other social needs and in advancing new ways in which city form and transportation systems may be integrated.”[10]

9 Krieger and Saunders, *Urban Design*, 2009.

10 Ibid.

At various points in our urban histories, architects have also laid a strong and contentious claim to the design of our mobility networks. As discussed in Chapter 2, the era of Modernism in our history was a particularly influential moment for architects and their urban plans that included the organization of transport systems. These large-scale visions range from Le Corbusier’s various urban proposals including *Ville Radieuse* and *Ville Contemporaine*, to Frank Lloyd Wright’s *Broadacre City* and many others. In particular, modernist utopias that enabled, and were enabled by, the advent of automobile technology lay comprehensive claim to not only urban systems and form, but also lifestyle and societal ambitions. They had wide-ranging effects on subsequent attempted implementations in cities, plans, and proposals, of which much criticism has been levied in contemporary conversations. This criticism is discussed in more depth in the first chapter of Book 2. Due in part to these critiques, the profession of architecture has largely withdrawn from laying claim to urban space, infrastructure, and mobility in recent years, preoccupying itself more with the defining of form and buildings. In the void that was left, the professions of urban design and landscape architecture have staked larger claims to that space.

The landscape discipline has always had a hand in the design of urban mobility and public space as well, and in particular our streets and associated landscapes. Ranging from influential landscape architect Frederick Law Olmsted who designed parkways and boulevards in the 19th century, to urban planner Ebenezer Howard who’s influential 1902 treatise *Garden Cities of Tomorrow* advocated for the reintegration of the garden landscapes into city design, landscape architecture has always played a critical role in the theories and practices of urban space and infrastructure. In recent years, the discipline under the framework of landscape urbanism, defined by figures including practitioner James Corner and academic Charles Waldheim, has staked even more expansive claims to the space, arguing that its ability to address ecological and environmental issues in a time of great environmental crises be fore-fronted.[11,12]

11 Corner, *Recovering Landscape*, 1999.

12 Waldheim, *The Landscape Urbanism Reader*, 2006.

Lastly, but arguably one of the most influential, the engineering professions have had one of the most prominent roles in dictating and designing urban mobility infrastructure, especially in recent years. The history of engineers’ prominent voice in this space can be traced to the tradition of requiring our military and its associated engineering lineages to design and build infrastructure. In the U.S., the Army Corps of Engineers is an influential group reporting directly to our military. Since its founding at the turn of the 19th century, it has been charged to deliver vital “public and military engineering services... to strengthen our nation’s security, energize the economy, and reduce risks from disasters.”[13] The Army Corp of Engineers has been responsible for not only a variety of flood control projects in

13 “A Brief History,” The U.S. Army Corps of Engineers, 2023.

our nation including dams, reservoirs, canals, levees, and sea walls but also other large-scale public works projects including the design and construction of bridges, roads, highways, and civic infrastructure. They have been granted vast disciplinary weight and power in designing and shaping cities in the 20th century, to much criticism in recent years. This is because applying *only* a techno-engineering lens to the complex and contested design of urban infrastructure means that concerns of efficiency, functionality, and technology become primary drivers. Meanwhile, social, cultural, ecological, and environmental concerns are secondarily relegated or even unaddressed altogether. This type of thinking, if not conducted in dialogue with other professions, can result in the proliferation of mono-functional infrastructures in our cities—infrastructures which serve a single use at the expense of all other uses. Their outcomes and impacts are a result of prioritizing one disciplinary voice over others that also lay claim to this space. This is a trait for which much criticism has been voiced towards as we contend with the modern-day environmental, ecological, and social ramifications of techno-engineering projects. The engineering professions, including the Army Corps of Engineers, continue to play a prominent role in contemporary city-wide infrastructural projects and urbanization at large.

Within these multi-disciplinary claims and contestations, the case of the street is a prime example of how disciplinary autonomies have torn its configuration and use apart. The historic urban street can holistically be thought of as being composed of three overlapping elements: movement, place, and form. Movement, of course, describes the street's vital function in moving people and goods. Place, describes the activities and open spaces that take place on a street including the destinations that people travel to. Form, describes the architecture, buildings, and frontages that give physical shape to the edge of streets, defining its aesthetic and spatial character while housing destinations or shaping open space. Stephen Marshall, in *Streets and Patterns*, describes that traditionally these three elements can be loosely correlated with the disciplinary concerns of the transport engineer (movement), the landscape architect (place or open space), and the architect (form).[14] These three elements and professions are always in symbiotic negotiation and theoretically work in tandem to create vibrant and active streetscapes.

14 Marshall, *Streets & Patterns*, 2005.

However, since the advent of the modern road in the early 1900s, the transport engineer was the discipline that was called in to "solve" the various safety issues that plagued them when high-speed automobiles were introduced. Their solutions—removing pedestrians and other uses from its primary use-right, introducing curbs, crosswalks, traffic signals, and pedestrian over- or under-passes—have primarily driven street design principles and the prioritization of "movement" over the other equally important elements of "place" and "form." Rather than being locked together, modern street paradigms have allowed roads to follow their own fluid geometries to prioritize automobile speed. Meanwhile, urban form was allowed to be expressed as divergent forms set back from engaging with the street edge, bereft of the responsibility to condition place and open space. Marshall describes how this essentially amounted to "a schism between the treatment of roads as movement channels,

The street as a confluence of three overlapping elements: (1) movement, (2) place, (3) form.

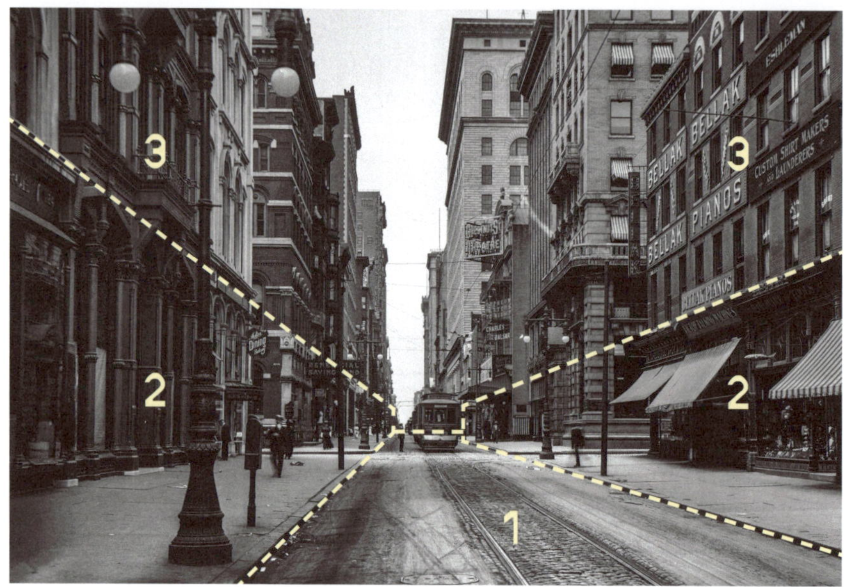

An urban highway, which only serves as a movement channel (1).

15 Ibid.

16 Ibid.

and the treatment of buildings and public space."[15] This schism also became apparent in the disciplinary roles and divisions of labor in modern street design. Road layouts, uses, efficiencies, and safety concerns became the primary domain of the transport engineer, while architects and landscape architects were designated to design only buildings and parks. Extrapolated to an extreme end, "optimizing" a road from a techno-engineering perspective means shedding it of intersections, direct frontage access, and place-making in order to segregate pedestrians and motor-traffic. This extrapolation precisely describes how our vast highway networks have come to exist in design and function—or as Marshall describes a street that has "dismembered it's body and evacuated it of its soul."[16]

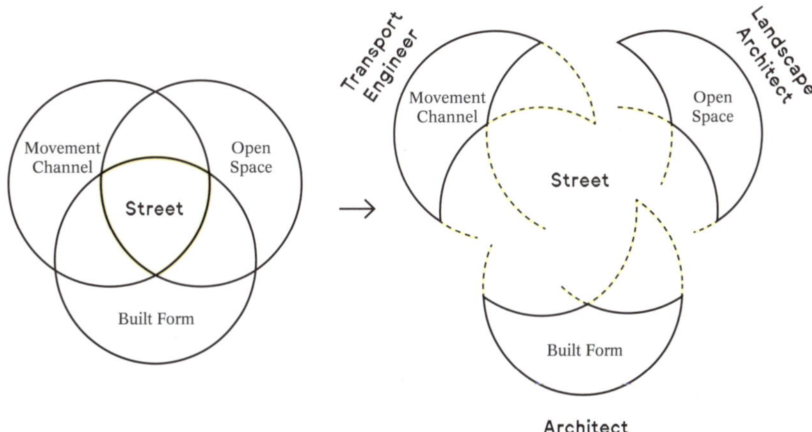

This schism has dictated not only mobility infrastructure design, but also the legal and social language defining its use. Today, most people will define a street only as a function for movement—measured by how many vehicles it can move per hour. Movement, however, is useless without placemaking to facilitate a variety of destinations not only at trip ends, but also along the way. And those placemaking destinations must be expressed by the design of the form and space that create their usability, aesthetic vibrancy, and success. The street—in all of its modern-day asphalted and impermeable material construction—must also forefront its critical environmental role in facilitating and filtering stormwater runoff, fostering urban greenery, and decreasing urban heat island effects. To think of the design of any mobility infrastructures, including ones that incorporate new technological disruptors, as the responsibility of a single discipline is to fail to understand the multifunctional purposes that our infrastructures must provide for cities. When our urban infrastructure and mobility can free itself from the disciplinary shackles that have been placed on it—when our land use, placemaking, transport, and open space design work in synergy—then better streets and urban spaces will result.

This summarized debate on design governance and disciplinary autonomy can continue indefinitely, existing as much in theory as it does in practice. What this book is more interested in, however, is looking forward and setting this discussion within the context of the new mobility disruptors landing on our streets today. The point of this dialogue is not to argue that one profession should lay dominant claim over another on urban space. Rather, moving forward with any technological innovations must require breaking down the siloed boundaries segregating professional responsibilities, creating productive results that will emerge from such interdisciplinary blurring.

When it comes to new mobility technologies like AVs, much of the current autonomy over their design and conversation is currently in the hands of the technology innovators, companies, and corporations charged with solving their remaining technological issues. As transport engineers have largely dictated street design in the contemporary city, they are likely to continue to have a prominent voice in influencing the impact of AVs on our current infrastructures. Disquietingly so, the various design professions—urban, landscape, architecture, planning—have been largely silent in shaping contemporary AV conversation. This is highly concerning, because it is precisely these design disciplines that should have an equal and influential voice at the table of not just public opinion, but of implementation, regulation, and design. Safety, efficiency, and functionality are certainly important considerations of driverless mobility, but no more so than the multiple other critical issues that moving around in cities elicits. Hence, this book calls on these design disciplines to take up this mantle, to insert themselves in these conversations, and to drive the decisions that result—using the tools that it has provided to do so.

Perhaps much of the reason that these interdisciplinary conversations have not taken place is that they are a product of the contemporary socioeconomic and sociopolitical systems that our cities and societies operate within. Sectors are classified into public and private on the basis of ownership, responsibility, and incentives. AV innovation is principally occurring in the private market, which has the overwhelming capital to fuel investment in a technology that has less-clear short-term fiscal incentives, but for which long-term market-share and economic payoffs will be vast. Our public sectors, on the other hand, lack fiscal capital but retain regulatory and governmental powers and responsibilities. While the incentives that must drive driverless technology oversight are less visibly clear on the regulatory side, they have been explicitly surfaced by the discussions in this book.

Again, the point of this dialogue is not to dissect these systems or the reasons why they exist but rather to argue for hybrid partnerships that leverage the strengths and tools of both. In the current neoliberal paradigm of technological development, it is shortsighted and naïve to think that public agencies can in-and-of-themselves drive AV innovation and implementation, just as the various examples of late-capitalism has illustrated that leaving the private market to determine that the public good is well-served and prioritized is a fool's errand. Some of the most successful contemporary examples of high-level urban mobility execution and service, exist in cities that have engaged in smart and synergistic partnerships between private-public entities. Private market competition and innovation can lead to higher qualities of service that are more efficient, responsive, and offer more options for mobility consumers. Meanwhile, public regulatory oversight can ensure that the public realm is protected and that equity and accessibility remain central tenets.

Mobility service and transportation systems in Singapore, Hong Kong, Curitiba, and Seoul—considered some of the best in the world—are cities, amongst others, that have hybrid transport ownership-operation models. These consist of a combination of private bus, rail and/or taxi operators with various roles in designing, financing, and constructing service lines. These operations are then organized under levels of public oversight, profit-sharing, and infrastructure maintenance. Singapore's state-owned Land Transport Authority (LTA), for example, finances the public infrastructure and capital assets of its railways and buses, while tendering the operation and maintenance of these system to private companies who bid on contracts. In Seoul, nearly 90 private bus companies operate its thousands of buses through different routes in the metropolitan areas, while its government regulates the bus companies as a public utility to ensure reliable and consistent service. In Curitiba, a similar partnership method is deployed. Although managed by the city's transit authority, services are contracted out to 22 private companies who operate buses and taxis, and are required to share revenues with the city to support road maintenance and infrastructure upkeep. In the unique case of Hong Kong, the city's Mass Transit Railway (MTR) is actually partially privatized, and listed as a company on the Hong Kong Stock Exchange. The city government owns about 75% of the company, while still providing the same support to transit operations and management as before it was privatized. Despite its partial ownership by public agencies, Hong Kong MTR is independently managed on commercial principles, fiscally independent from any subsidies from the government, and answers to its shareholders. These varying levels and methods of public-private hybridization are less common in the U.S., but as these successful global cases have shown, their lessons can be transplanted and must be taken advantage of.

Within these structures, this book also calls on public sector agencies to work across their siloed departmental branches. Larger metropolitan cities are often run by intra-agencies that divide the administration of public services between departments of housing, transportation, planning, economic development, building and construction, parks and recreation, and more. Doubtless, these departmental delineations have come to exist for a reason, but a result of such divisions of work is that it lends itself to departmental silos that can often impede cross-platform, comprehensive solutions to complex urban issues. When it comes to the holistic nature of transportation, public agencies must be far more agile in their ability to work together to regulate AVs already operating on city roads. Mobility, urban form, natural systems, economic development, land use, and zoning are intertwined, and thus our policies and plans put forth must be equally interdependent and integrated. Each of these departments must use the particular tools available to them in more synergy, and therefore to greater effect, and they must be far more proactive rather than reactive to new technological disruptors.

Lastly, but importantly, what role should the citizen play in the bottom-up design governance of urban space and mobility? As the constituent of all the aforementioned agencies and parties, the general public is often left out of the conversation and design decision-making processes. They are then left to contend with the

Tactical urbanism transformation of the intersection of Spoleto and Venini streets in Milan, Italy, through its Piazze Aperte initiative.

outcomes and decisions that result, being the primary inhabitants, users, and consumers of those spaces and systems. In our urban histories, this neglect has time and time again led to the various spatial, social, racial, and environmental inequalities permeating our urban geographies and built environments. Who better to champion for their own needs and preferences than those very urban citizens themselves? In recent years, movements like tactical urbanism have gained operational power to great effect.[17] Referred to colloquially as guerilla or DIY (do-it-yourself) urbanism, tactical urbanism describes organizational, often citizen-led, approaches to neighborhood building using short-term, low-cost, and scalable interventions to catalyze long-term change. Tactical urbanism offers more agile, quicker, and participatory alternatives to the conventional routes of creating change within the built environment and public realm. It arose in response to the shortcomings of traditional city planning, criticized as being frustratingly slow at implementing changes on a more localized scale or not meeting the needs of particular communities. Tools used by tactical urbanist approaches and citizens often include deploying paint, temporary signs or bollards, pop-up bike lanes, and urban furniture to reclaim streets and lanes for alternative uses. Some of the examples mentioned in this book, including New York City's pedestrianization of major thoroughfares under the leadership of DoT Commissioner Janette Sadik-Khan, began as tactical urbanist projects that due to their success became permanent fixtures for cities. They demonstrate how the everyday citizen, in coordination with larger organizations, can take control of urban space themselves to question assumptions, drive narratives, and produce results. Their successes demonstrate the important role they must play in any discussion on shared responsibilities and governance.

17 "What is Tactical Urbanism?" Tactical Urbanist's Guide, 2023.

Altogether, this discussion serves to remind us that we as a society, are in this together. Therefore, those choices must also be made together, leveraging the expertise, knowledge, advantages, and incentives provided by all parties in pursuit of a common goal. Without this interdisciplinary, interagency, and inter-participatory approach, then we risk repeating the failures of 20th century urban infrastructure and transportation design as driverless vehicles proliferate into our urban environments.

Reprogramming Mobility: Negotiating Digital-Physical Spatial Overlaps

The contentious governance over the physical realm of our cities extends to its digital space as well. With disruptive driverless technologies, this digital governance, negotiation, and control has far-reaching impacts on determining how we use and move around urban space. This digital "overlay," if you will, is often not immediately visible to the naked eye. While this conversation will continue to evolve quite rapidly as new technological overlays arrive in our future, there are several of these "negotiations" that have already begun playing themselves out in the physical space of cities.

The first and most explicit, is the fight over who has access to the vast quantities of mobility data that driverless mobility will produce. In particular, the advent of GPS (Global Positioning Systems) opened up a treasure trove of data in the transportation space. Used widely by mobility service providers, including transport network companies (TNCs) like Uber and Lyft, hard-fought negotiations have transpired over who should have access to mobility data and to what extent. Data ranging from precise pick-up/drop-off locations, average vehicle speeds, traffic congestion, ride costs, route decisions, and geographic coverage are gathered for every single ride-hail trip taken. With the arrival of various shared micro-mobility service (bike-share, e-scooter-share, etc.), further mobility data at these scales of travel are also captured. With AVs, even more data will be identified and gathered by its neural networks. The array of sensors and cameras in AVs will capture real-time monitoring of the physical environment around it, including the weather, the number of vehicles and the distances between, curb conditions, trees, street-signs, pedestrian behaviors, and much more. Multiplied by the number of future AVs potentially operating on streets simultaneously, and one could conceivably imagine that real-time imaging of entire cities could be done at all times of day—as if Google Earth's Street View function was live.

Ownership of this data becomes an incredibly powerful tool that is used to not only understand current mobility patterns and behaviors, but influence and projectively change them. This data is something that TNCs already feed into their machine algorithms to impact vehicle redistribution, traffic and accident avoidance, and service coverage. They also use this data to incentivize travel behavior. Features like dynamic surge pricing levies a fee on passengers to constrain TNC use during periods of high demand, while other bonus fiscal incentives are offered to TNC drivers to get them onto roads to match supply to demand.

When extrapolated to a city-wide scale, comprehensive network reorganization could occur. One example of such network disruption occurred with the Israeli start-up app Waze, which is a mobility app that provides real-time driving directions based on live, user crowd-sourced traffic updates. In 2011, Waze had become an indispensable and widely used app in the car-dependent city of Tel Aviv, the start-up having first rolled-out in the Israeli capital a few years prior. In August of that year, a series of outages in Amazon's Web Services, for which Waze had been renting virtual servers from, caused the app to go down. Subsequently, rush hour in Tel Aviv on

August 7th was catastrophic. Drivers who had come to rely solely on Waze for directions and traffic had to suddenly contend with saturated, traffic-ridden street networks. The app, and its ability to redistribute and reroute traffic flows, had become so integrated into the Tel Aviv transportation network that its service interruption threw the city's mobility into commotion. Until that day, this virtual permeation and digital dependency had become so seamless that city officials and citizens hadn't even realized it. Google certainly did after this incident, and acquired the company for a record $1.3 billion not long after.

This pervasive *reprogramming* of our mobility systems is playing itself out all over our cities in front of our eyes, often without us even explicitly realizing it. In Chapter 3, we already discussed the lengths to which TNCs use has drastically increased traffic congestion in cities. While touting itself as a solution to car ownership, TNCs apps and services unintentionally increased congestion on urban streets—an effect which was a symptom of excess car ownership in the first place. These examples illustrate the powerful ways that our digital technologies often have unintentional but distinct impacts on the use of physical urban space in cities. Those who gain control over these tools wield significant influence in impacting these spaces and systems.

There are several other digital tools that cities can use at their disposal. One of the most powerful is the concept of geo-fencing, which describes a *virtual* perimeter that is overlaid on a real-world geographic area. Generated dynamically in a radius, or to match real-world boundaries like streets, its implementation can dictate the use of mobility technologies in particular zones or areas of cities. Geo-fencing requires location-aware devices and services, like GPS in cars or a phone-based app, that are used to limit where these devices can travel. Integrated with AVs, it can dictate where they are allowed to pick-up/drop-off passengers on curb space, or virtually close off certain streets and urban areas during certain hours of the day. Geo-fencing technology has already been used in Los Angeles to limit where users can operate and park the various on-demand shared e-scooter services that have proliferated there, preventing them from traveling within or piling up in pedestrian-only sidewalks and areas.

This symbiosis between digital and physical space also begins to conflate in the physical infrastructure of cities. Our urban curb space has become increasingly digital with the replacement of single-space meters by parking kiosks and even virtually through pay-by-app services. Digital management of curb space will need to be increasingly monitored as AVs continue to add congestion and demand to its use. The digitalization of our street signs and signalization could also be impacted and deployed as tools. For example, the duration of a "walk" signal at an intersection could be extended right after a nearby driverless bus has just dropped off a large number of passengers through vehicle-to-infrastructure communication. Transit signal priority (TSP) could also be expanded, which is a technology in which transit vehicles and traffic signals communicate prior to its imminent arrival to grant a green light. This would serve to prioritize efficient and fast passage of transit vehicles at an intersection.

The exciting potential of dynamic lane management also demonstrates the need for cities to be more agile in overseeing digitally and physically intertwined space. While cities typically conceive of urban infrastructure in fixed spatial states, the lane uses of a street could be flexibly configured and assisted by technological overlays—converted for driverless buses during rush hours, urban freight in overnight hours, or close completely for pedestrian use during weekend hours, for example. AVs could be dynamically designated to avoid certain lanes or streets when uses have changed, while urban street signage, lighting, and bollards could be digitally coordinated to indicate pedestrian's access to those changing use designations.

In a more digitally seamless future, coordinated interfaces between various mobility applications and services must also be offered to consumers. Rather than having separate phone-apps for transit, regional rail, ride-hailing, and bike-sharing, our mobility services should be integrated into a common-standard app. This would not only mirror the holistic way that transportation services should be planned out and operate but also expand a consumer's ability to access, compare, and choose between the diverse options available to them. This digital gateway would allow users to compare travel mode by time, cost, route, and other potential parameters including even carbon footprint by trip. Incentive ticket discounts or surge-pricing fees could be integrated into this unified interface, leveraging service competition to induce travel behaviors based on shifting demand and congestion by location or time of day. This unified platform would also list arrival times for buses and other travel modes, passenger use-capacity of that mode (how full that vehicle is), and passengers could even have options to reserve seats. Purchasing tickets or hailing vehicles for all modes could be done directly from within the app. This would allow trips to be more easily and seamlessly chained together.[18] Various mobility service payment models could also be introduced, including monthly or yearly subscriptions that include a set number of modal options, depending on consumer lifestyle or need. In totality, this seamless digital overlay would make our mobility choices through urban space more accessible and comparable, which is exactly what the

18 Trip-chaining refers to the grouping of multiple legs of a trip between transfers of varying modes (from micromobility to rapid transit) under a single ticket fare.

A sample of all the mobility apps and services available to consumers, offered on separate platforms.

concept of mobility-as-a-service envisions. Along with automation, integrating our mobility in this digitally holistic manner could improve network efficiency, decrease costs for the user, improve transit ridership, and reduce congestion.

As all of these examples illustrate, it is vital that our cities gain autonomy over digital space and the tools that emerge from doing so. This starts first with the important requirement that any mobility data collected be shared with all stakeholders. Data-sharing must occur before these technologies roll-out mass market through the private sector, as it will be very difficult to safeguard digital space once it has become privatized, accepted, and consumed by the public. This lack of regulatory foresight has already occurred with TNCs and the massive mobility data that they own and use. TNCs have mostly rebuffed the many attempts by regulatory agencies to access that data, instead fencing it in walled gardens.[19] This severely impacts a city's ability to understand and work in partnership with them to regulate their transportation impacts in a comprehensive manner.[20] Some progressive cities, however, have begun implementing data-sharing standards and requirements (like MDS in Los Angeles, more on this in Chapter 6) to great effect.[21] These efforts must expand, with the eventual goal of standardizing all data on an open mobility ecosystem with shared-access platforms. Who has access to mobility data? How can cities ensure that individual privacy is anonymized and protected, while larger data trends are extrapolated? What other digital levers can be used to shape mobility patterns and networks? How can digital space be projectively leveraged for the public good and to further safeguard the physical space of cities? These are not easy questions to answer but ones that cities must as we move into an era where our digital and physical mobility become impossible to separate.

19 A walled garden in digital technology refers to a closed platform or ecosystem where the platform owner has total control over the data and its contents, restricting external access with the end goal of creating a monopoly.

20 Dobush, "Uber has troves of data on how people navigate cities... will [cities] ever get it?" *Medium*, 2019.

21 "What is MDS," LADoT, 2023.

Another important "negotiation" that plays itself out in urban space is the implicit but imperative one that takes place between street users. These unspoken exchanges often occur between human drivers in cars and with pedestrians or cyclists. Traveling through streets involves not only measured movement through physical space but also psychological and social negotiation amongst actors respecting various social cues or informal rules of the road. A nod of the head, a wave of the hand, a forward nudge of a car, and eye contact, are all subtle cues that pedestrians and drivers make and respond to when moving around in cities. They are delicate but essential to human safety as one crosses an intersection or makes a left-turn.

In a fully autonomous future, in which all cars have become driverless, direct vehicle-to-vehicle communication, which AV technology will usher in, may make human driver to driver communications obsolete. However, not only is this future—a complete permeation of our mobility systems by automation—one that is far off, it may also never fully occur, meaning that there will always be a small subset of mobility consumers who refuse to relinquish personal driving to AVs. The majority of intervening decades will also involve both human and non-human operated vehicles sharing the roads. Additionally, driverless vehicles will *always* need to share use-right of streets with pedestrians and bicyclists. This means that communicating through and negotiating in space will always be in contestation. As a result, the unspoken nonverbal cues that happen between vehicles and pedestrians will need to be reexamined as humans are removed from the driving equation.

For several years, AV companies have been working on ways for their vehicles to communicate with humans.[22] Self-driving company Waymo has recently begun implementing LED screen displays into roof domes on their vehicles in order to digitally communicate to pedestrians that the vehicle is yielding to them or to display the initials of the passengers who had ride-hailed the vehicle. The company has also explored using audial communication methods with pedestrians. Just like brake and signal-turning lights in our cars today, Waymo calls these recordings and cues "tertiary communication" and they have the potential to not only replicate but also introduce new ways for digital and physical negotiations to occur. What must be prioritized in these new forms of digital communication, however, is the *right* of the pedestrian to move about in urban space. With any methods of one-way messaging, it is critical that they are in service of making streets safer for humans, and not the other way around. Instead of warning pedestrians to move out of the way, or forcing cyclists to acquiesce to the driving maneuvers of the AV, this digital communication must be primarily beneficial to the human, their safety, and their mobility. Regulatory oversight must be established to ensure that this occurs, otherwise each new digital innovation will only further solidify a moving vehicle's claim to the use-rights of streets—as has already been witnessed in the language and media wars won by Motordom in history.

22 Hawkins, "How will driverless cars 'talk' to pedestrians?" *The Verge*, 2023.

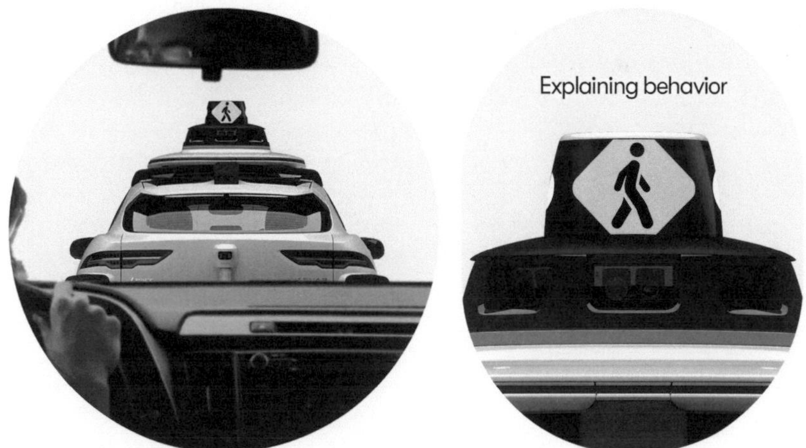

Waymo self-driving vehicle with integrated visual communication displays.

This communication and behavioral negotiation over urban space also plays itself out in other ways as well. If you live in a city in which AVs have already begun rolling out, you may have had the opportunity to ride in a driverless vehicle. The experience of riding in an AV today is described by many as similar to riding in a car "driven by your grandma." AVs are never the first vehicle off the line at a stop light or intersection, they don't speed nor accelerate too quickly, they are cautious when executing riskier driving maneuvers like lane changes, and they are especially courteous with pedestrians and cyclists. Certainly, the technology is advancing rapidly and will eventually reach the point where the neural and behavioral aspects of AVs will become indistinguishable from a standard human driver, but this is where the technology stands these days—overly cautious and careful.

Might the "behavioral tendencies" of AVs assist this book's call to reclaim our urban space for more human-centric uses? Academic Adam Millard-Ball uses game theory in a 2016 study to speculate that AV behavior may actually facilitate a shift towards pedestrian-oriented urban neighborhoods as the dynamics between vehicle and pedestrian negotiation plays out.[23] This is because AVs are programmed to be risk-averse and law-abiding. Instead of aggressively skirting road laws like human drivers—speeding, pausing for less than 3 seconds at a stop sign, or encroaching into crosswalk space—AVs will always obey road rules "to a T." Millard-Ball speculates that this may lead to pedestrians that intentionally begin stepping into the space of roads with confidence, knowing that an AV will always acquiesce to them. At an intersection crosswalk, for example, pedestrians today must play a "game of chicken" with an arriving vehicle as Millard-Ball describes—i.e. stepping tentatively out first and then hesitating to make sure that a human driver will yield before proceeding across. In a driverless future, this negotiation will shift in favor of pedestrians, who will be confident that all AVs will defer to the human.

23 Millard-Ball, "Pedestrians, Autonomous Vehicles, and Cities," *Journal of Planning Education and Research*, 2016.

These new behavioral equilibriums could allow certain areas or streets of our cities to be designed in favor of pedestrian's claim to its space. This shared use-right has been exhibited to a small degree in the design of "living streets" or *woonerfs*, which are streets designed to rely on visual and social interactions between street users. Common in the Netherlands and Belgium, *woonerfs* allow street users to negotiate a use equilibrium through social norms and deliberate uncertainty via democratic street design, rather than physical barriers, curbs, paving materials, and legal rules.[24] In a *woonerf*, vehicles, pedestrians, and cyclists all share the same space and they have been successfully shown to reduce car speed and encourage walkability, sociability, and safety when designed correctly.

24 Gooden, "Woonerf: A Living Street Concept," *Citygreen*, 2020.

With AVs, these democratic streets become even more practical and implementable. They would allow cities to expand shared street concepts in residential and others slow-speed zones, safe in the knowledge that AVs will always yield for pedestrian safety. In denser areas of cities, this may also result in AV algorithms that begin to avoid these streets, or at least allow their passengers the choice to do so if speed of travel is a priority. As increased pedestrian activity slows down AVs, they could drop off passengers on adjacent arterial roads, incentivizing them to walk the final block of a trip if they so choose. This would further decrease traffic congestion, increase walkability, and inevitably increase urban density in these areas.

Of course, the opposite could also very well occur if Motordom were to drive this conversation. Reacting to pedestrian's newfound impunity, policies and laws could be enacted to instead reinforce the legal use-right of AVs to our roads through sidewalk barriers, fines, and penalties for impeding the efficient and fast travel of a driverless vehicle. Pedestrians could even be required to carry around transponder beacons that prioritize the digital communication of AVs, and their technological limitations, over human communication (recall our previous discussion on beaconization in Chapter 2). The claim to urban space by a new mobility technology could be further advanced, as it was in the 20th century when automobiles

A *woonerf* shared street designed without curbs and with continuous paving material in Dordrecht, Netherlands.

Rendering of a future commercial shared street design with AVs.

Rendering of a future residential shared street design with AVs.

proliferated. This is why, in the context of this chapter, the critical values, principles, and stakeholders that we apply to these negotiations will go a long way in determining which future we want and can therefore result.

The last "negotiation" over our digital-physical space we will discuss is the use-occupancy or program of an AV in-and-of-itself. Some AV-utopists have suggested that in addition to transporting people, AVs could also transport goods or even services directly to a person's location. In the matter of transporting goods (i.e. urban delivery of food and packages), this is already happening at a large scale in cities. We have previously touched upon how urban freight may be impacted by automation earlier in this chapter. While first/last mile deliveries may still require humans, the long-distance intercity freight industry will be dramatically revolutionized by AVs as the technology is already adept at traveling on highways which are a far more predictable driving environment than city streets. Urban delivery within cities could be impacted as well. Dramatically accelerated during the height Covid-19 pandemic, our cities saw a rapid increase in the use of digital food delivery apps and cars that delivered those meals. The continued use of increased food delivery services even as urban mobility returned to pre-Covid states in intervening years, meant that more traffic congestion was added to our city streets. A few studies that have been conducted have shown that the impact per trip of rapid food deliveries may be even greater than that of ride-hailing. Their negative externalities include not only increased road congestion through added vehicles miles, single passenger occupancy, and deadheading, but also increased curbside use and congestion. In some cities, food delivery drivers spend on average 10 minutes idling at a curb waiting for restaurants to finish food prep, while passenger pick-up/drop-off times in ride-hailing services range from 30 seconds to 5 minutes on average.[25] Food delivery serves a vital role in supporting restaurant businesses but their increased use has had unintended impacts on urban mobility and space. They also operate relatively independently of ride-hailing services. All of these trends could further exacerbate with driverless technologies. In 2022, Uber Eats, the food delivery service arm of Uber, announced that it would begin testing driverless food delivery-only vehicles and services in two cities in California and Texas.[26] Other self-driving companies like Nuro, have begun testing

25 Felix and Pollack, *From App to Table*, Metropolitan Area Planning Council, 2022.

26 Bushard, "Driverless Food Deliveries Grow," *Forbes*, 2022.

Driverless Uber Eats food delivery vehicle operating on city streets.

their driverless grocery and package delivery-only vehicles in partnership with Uber Eats, FedEx, Walmart, Dominoes, and other corporations. If successful, they will undoubtedly add further congestion to our streets and curbs.

With all of these parallel technologies, the movement system for goods delivery and for passengers have the potential to be rethought more holistically and efficiently. Food deliveries, packages, and people traveling from similar locations and times to similar destinations could be grouped together by mobility algorithms if these independent systems were to work in coordination. Certainly, there are many issues that must be solved, which include economic, logistic, and operational challenges. Even vehicle design must be rethought in order to compartmentalize food or delivery packages from passengers. If done, however, this could result in a more efficient utilization of urban space and time in order to reduce the duplicate negative externalities caused by both mobility systems operating simultaneously.

In the matter of transporting services directly to a resident's home or location, some have suggested that our AV urban futures should involve "shops-on-wheels" or "moving buildings." This notion has been deployed on a very small scale in cities—with the use of restaurants-on-wheels (i.e. food trucks) or even mobile services like pet grooming vans that provide their services directly in front of your home. Cities already regulate the curb use and deployment of these businesses through required permits. In a driverless future, mobile food and other service trucks could coordinate to arrive and service a dynamic lane-managed street during lunch hour, when it has closed for pedestrian-only use, for example, and then depart when lunch-time demand has passed. However, increasing the categories of these mobile services to an economy of scale is probably less feasible, given the limit to the range and type of services that mobile vehicles sizes will be able to accommodate. If they were to proliferate, however, they bring up key lifestyle questions. Do we as a society value the experiences that come with traveling to destinations and seeing the diverse spaces and land uses one encounters on the way, or is a future promoting more sedentary and secluded lifestyles—in which more and more services are delivered directly to our front doors without requiring us to leave our homes—good for cities? These discussions demonstrate the continued need for cities to regulate curb uses and monitor congestion in urban space. One can easily imagine, as ride-hailing and food-delivery apps have demonstrated, that services on wheels will only add to the spatial congestion and contestations already occurring in our physical and digital realms.

As all of the varied discussions in this section have illustrated, we must be highly attuned to the ramifications of embracing today's technological tools, including understanding their unintended externalities. This book does not reject technology, and if anything, it welcomes it as a useful tool for helping us define what we value in our cities and assisting in the efforts to achieve those goals. However, prioritizing technological efficiency above all else, as 20th century highway construction did, results in what can be devastating negative externalities that are highly difficult to reverse. Others are already in motion—ride-hailing, food delivery apps, AV algorithm behaviors, and data collection—and we must continue to wrestle with their externalities as they unfold. Yet, others that

Soquimich lithium extraction field in Atacama, Chile.

Mutanda cobalt mine in Katanha, Democratic Republic of Congo.

are the technology's supposed great advantage—like the electrification of fossil-fuel based mobility—have already begun to cause negative environmental impacts elsewhere.[27] This includes the damaging land-water pollution and labor exploitation involved in the large-scale extraction and mining of cobalt and lithium, occurring largely in the global south for the global production of electric vehicle batteries. They all demonstrate how our technologies and digital spaces are implicated in the social, physical, and environmental crises that urban economies of scale cause. How can we make sure that the proliferation of AVs and mobility services limit their negative externalities? It starts with carefully defining our shared values and spatial principles and then continually evaluating whether those technological intentions are holistically met. The solutions of today must not create the problems of tomorrow.

27 Zheng, "The Environmental Impacts of Lithium and Cobalt Mining," *Earth Org*, 2023.

5 Transportation Policies

5 Transportation Policies

In this chapter, our focus shifts to a variety of key transportation policies that frame the modal shift towards the new Transitopia that this book argues for. The potential infrastructural changes and spatial design transformations described in this book cannot exist without concrete policy frameworks that enable their implementation and use. Spatial design changes must go hand-in-hand with economic and fiscal levers, as both play important synergistic roles in inducing mobility behavioral changes. They are all part of the critical toolbox that transportation regulatory agencies must take advantage of and utilize.

These transportation policy concepts are drawn from global transit-rich cities, all of them having proved effective in curbing automobile dependency and incentivizing multi-modal transportation in varied contexts and cultures. To cities in the U.S. in particular—shaped so spatially, geographically, and politically by the automobile—some of these concepts may seem far-reaching and progressive. They also necessitate a measure of political appetite by government officials and public agencies in charge of funding and enacting the regulatory bills, ordinances, and measures behind them. These policies must also be supported by the stakeholders and the public, for which these policymakers answer to. However, this measure of regulatory public oversight is critical to ensure that both the public realm and the public good remain at the forefront of any new technological advancements that land in cities. With the right leadership, able to negotiate the many contentious stakeholders and political challenges that these policies may confront, these transportation concepts are achievable in the cultures and contexts of even the most automobile-dependent cities. Collated together, their effect on cities could be transformational. If implemented correctly, they could unlock the exciting spatial and design opportunities illustrated typologically in the second half of this book. Along with spatial and infrastructural design changes, far greater mobility options, access, and value for the urban inhabitant could be created.

Transportation policies in this chapter are structured into two categories. Incentive policies are meant to motivate or "incentivize" mobility behavior towards alternative multi-modal ways of moving around by providing discounts, time-savings, convenience, and better alternatives to private single-occupancy vehicular use. Penalty policies, on the other hand, are meant to discourage or "penalize" personal single-occupancy automobile use, often through fines, fees, or other restrictions. Conceptually, these two ways of thinking about policy implementation follows the "carrots and sticks" approach to inducing desired mobility behavioral shifts.[1] Importantly, both must be used in conjunction. Incentives such as high-occupancy vehicle (HOV) or express transit lanes if used alone may not inspire drivers enough to voluntarily shift their behaviors en masse. In a similar vein, disincentives, such as congestion pricing, cannot be enforced alone without offering an appealing and effective alternative to individual car use. Both policy categories involve a combination of financial, social, and psychological levers. Once legally in place, these levers can be adjusted in different combinations and effects over time for the context of their locales, depending on how effective they are in catalyzing the mobility paradigm shift that this book envisions.

1 Metaphorically, "carrots" are rewards while "sticks" are punishments. The "carrot and stick" method is a metaphor for the use of both to induce a desired behavior.

Each policy concept is described through the following structure. Firstly, the introduction of AVs brings potential threats to the mobility and urban environment of cities, thereby demonstrating the need for these policies to be implemented to mitigate or address these automation fears. Secondly, AVs also bring potential opportunities, guided by values and principles that forefront the public good. Lastly, these policies have precedent in implementation, impact, and use in cities around the globe which are referenced here. Let's examine these policy concepts one by one next.

Penalty Policies

- **Congestion Pricing**
  Fixed or dynamic fees levied for personal AV and automobile use in high-traffic congestion areas of the city

- **Transport Zone Fees**
  Fees levied on personal or low-occupancy AVs and automobiles that travel long distances, crossing multiple city zones

- **Certificate of Entitlements**
  Quota license or permit system that limits the ownership and use of private AVs by individual households

- **Curbside Management Fees and Geo-fencing**
  Street curbside queuing fees based on waiting time and curbside demand levied on ride-hailed AVs to deter chaotic free-for-all drop-offs and pick-ups

- **Parking Mandates and Street Parking Fees**
  Removal of arcane parking minimum mandates from zoning regulations along with implementation of parking maximums; dynamically priced street parking fees and even the removal of street parking lanes for alternative uses

- **Required Data-Sharing Systems**
  Mobility data from private AV-operators—including number, use, and location of vehicles—must be provided real-time to regulatory public agencies

Incentive Policies

- **AV-Incorporated Transit Network**
  Comprehensive automated transit network incorporating a diverse range of mobility options at multiple scales and hierarchies, from micro-mobility and on-demand ride-hailing to driverless bus rapid transit and rail

- **Multi-Modal Fare Trip-Chaining**
  Grouping multiple trip legs between transfers of varying mobility modes, from micro-mobility services to rapid transit, under a single ticket fare

- **High-Occupancy Travel Discounts and Surcharges**
  Travel fare discounts for on-demand AVs when occupancy is greater than one or two passengers; surcharge on operators when AV occupancy is at zero to discourage "zombie cars"

- **Dedicated Bus and High-Occupancy Vehicle (HOV) Lanes**
  Mobility infrastructures—street right-of-ways and highways—dedicate multiple lanes to express and local bus transit use-only, as well as dedicating lanes to HOV use

- **Transit-Oriented Zoning and Land Use Development Policies**
  Zoning and land use policies that encourage denser development around major transit corridors and stations; affordable housing must be prioritized in new developments

- **Fare Subsidy Programs and Services**
  Fare subsidy programs for low-income households in which transit use still remains cost-prohibitive; para-transit service programs for seniors and mobility-challenged individuals

Congestion Pricing

Congestion pricing is a system of surcharging users a fixed or dynamic fee levied for personal car use during times of high-traffic and high-demand, in congested urban areas. Its objective is to use price mechanisms to make users conscious of the costs that they impose upon one another when using private modes of transport, and therefore that they should pay for the additional congestion they create. General types of congestion systems include cordoning and pricing an area, a city center toll ring, or corridor pricing where access to a lane or transportation corridor is levied a fee.

Automation Threats

Removing drivers from vehicles could lead to significant decreases in vehicle-occupancy—down to even zero-occupancy "zombie cars"—as well as increased vehicle use and miles traveled on city roads and highways. Already congested areas of the city could see traffic skyrocket, especially during high-demand, rush hour times of day. Other urban areas that are reaching vehicle traffic capacity could also begin to see gridlocks.

Automation Opportunities

Congestion pricing is one of the most important tools that cities have at their disposal to directly influence modal shifts. It levies a cost on the use of a shared urban resources—the right-of-way of the street—something that most people take for granted and thus neglect to understand how that use comes with spatial costs. It also encourages users to shift into other modes of transportation that could be unaffected by congestion, like transit, or shift individual vehicle use to non-peak periods of time. The arrival of AVs is an opportunity for cities to roll out congestion pricing systems in coordination with monitoring operations and overseeing management of the technology and MaaS at large. Cities could take advantage of the integrated system communications that AVs will require, and congestion could become dynamically priced, fluctuating based on traffic, demand, and by geography. Funds from congestion fees should then be reinvested into transit services and operations.

Policy Precedent

Congestion pricing is often viewed as a political non-starter in car-dependent cities. However, certain progressive cities have already implemented this policy to great effect. Singapore was the first country to introduce it in 1975, and it has been refined since to incorporate real-time variable pricing through their Electronic Road Pricing (ERP) system.[2] ERP can predict congestion levels up to an hour in advance and price accordingly. Since then, other cities, like London and Stockholm, have implemented the system successfully through area-wide cordons and toll rings. In the U.S., New York City has enacted congestion pricing in Manhattan, while even Los Angeles has initiated pilot studies to explore pricing certain major arterials. Private market variations of congestion pricing include Uber's surge pricing, which levies additional charges when demand is high. This type of system could be adopted by public agencies at a far greater geographic extent and scale.

2 "Electronic Road Pricing (ERP)," Singapore LTA, 2023.

Transport Zone Fees

Transport zone fees can be levied on personal, low-occupancy AVs and automobiles that travel long distances and cross multiple city zones. These long distance travel fees encourage the use of more efficient forms of shared rapid transit—which would not incur these fees—while limiting private, individual, and low-occupancy AV use to urban areas within localized geographic zones.

Automation Threats

The convenient point-to-point mobility that private automobiles and transportation network companies (TNCs) provide today—both of which often service single passengers—puts them in direct competition with shared, efficient, mass transit. While TNCs are priced by distance, the cost to use individual automobiles (more gas) is far less financially punitive in comparison. AVs could exacerbate these trends, particularly if they replace individual automobile use and if they travel even greater geographic distances due to the ability for passengers to rest and work instead of having to drive. A three-hour drive becomes a lot less daunting if it becomes a three-hour nap or work session.

Automation Opportunities

Transport zone fees encourage rapid mass transit for longer distance travel, while limiting individual mobility uses to shorter distances. In concept, the farther one travels in an individual AV, the higher the fee a user pays because of the greater spatial and traffic congestion costs incurred on the shared resource of the street. With automation, this system could be supported and expanded in coordination with geo-location technologies that are an integral part of AV software and hardware design. This would incentivize and limit individual AV use to local areas within city-designated transport zones, since crossing between geographic zones would incur a fee. Funds from transport zone fees should be reinvested in automated transit that is more efficient at moving people for long distance commutes, incurring less spatial costs per passenger-trip.

Policy Precedent

The transport zone concept applies to certain transit fare pricing systems that charge more for a ticket when traveling to farther destinations. Copenhagen uses a zone-system for its transit service that allows free transfers *within* one of its designated transport zones, but charges additional fares when a zone is crossed.[3] However, automobile use enjoys far greater economic liberties and doesn't typically incur a punitive cost for longer distance uses despite the expanded spatial costs they incur. Certain highway tolling systems are a simplified version of this idea, charging highway users variable tolls based on distance traveled per mile or per area. TNC services are also priced by distance traveled. These concepts need to be adopted to a far greater geographic extent and scale for all private and individual AV use.

3 "Zone Line Maps for Buses, Trains, and the Metro," Copenhagen Metro, 2023.

Certificate of Entitlements

Certificate of entitlements (COEs) are a quota license or permit system that limits the use and ownership of private vehicles by individual households. The license is obtained by purchase or winning bid that grants the legal right of the holder to register a vehicle and use it on city streets. When demand is high, the cost of a COE can exceed the value of the car itself. This system is implemented to restrict the number of vehicles on the road in order to regulate traffic congestion, especially in cities where space is contested and limited.

Automation Threats

AVs used primarily as individually-owned vehicles could lead to a sharp increase in vehicle miles traveled and more total vehicles owned as household transition to purchasing new vehicles of the automated variety. Furthermore, ride-hailing companies that currently depend on each of their cars to be owned and operated by drivers themselves, would shift towards fleet ownership models in a driverless world. This would require them to purchase AVs for ride-hailing operations, adding further total vehicles to city roads.

Automation Opportunities

The catalytic arrival of driverless vehicles must be used by regulatory agencies to establish a precedent that limits the total number of private AVs that can be owned and used. This is something that is likely already untenable with traditional automobiles that have already become widespread in private use, ownership, and acceptance in cities today. Establishing AV COEs would importantly encourage driverless technology to be deployed in mobility-as-a-service models and automated transit service instead.

Policy Precedent

A smaller-scale version of COEs is the taxicab medallion system that has been in use in several major U.S. cities like New York City.[4] Medallions are permits that allow taxicab drivers to operate their ride-hailing vehicles to run service in the city. The system intentionally constrains supply in order to protect the economic livelihood of its owners and operators by preventing an oversaturation of the market. The value of medallions has significantly plummeted with the arrival of TNCs in the early 2010s. As a result, certain cities like New York have implemented and required Uber and other TNC operators to apply for individual TLC licenses from the Taxi and Limousine Commission (TLC) to operate in city limits.

COEs at a city-wide scale have been pioneered by Singapore transportation agencies, which require all of its citizens to obtain a COE if they wish to own and operate a personal automobile in the city.[5] The city limits the number of certificates awarded each year, which are valid for 10 years, and they have become widely successful in promoting a car-light, transit-rich society. In China, cities like Shanghai and Beijing have implemented a license auctioning system for limiting vehicle quotas on the road. Some of their policies include limiting household use of vehicles to odd- or even-numbered calendar days in order to restrict car use and congestion.

4 "Taxicab Medallion," NYC Taxicab and Limousine Commission, 2023.

5 "Certificate of Entitlement (COE)," Singapore LTA, 2023.

Curbside Management Fees and Geo-fencing

Curbside queuing fees or other space management strategies like geo-fencing must be implemented on ride-hailing AVs in order to deter chaotic, free-for-all drop-offs and pick-ups. Geo-fencing refers to a digital or virtual perimeter that is established in geographic space in order to prevent or limit geo-linked vehicles from accessing, crossing, or parking in a particular boundary. Additionally, curbside access fees could be levied on private or ride-hailed AVs that use in-demand streetside curb right-of-ways based on demand, waiting times, or queuing lines.

Automation Threats
Automation will lead to a sharp decline in the need for parking in cities. Inversely, cities could see a rise in curb-space demand. Increased use of AV ride-hailing means more vehicles will need to stop to execute pick-ups and drop-offs instead of parking. Depending on the number of vehicles at the curbside, automated ride-hailing vehicles could stop two to three lanes deep from the curb, or otherwise double- or triple-park to create traffic blockages and exacerbate congestion. This behavior and its associated congestion effects have already been exhibited with increased TNC use on contested, in-demand, curbs and sidewalks of cities.

Automation Opportunities
To address these concerns, cities should establish regulations dictating the right to access pedestrian curbs and sidewalks, charging fees when long queues form and prohibiting double or triple parking that impede street right-of-ways. AVs must be geo-located, obeying dynamic geo-fencing perimeters deployed to limit or prevent certain in-demand curb areas from being used during high-traffic times of day. Dynamic curbside fees could be implemented that fluctuate according to demand and available curbside space. Additionally, sidewalks, curbs, and street right-of-ways could be infrastructurally redesigned to designate specific zones where ride-hailed AVs are allowed to drop off or pick up passengers. Fees generated from these policies could substitute for lost parking revenue.

Policy Precedent
Regulations around curbside use have already been implemented in urban locations that see a large influx of pedestrians using ride-hailing services, including airports and large entertainment venues like sports stadiums. However, more widespread implementation of curbside fees and infrastructural adaptations remains rare. Geo-fencing has been used by transportation agencies in cities like Los Angeles, and others, to limit certain in-demand areas of the city from parked on-demand e-scooters or e-bikes, which have been known to pile up on sidewalks leaving no space for pedestrians.[6] These preliminary examples should be expanded to encompass all new driverless technologies.

6 "What is MDS," LADoT, 2023.

Parking Mandates and Street Parking Fees

City zoning ordinances typically require new developments to provide a minimum number of off-street parking spots in new construction. These arcane parking minimum mandates must be eliminated. Furthermore, the assumed right to park free-of-charge on many city streets—a shared resource—must be re-evaluated by either expanding street parking fees that incur a financial cost for the spatial cost of using curbside space or by removing street parking from street right-of-ways. Both assumptions have been a result of decades of catering to and prioritizing individual car use in cities.

Automation Threats
Municipal adoption of parking minimums has had a substantial effect on urban form, including reducing density of built form in favor of disconnected buildings with expansive parking lots which is the easiest way to meet parking requirements. Sometimes, these requirements have forced businesses to build lots that are never full even on their busiest days. Automation will decrease parking demands including on-street parking. However, these lanes could be concerningly reallocated to expand the street's right-of-ways for more vehicular use through privately owned and individually operated AVs.

Automation Opportunities
Instead, cities must establish priorities for what reclaimed parking space should be used for. Removal of this vast dedicated space opens up transformative opportunities to redesign our street right-of-ways by expanding the city's public realm for pedestrian-oriented sidewalks, multi-modal bikeways, transit lanes, and urban landscaping. At minimum, dynamic-pricing fees for street parking must be implemented so that car owners contribute to the spatial cost they incur on a shared public resource. The removal of parking minimums in zoning and development could incentivize building typologies that are more oriented to walkable, vibrant, and human-scale urbanism. Parking maximums could be established to limit the construction of new parking stock. The potential loss in revenue provided by city-managed street parking tickets could be substituted by implementing the other fee policies outlined in this chapter.

Policy Precedent
In recognition of these issues, many cities have already begun overhauling or repealing their parking minimum laws. Parking maximums are more common in European cities, where many new developments calculate parking requirements to incorporate a minimum percentage of visitors arriving by transit. New York City, under Janette Sadik-Khan's leadership, reclaimed street parking in areas of the city for alternative uses. These initiatives have continued under her Global Designing Cities Initiative (GDCI), a nonprofit that advises cities around the world looking to reimagine their streets.[7] Additionally, expanded dynamic curb-pricing systems include San Francisco's Demand-Responsive Parking system that adjusts street parking meter prices in real-time to match demand, which works to redistribute demand to alleviate parking congestion.[8]

7 "Change Streets, Change the World," GDCI, 2023.

8 "Demand Responsive Parking Pricing," SFMTA, 2023.

Required Data-Sharing Systems

Mobility data from private AV-operators—including number, use, and location of vehicles—must be provided real-time to regulatory public agencies through a common standard, shared data vocabulary. This would allow cities to communicate directly with mobility operators in real-time and collect, analyze, and manage the use of mobility technologies and their impact on broader transportation systems, services, infrastructures, and the public right-of-way.

Automation Threats
TNCs like Uber and Lyft proliferated on city streets without any data-sharing regulatory requirements by cities. Because of this, the impacts they had on increasing traffic congestion were not understood by public agencies until it was too late, after their acceptance by the mass market. This made any data-sharing regulations post-TNC proliferation much more difficult to implement and obtain. If AVs become widely used on our city streets without any data-sharing standards, these trends could exacerbate. It will be extremely difficult for cities to not only evaluate but to manage their transportation network impacts and any negative externalities they may bring—like increased traffic congestion or curbside use demands.

Automation Opportunities
Transportation agencies must implement data-regulatory standards *before* AVs become widespread in use and adoption. If implemented correctly, both public and private transportation operators could work in far greater tandem to ensure that their mobility services support each other and the mobility consumer. Furthermore, regulatory agencies should establish a larger role in monitoring and managing system wide mobility in both digital and physical space. This digital management should be used to induce modal shifts, redistribute traffic bottlenecks in real-time, dynamically manage the right-of-ways of street and highway lanes, implement dynamic pricing fees, and other important transportation systems and policies.

Policy Precedent
Los Angeles' Department of Transportation has pioneered a data-standard system called MDS (Mobility Data Specification) that requires all of its ride-shared bike and e-scooter companies to connect to a city-run application programming interface (API) in order to operate in the city.[9] This has allowed the city to manage the use and parking of these micro-mobility modes on urban sidewalks and curbs, prohibiting use or inhibiting high speeds in certain designated areas. As part of Singapore's next-gen Electronic Road Pricing (ERP) congestion pricing system, all vehicles are required to be equipped with on-board devices that leverage GPS to transmit mobility data including location and speed.[10] ERP is also able to alert drivers to real-time congestion and suggest alternative routes. These precedents should be established for all new AV companies wanting to operate on city streets.

9 "What is MDS," LADoT, 2023.

10 "Electronic Road Pricing (ERP)," Singapore LTA, 2023.

AV-Incorporated Transit Network

Driverless technology must be integrated into a city's transit network in order to expand its geographic reach and support its service operations. Mass transit networks must expand in scale and hierarchy to incorporate a diverse range of mobility options—from micro-mobility and on-demand ride-hailing to automated bus rapid transit and rail. This must be done both through policy measures as well as physical transportation network planning.

Automation Threats

As previously discussed in expanded detail, the arrival of driverless technology could serve to expand the use, reach, and affordability of TNCs like Uber and Lyft. TNC proliferation has been shown to increase traffic congestion in cities, as well as shift passengers away from other more sustainable and efficient mobility modes, including primarily mass transit. If this model and method of AV use were to be adopted leading to zero-occupancy "zombie cars," traffic congestion will magnify and mass transit use could be significantly diminished.

Automation Opportunities

Driverless technology could be used to revolutionize bus transit services at local and express scales. Automating bus transit by reducing driver costs could expand service frequency and network reach. Additionally, driverless transit vehicles at smaller scales and service types, like on-demand mini-shuttles, could act as first/last mile feeder services to major transit hubs and corridors that support long-distance rail or bus rapid transit. Vehicle platooning and other compounded spatial efficiencies could allow transit service networks to be far more agile, flexible, and responsive to network and passengers demands.

Policy Precedent

Transit-rich cities exist widely around the world with varying network types and structures. A few have begun piloting automated buses and feeder services to support their wider transit operations. The first commercial driverless bus service arrived in Singapore in 2021, developed in partnership between the city's public Land Transport Authority and its two largest bus and rail operators: SBS Transit and the SMRT Corporation.[11] Other city transit agencies, including Valley Metro in Phoenix and Capital Metro in Austin, have launched pilot demonstration programs to smaller degrees, testing on-demand autonomous feeder services to transit hubs.[12,13]

11 "Autonomous Vehicles," Singapore LTA, 2023.

12 "Waymo AV Partnership," Phoenix Valley Metro, 2023.

13 "CapMetro, City of Austin, testing Autonomous Transit," CapMetro, 2023.

Multi-Modal Fare Trip-Chaining

Fare trip-chaining refers to grouping multiple trip legs between transfers under a single ticket fare. This concept could be applied to encompass all trips that include multiple modes of transportation, from micro-mobility services like shared e-bikes to regional commuter rail. Doing so removes barriers from multi-modal transportation and transit use. Trip-chaining also works to provide a more integrated service experience between modes and providers for the passenger.

Automation Threats
Wait times associated with transferring as well as compounding additional fares between transit modes are often cited as a barrier to the use of transit and multi-modal mobility. AVs could further exacerbate these barriers by making on-demand, point-to-point mobility even more convenient and affordable than other alternatives. If AVs proliferate primarily as individually owned or operated vehicles, they will cause an increase in vehicles miles traveled at the cost of decreased use of multi-modal transportation.

Automation Opportunities
Driverless technologies spark an opportunity to consolidate the diversity and range of transportation options under the comprehensively integrated digital management systems needed to support multi-modal fare trip-chaining. For example, in the context of first/last mile scenarios, trip-chaining an AV mini-shuttle or bike-share ride to light rail could incentivize the use of multi-modal transit over ride-hailing an AV from a trip's start to finish. The price of the first/last milc lcg of that trip would be chained to the rail transit fare, which would be cheaper than alternative transportation options while still providing service from the trip's beginning to end.

Policy Precedent
Fare trip-chaining is already a common standard within transit systems on one mode, as transferring between two subway lines typically does not incur additional fare charges. This has been expanded in some cities to encompass transfers between two mobility modes. In New York City, for example, passenger transfers between the city's bus and rail network within a two-hour window will not be charged additional fares.[14] These transportation chains could be expanded to encompass more mobility modes—particularly in the first/last mile gap—if cities are progressive about implementing these standards and systems.

14 "Everything you need to know about Fares," NYC MTA, 2023.

High-Occupancy Travel Discounts and Surcharges

Travel fare discounts can be provided to on-demand AV passengers when the occupancy of that individual vehicle is greater than one or two passengers. Additionally, surcharges can be applied to AV operators when the occupancy of a vehicle is zero. Both pricing adjustments work in tandem to encourage on-demand, ride-hailed AVs to carry multiple passengers at greater efficiencies and space utilizations.

Automation Threats

Current on-demand, ride-hailing mobility services remain primarily single-occupancy, meaning that one or fewer passengers are serviced in addition to the vehicle driver. Average vehicle occupancy in automobile travel is 1.5 persons per vehicle, and even less in ride-hailing services at 0.8 when accounting deadheading (which describes the time the vehicle travels between passenger pickups). With AVs, this figure could plummet. Driverless vehicles without any human occupants ("zombie cars") could proliferate, roaming city streets in search of passengers while contributing to traffic congestion and street use while vehicle miles traveled are added to roads.

Automation Opportunities

To counter this threat, city transportation agencies must utilize these fare pricing, discount, and surcharge levers in their AV transit operations. They must also work closely with private ride-hailing operators to regulate and implement incentives and penalties for vehicle occupancy metrics in their service operations. Both must be done in order to shift single-passenger vehicle use towards shared and multi-occupancy rides in not just transit but all driverless vehicles.

Policy Precedent

Today, ride-hailing companies like Uber and Lyft offer "pooled" or "shared" rides that give a fare discount to passengers willing to share their on-demand ride with other passengers traveling in similar directions. However, use of these options remains low, and when the Covid-19 pandemic arrived it virtually eliminated their utilization. Currently, ride-hailing services also do not account for multiple riders from the same source in their fare charges. Surcharges on zero-occupancy driverless vehicles have not yet been considered nor implemented even though commercial AV services are becoming available. Fare surcharges must be expanded, fine-tuned, and made punitive so that they become a useful financial lever to incentivize mobility behavioral shifts.

Dedicated Bus and High-Occupancy Vehicle (HOV) Lanes

Our city streets and highways at a variety of scales must have laneways dedicated to express and local bus transit use-only. Additionally, high-occupancy (HOV) lanes, often referred to as "carpool" lanes, must be implemented in highly in-demand streets and highways for AVs that carry two, three, or more passengers. Transit Signal Priority (TSP), in which traffic signals communicate with transit vehicles to grant green lights upon their arrival, must be implemented to prioritize the efficiency and speed of mass transit.

Automation Threats

As previously elaborated, AVs will almost certainly reduce costs per mile traveled. This could lead to increased vehicular use, causing further traffic deadlock and congestion on our city right-of-ways, spaces that already face much contestation in their demand. In particular, buses and other transit vehicles that already have to fight for space in cities, and that are often hampered by car traffic, could face further exacerbations of these issues.

Automation Opportunities

To address such concerns, transportation agencies could seize AV-rollouts to prioritize the implementation of higher-speed laneways and TSPs that service automated transit and multi-occupancy vehicles. The conversion of existing automobile laneways for HOVs is relatively inexpensive to implement, requiring only paint and signage, especially in comparison to the construction of rail infrastructure. The ability for AVs to platoon means that automated bus rapid transit—if serviced by dedicated laneways facilitating their fast passage—could begin to approach and even rival the carrying capacity of rail carriages. Altogether, these relatively inexpensive right-of-way transformations could help shape the perception of bus transit as one that is faster and more efficient than a traffic-ridden car, in addition to actually reducing the congestion and service delays frequently encountered by them.

Policy Precedent

Minneapolis' Metro BRT lines have rolled out Traffic Signal Priority technology in many of its intersections and routes to great success.[15] Widely successful precedents for BRT-dedicated express lanes and bus-priority lanes have been pioneered and implemented in transit-rich cities like Curitiba and Seoul, amongst others. These cities have—through political willpower, infrastructural transformations, and eventually cultural embrace—prioritized the passage of more efficient surface transit on their city streets. Bus rapid transit in the city of Curitiba was so well-used and successful that even the city's urban development began densifying along and responding to major BRT corridors. In the American context, HOV lanes are typically restricted to expressways and highways, and often remain widely under-capacity in use. In Los Angeles, for example, the HOV lane on its I-110 highway was converted to paid toll lanes in 2012 due to underuse. Meanwhile, dedicated bus lanes on local streets remain rare in most U.S. cities. Both must be expanded and their uses incentivized in a driverless future.

15 "Speed & Reliability Program," Minneapolis MetroTransit, 2023.

Transit-Oriented Zoning and Land Use Development Policies

Zoning and land use policies must be coordinated with transit planning in order to encourage higher-density development and mixed land uses around major transit corridors and stations. New transit-oriented development must prioritize the construction of affordable housing stock to address the housing crises facing many cities, as well as provide affordable housing for populations that use transit services. This work must be done in close coordination with multiple agencies, collaboratively between transportation, city planning, housing, economic development, and others so that all of the city's urban systems work supportively together.

Automation Threats
If automation were to proliferate through personal vehicle ownership, it could continue to fuel the lower-density suburban sprawl that automobiles encouraged in cities during the 20th century. These patterns of suburban development are far more car-dependent, unsustainable, and monotone in their built environments than their denser urban counterparts. Even if automation were to be instrumentalized in transit vehicles, increasing transit ridership must still be supported by higher-density development, zoning, and mixed land uses, or they may lead to the proliferation of auto-oriented transit typologies like park-and-rides, which describe commuter transit stations that are surrounded by parking garages and surface lots.

Automation Opportunities
Smaller scale AVs could unlock and expand the first/last mile radius around transit hubs and corridors. This would lead to a larger area radius for new development opportunities and increased densities that support rapid transit as the backbone of long distance travel. Developing these expanded urban centers creates a blueprint for the structural evolution of a city's future growth. This structure would become one that is more concentrated, sustainable, pedestrian-oriented, and vibrant—one that provides increased opportunities for jobs, homes, and services to these neighborhoods around the clock.

Policy Precedent
Transit-oriented development (TOD) and transit-oriented community (TOC) programs and policies have been widely developed in cities around the world, in recognition of the benefits that come with developing compact, mixed-use neighborhoods near transit. One of the earliest and most successful examples of TOD is in Curitiba, Brazil, which influenced many other cities around the globe to implement official TOD policies and ordinances to support their transit systems.

Fare Subsidy Programs and Services

Fare subsidy programs must be implemented for low-income households in which automated transit or other transportation use still remains cost-prohibitive. Para-transit service programs and wheelchair accessible vehicles (WAVs) for seniors and mobility-challenged individuals must be introduced. Driverless technologies must bring benefits not only to the wealthy who can afford their services.

Automation Threats

As with the arrival of many new technologies, the mass adoption of driverless technology could disproportionately benefit higher-income residents of the city who can afford to pay higher fees for low-occupancy AVs or point-to-point ride-hailing services. Meanwhile, lower-income residents of the city who cannot afford these services could be left out. Additionally, if market forces dominate the use and implementation of AV technology to the detriment of public transit, lower-income urban residents who are often the predominant users of transit could be even further disproportionately and unequally impacted.

Automation Opportunities

To address potential access inequities, public agencies must expand and increase fare subsidy programs to ensure that low-income residents benefit from AV technology. Driverless, on-demand para-transit and WAV services for seniors and the mobility-challenged must also be introduced. Furthermore, all public transit could even be provided free of charge, fare-free for all users. Transit in most cities is already heavily subsidized, funded primarily by taxes and means other than fares which make up a minority percentage of its service and operational costs. Additional costs for low-income, mobility-challenged, or fare-free/zero-fare transit could be subsidized by new congestion pricing, transport zone, and curbside drop-off/pick-up fees that work in conjunction to redistribute affordable transit access to those who are in most need of it.

Policy Precedent

Various cities have already implemented reduced fare programs for low-income riders. For example, Los Angeles Metro has a rider relief program called LIFE (Low-Income Fare is Easy).[16] Other cities run on-demand, para-transit services for seniors and mobility-challenged individuals. These include Switzerland's Betax service and New York City's Access-A-Ride service.[17,18] The concept of fare-free/zero-fare public transit has also already been tested and implemented to great success. Entire countries, like Luxembourg and Malta, have made their public transport systems free for all residents, subsidized by a combination of taxation, commercial sponsorship, and advertising rights.

16 "Low-Income Fare is Easy (LIFE)," LA Metro, 2023.

17 "With Betax, you can move forward," Betax, 2023.

18 "Welcome to Access-A-Ride Paratransit Service," NYC MTA, 2023.

Los Angeles as a Testbed City

Many of the discussions and concepts explored so far in this book have been done so in broad strokes. The impacts of driverless technologies will be distinctly felt in all cities at large and require these conceptual frameworks and transportation policies to guide their deployment. However, it is equally important to explore what happens when we land these ideas in the specificities of real lived cities—cities that, by their very nature, have their own cultural peculiarities, transportation histories, regulatory contexts, mobility services, and local inhabitants that use these existing systems and services. By testing these concepts in a real city, we are forced to grapple with the urban environments, built forms, transportation systems, and mobility infrastructures of that specific locale, contextualizing how driverless technologies will be deployed and absorbed. Confronting a real city also forces one to wrestle with the diverse populations who occupy it, and whose travel behaviors and urban experiences will be distinctly impacted by new technologies landing in their lives.

Situating itself in a real city, as the latter part of this book and as Book 2 (*The Experience*) does, also allows the design visioning that takes place in both to explore the spatial and formal impacts of new driverless technologies on real contexts and neighborhoods. Rather than hover in contextless "ideal space," these design implications on the future urban environment of our cities could very well take place. In parallel, the typological nature of many of the spaces and systems explored also act as a blueprint for other cities grappling with the arrival of new mobility technologies on their own city streets.

The city of Los Angeles is chosen as this prototype city for several important reasons. Firstly, the city has in its recent history, been known for its love affair with the automobile. This recent attachment to and subsequent reliance on the car, has been a fundamental shaper of the city's built environment and urban structure. It has fueled Los Angeles' urban expansion, functional segregation of land uses, redistribution of land values and real estate, and the demographic segregation of race and class groups. In recent eras, it has also given rise to a whole suite of building and infrastructure typologies: drive-through businesses and strip malls that have retreated from the street wall, residential forms calibrated to parking requirements, and expansive street widths and highway infrastructures dedicating generous space to the car. It is hard to think of another city in which the influence of the car has led to such a broad transformation of its built environment than in the city of Los Angeles. According to influential architecture critic Reyner Banham, "the less densely built-up urban structure of the Los Angeles basin has permitted more conspicuous adaptations to be made for motor transport than would be possible elsewhere without wrecking the city."[1] Additionally, the media and entertainment industry of the city has, in its recent history, further fueled the city's association with the automobile over other mobility modes through Hollywood movies and advertisements that have proliferated and been embraced world-wide. Exploring a future mobility paradigm for the city—one that incorporates driverless technologies to expand its transit systems and multi-modal transportation—offers an opportunity to not

1 Banham, *Los Angeles: The Architecture of Four Ecologies*, 1971.

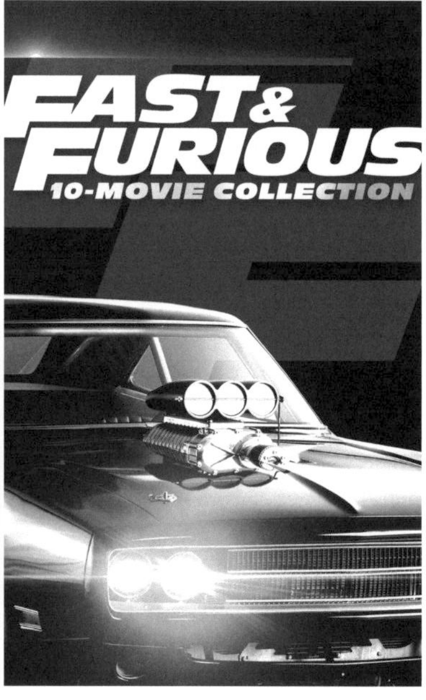

Various Hollywood-produced movies that celebrate and proliferate car culture.

only question historical status quos but also shape the evolution of its culture and identity. The city also offers an important context that frames the design-visioning medium and world-building method that Book 2 undertakes. Deployed through the format of an architectural graphic novel, the book utilizes these novel representation tools with the understanding that language and media choices critically shape the acceptance and proliferation of new technologies in the urban imaginary.

Secondly, Los Angeles is emblematic of larger cultural shifts and mobility trends in cities today. It is a city that, in its modern rendition, is undergoing rapid urban change in its mobility systems and infrastructures. It is a city that is confronting how its own transportation history has shaped the evolution of its urban structure and fabric. It is a city that is investing heavily into expanding its transit system in recent years, in recognition of the fraught externalities that the automobile has levied. It is a city in which its own residents, frustrated by decades spent in worsening traffic, pollution, and freeway expansions have begun clamoring for alternative ways to move around. As a result of decades of urban expansion fueled by the automobile, it is a city like many others, that is grappling with inward densification rather than the outward sprawl that marked its previous eras. It is a city whose newer generation of inhabitants increasingly prefer more urban lifestyles in higher-density, mixed-use built environments—rather than the single-family, suburban lifestyles that characterized the second half of the 20th century American city. These larger cultural and demographic shifts demand cities to re-examine their frameworks for future growth and the transportation decisions that they are inextricably tied to.

Lastly, Los Angeles has a robust legacy of transit-oriented development in its DNA to build upon, a history that will be briefly unearthed in the next section of this chapter. This historical DNA provides a blueprint for the city to reflect and build on as it charts a path forward in light of new mobility technologies landing quickly on its city streets. It is a city that has become a hot battleground in recent years for private-sector mobility technologies, including ride-hailing apps and on-demand micro-transit services like electric, dock-less, scooter- and bike-sharing companies. It is a city whose cultural appetite and public demand for alternative and new ways of moving around the city apart from the private automobile, is increasingly substantial.

Altogether, these various reasons and recent trends make Los Angeles ripe for examination as a prototype city that has the cultural context and public appetite to incorporate driverless vehicles into its transit systems and urban structure. Given all of this, if there were to be a model city able to pioneer such a transportation paradigm shift, and one that urgently needed such a paradigm shift, Los Angeles would be the ideal candidate. While certainly unique in its history and culture, the city is one in which its lessons can be extracted and applied to many cities grappling with similar transportation challenges and technological impetuses. In the end, if it can be done in city like Los Angeles—known historically for its love affair with the car—it can very possibly be done in other cities as well.

The "Three" L.A. Paradigm

The transportation history and context of Los Angeles is extensive and rich—and far less one-dimensional than it might seem at first glance to those unfamiliar with the city. This history, capturing the evolving mobility patterns in the city from the late 1800s until today, can be understood within a tripartite framework of the "First", "Second", and "Third" L.A.s, which are terms coined by former L.A. Chief Design Officer and architecture critic Christopher Hawthorne.[2] While the depth and nuance of this transportation history is beyond the scope of this chapter and book, this section will highlight its relevant themes in order to contextualize the city's present transportation challenges and ambitions for its future—setting the stage for the future integration of driverless vehicles.

2 Artbound, "Third L.A. with Architectural Critic Christopher Hawthorne," *KCET*, 2016.

As a testament to how widespread the proliferation of the automobile has been in its recent eras, as well as how distinctly it has been associated with the image of the city, the casual reader unfamiliar with Angeleno history may be surprised to learn that its urban structure was actually developed off of the backbone of a transit system. While it can be argued that no city in the U.S. is more identifiable with the personal automobile than Los Angeles today, the city in the late 1800s and early 1900s was in fact a global leader and at the forefront in mass transit. This is the "First L.A." that Hawthorne describes, which between 1880 and the mid-1940s, enjoyed a distinct civic and historic character that was relatively independent of the automobile, freeways, and their ever-present smog.

One of the first mobility technologies to land in the city was the arrival of electric trolleys, which initially appeared in the city in the late 1880s. By 1901, the Pacific Electric Railway Company was created, and with it, a privately owned mass transit system was established consisting of electrically powered streetcars and trolleys. This model of transportation expansion and urban development was predicated not on revenue generated from passenger transport, which rarely generated a profit, but rather on supplying electric power to new areas of the city now accessible by transit and the land speculation and real estate development that these new lines enabled.

The Pacific Electric Railway company established five rail lines that extended from the downtown of L.A.'s core outward to connect the greater counties including Orange County, San Bernardino, and Riverside. This speculative real estate expansion, fueled by private mass transit, enabled the company to become the largest electric railway system in the world by the 1920s. The common misperception of the city is that the urban sprawl of its metropolitan structure was caused by the private automobile but that's simply untrue. Rather, the wide metropolitan reach of it's urban geography was actually enabled by this electric streetcar system which, fueled by land speculation, drove the vast expansion and development of the streetcar suburbs of the city.[3] It was this backbone established by these major rail development corridors that produced the region we've come to understand today as constituting the metropolis of Los Angeles.

3 Streetcar suburbs are residential communities whose growth and development were strongly shaped by the use of streetcar lines as a primary means of transportation. Such suburbs developed in the U.S. in the years before the automobile, which allowed a city's growing middle class to move beyond its downtown cores.

Henry Huntington, the owner of the Pacific Electric Railway, profited greatly from the buying and selling of new real estate in which his railway lines unleashed. However, once the company's land holdings had been developed, its major source of income began

1926 Pacific Electric Railway System Map of Los Angeles.

LINES OF THE PACIFIC ELECTRIC RAILWAY IN SOUTHERN CALIFORNIA

PACIFIC ELECTRIC RAILWAY
WELLS-FARGO & CO. EXPRESS
WORLD'S GREATEST ELECTRIC RAILWAY SYSTEM
1000 Miles of Standard Trolley Lines
To All Points of Greatest Interest in the Heart of SOUTHERN CALIFORNIA and Traversed by
2700 SCHEDULED TRAINS DAILY
Including 5 Trains at Convenient Periods to
WORLD FAMOUS MOUNT LOWE
A Climb from SEA LEVEL to CLOUDLAND
By Trolley Through
America's GREATEST SCENIC WONDERLAND

PACIFIC OCEAN

PACIFIC ELECTRIC COMFORT · SPEED · SAFETY

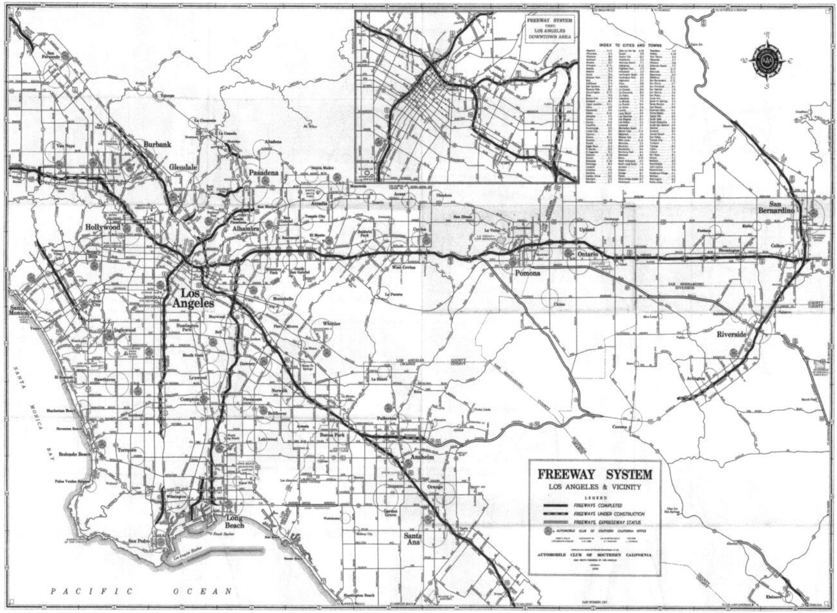

1960 Freeway System Map of Los Angeles, showing freeways completed and under construction.

to diminish. Electric transit lines that had their service operations and maintenance subsidized by their land development, could no longer be subsidized once those land holdings were all sold. This resulted in the lines being slowly converted to cheaper bus routes, starting as early as 1925.

By this time, the impacts of new automobile technologies were also beginning to be felt. In downtown Los Angeles, for example, mass electric transit tracks had been previously laid at the center of roads. As a result, electric trams, new automobiles, trucks, and pedestrians had to share these contested thoroughfares. By the 1930s, increasing automobile use led to increased congestion on street right-of-ways, causing slower mass transit speeds. While the city favored and upgraded its rail transit system that was focused on its central downtown, Motordom was convinced that a freeway

Aerial view of the Harbor and Century Freeway interchange facing north towards downtown Los Angeles.

system could solve the region's transportation ills. In response to this congestion, and mirroring the trend that was happening all around the country at the time, the Automobile Club of Southern California lobbied for the construction of a sprawling freeway system in the city. Originally planned, interestingly enough, with tracks in the center margin of each freeway for the passage of inter-urban mass transit, these aspects of those highway designs were never implemented. Fueled by federal funding, Fordism, and the media/culture wars won by Motordom championing automobiles nation-wide, the foundational network of the freeway system in Los Angles was constructed over the original five Pacific Electric Railway lines that had been laid out decades prior. The underlying structure of its original rail corridors that connected the many outlying parts of its metropolitan areas is still legible today in the form of multi-lane freeways that replaced and formed the urban structure of the city in its next era.

A combination of all these factors led to the majority of electric mass transit lines decline in use and revenue by the 1930s and '40s, eventually resulting in many lines being eventually shut down. By the early 1960s, the last few remaining rail lines in the city were removed and replaced with bus transit while the city's inhabitants and urban development embraced the proliferation of the automobile.

This embrace defines what Hawthorne terms the "Second L.A." It is a period in which the city became synonymous with the personal automobile, expanding freeway construction, ever-worsening traffic congestion, and a shrunken civic realm. Following the dismantling of its original streetcar transit lines, this period of the city runs roughly from the mid-20th century to the early 21st. This is the era that produced all the familiar stereotypes that one commonly associates with the city's image today.

1961 Construction of the Santa Monica Freeway.

1987 Construction of the Century Freeway.

1958 Construction of the Harbor Freeway.

This image, and its cultural associations in the urban imagination, is described famously by Reyner Banham in one of his seminal works, *Los Angeles: the Architecture of Four Ecologies.* One of these ecologies he famously coins "Autopia", which describes the city's extensive freeway network as not only physical infrastructure but as a lived place and a space for publicness as well. He dubs the city's infamous freeways as a "single comprehensible place, a coherent state of mind" in which Angelenos "live a large part of their lives."[4] It is a place in which its inhabitants treat subconsciously as an essential form of public expression, one in which its citizens venture out to see and be seen. It is one in which the public space of the city permeates through the windows of a car into its interior and where the automobile itself is treated as an artistic form of personal expression, identity, and character. The development and use of the automobile in the city served not only as the primary means of transportation but also as a protean form of identity expression for particular ethnic cultures and geographic areas. This is exemplified by the sleek and whimsical "Custom Cars" popularized by George Barris in the 1950s and '60s, the "Lowrider" cultural movement of

4 Banham, *Los Angeles: The Architecture of Four Ecologies*, 1971.

Auto-meet at the Petersen Automotive Museum's annual car cruise-in.

working-class communities in East Los Angeles, and the frequent car clubs, auto-shows, and auto-meet occurrences that draw hundreds of motor-enthusiasts and specialty cars from around the city.[5]

In this era, the automobile and its freeways became not only synonymous with Angeleno culture and public identity, but its arrival also expanded and solidified the sprawling geographic reach and structure of the city. This second era of L.A., post-World War II, witnessed a large influx of populations from soldiers and citizens who had migrated to the city in response to wartime efforts. These populations dispersed and scattered throughout various geographic areas, spurring what was previously a mono-centric (or mono-nucleated) city—that is, a city with a concentrated and strongly developed single urban core—into one that developed distinct poly-centric characteristics. This poly-centricity (or poly-nucleation) describes a city that has developed around multiple urban centers of economic activity without a strongly defined center core. While Los Angeles still retains its traditional downtown, this center is far less dense in form and population than other cities of comparable size. Instead, Angeleno inhabitants are spread throughout a variety of poly-nucleated

5 A lowrider is a car with a lowered body often fitted with hydraulic suspension systems, generally highly customized as artworks expressing Chicano culture. Lowrider culture proliferated among Mexican-American youth in the 1970s and beyond.

employment centers of sizeable density that also begin to take on their own distinct cultural characteristics. Some of these poly-centers, like Century City or Culver City, have developed into prominent employment and commercial centers that were formed off the backbone of their film and entertainment industries. Others, like Hollywood or Santa Monica, developed into prominent entertainment and commercial centers fueled by the backbone of tourism economy, art and culture, and other important employment industries.[6] The distinct poly-centric evolution of Los Angles was also fueled in part by the city's expansive freeway construction, a network which grew with and allowed the city's multiple spread-out centers to be reconnected through efficient infrastructure. Hand-in-hand with the proliferation of affordable automobiles enabling households to move outside of its traditional urban core, the poly-nucleated structure of the city, in concert with its mobility systems, defined L.A. in its second era. This poly-centricity is a key characteristic that has become distinctly associated with the city, and must therefore be grappled with in any transportation planning or visioning that is undertaken.

6 Culver City and Santa Monica, among others like Beverly Hills, are technically separate incorporated cities that are independent from the city of Los Angeles. While they operate with their own city government services, they altogether form the urban structure of the entire metropolitan region of L.A. as a whole.

Along with the urban evolution and larger restructuring that occurred during this era of Los Angeles, the proliferation of the automobile also gave rise to a whole suite of architectural building and infrastructural typologies that are distinct to the context and culture of the city. Amongst the smorgasbord of Southern Californian architectural styles that evolved, the "Googie" architectural style was one that was distinctly influenced by automobile technology, car culture, and other future-oriented cultural references including the 20th century Space and Atomic Ages. Characterized by curvilinear, geometric, free-form shapes, bright colors, and modern materials that referenced the future, Googie-themed architecture proliferated amongst roadside automobile-oriented businesses in L.A., including motels, coffee houses, and gas stations. Boldly extravagant and eye-catching, Googie-style buildings were meant to advertise to and attract the eyes of people driving by in cars.

1949 Googie's Coffee Shop designed by architect John Lautner, which lent it's name and sparked the Googie architectural style in mid-century Los Angeles.

Even more so than just influencing the aesthetic image of buildings though, the automobile also contributed directly to the development of distinct typological forms that arranged themselves around the spatial constraints and dimensions of the car. The Southern Californian dingbat typology, for example, is a commonly found residential apartment typically formed by a two to three story inexpensive housing box holding 6 to 12 units that is elevated along the face of the street to allow cars to park underneath in a soft story. This typological form has evolved in careful calibration to meet the parking requirements and density of its inhabitants, creating a full-width curb cut that runs the full face of the building so that the entire width of the lot is given over essentially as a driveway. Other car-centric typologies like the drive-through restaurant, retail strip mall, motel, gas station, or the porte-cochère while first invented elsewhere, enjoyed widespread proliferation and use in the built environment of Los Angeles. These are buildings that exhibit a common typological trait of standing isolated from the block's edge in order to accommodate a swath of surface parking needed to service their programs. Some, like the retail strip mall or the motel (also known as a motor-hotel), have business or service densities that are in precise calibration with the number of cars needed to park and access them. Others, like the drive-through restaurant and gas-station, are designed precisely around the spatial queuing lines of cars. In proliferation, these car-centric building typologies form a pattern in the urban fabric of the city, one that suffers heavily from buildings that retreat from the traditional street wall in order to accommodate for vast parking lots. In almost all of these typologies, building signage is over-scaled to advertise to the speed of the passing car, rather than sized to the scale of the meandering pedestrian on-foot. This results in a built environment where even its visual urban communication cater to the automobile. Many of these spatial traits are exhibited in the street perspective views documented in the second half of this book.

Street right-of-ways and freeway widths in Los Angeles are also much wider than their east coast counterparts, both in lane dimensions and in number of lanes, which is in part a result of the cities

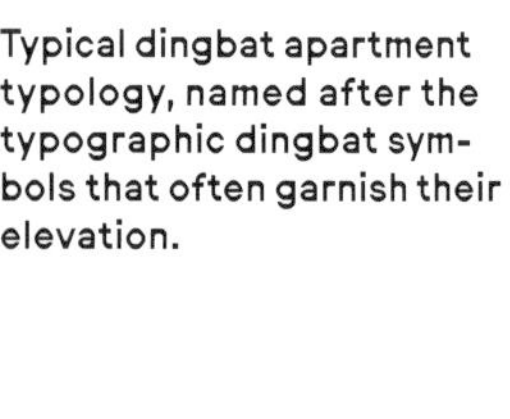

Typical dingbat apartment typology, named after the typographic dingbat symbols that often garnish their elevation.

Expansive surface parking lot fronting the beach.

generous embrace of the car in its second era. Up until even modern day, the city has invested in plans to further expand and widen its freeways, including the I-405 freeway widening project that adds multiple lanes in a 10 mile stretch between the I-10 and I-101 freeways. Other freeway expansion plans, like the I-710 expansion which was years in the making, was just recently canceled in 2022 due to backlash from residents and those communities it would impact.

As previously discussed, one of the largest accommodations our cities have made for automobiles is dedicating huge swaths of land to store them when they are not in use. In Los Angeles, this trait is on full display, even more so visible than many of its metropolitan counterparts. A 2015 study documented that the overall number of automobile parking spaces in L.A. County tripled between 1950 and 2010, primarily through the expansion of off-street non-residential parking, which multiplied by a factor of almost five times in amount.[7] Taking the form of multistory parking structures, street parking, and surface lots, the latter represents one of the most defining features of the urban environment in the city. Another recent study estimated that there are approximately 25.4 square miles of surface parking in the core of the city's metropolitan area, an astonishing figure that is comparatively larger than the entire land area of Manhattan in New York City.[8] Due the availability of land in its previous eras, surface lots were an opportunistic way to repurpose underdeveloped or derelict land for consistent, modest revenue to public and private lot owners at very minimal costs. They also represent one of the least valuable land uses that contribute to activating an urban environment, not to mention acting as temperature-boosting heat islands with impermeable surfaces dumping motor oil and stormwater runoff into city infrastructure.

7 Chester et al., "Parking Infrastructure," *Journal of the American Planning Association*, 2015.

8 Sanders, *Renewing the Dream*, 2023.

These cumulative effects result in an Angeleno public realm that is often criticized for catering its spatial orientations and densities to the automobile, constituted by islands of urban form separated by parking lots and wide streets. The result is a shrunken civic realm that is friendly and accommodating to the scale of the passing vehicle, and supremely unfriendly to the scale of a walking or biking person. These automobile-oriented adaptations might have "wrecked" other cities, as Banham playfully suggested. However, in Los Angeles, they have instead become culturally assimilated, and one might even say a point of cultural pride amongst some of its citizens, contributing distinctly to the image and urban experience of driving in the city.

Lastly, it is important to mention that the mobility choices that the "Second L.A." made are not only distinctly tied to the evolution of its urban structure and form, but also to its deep and fraught history of race and class struggles. It is a city that, like many other American cities at the time, bore witness to a series of highway revolts that in the case of Los Angles mostly failed. This left residents of less socioeconomic means to contend with the effects of roads and infrastructures that severed and in some cases completely destroyed their communities. This is a story in which the building of the city's mobility infrastructures was tied to economic and political privilege and class. Historian and Angeleno academic Eric Avila describes in his book, *The Folklore of the Freeway: Race and Revolt in the Modernist City*, the diverging fates of Beverly Hills and Boyle Heights in the city.[9] This was as an example of one highly wealthy and politically influential community in 1975 that was able to lobby and scrap the city's plan for the Beverly Hills Freeway, which would have severed it in two. Meanwhile in East L.A., the ethnically Hispanic neighborhood of Boyle Heights watched six freeways slice through it over a period of two years in 1960, quarantining and segregating it through two massive interchanges that displaced thousands of residents despite their challenges and protests. Highways like the I-10 went so far as to contributing to the complete erasure of certain communities, like the predominantly Black neighborhood of Sugar Hill in 1963. Los Angeles is a city that today is also still reeling from the geographic effects of racist redlining practices in which financial services, including the denial of credit, insurance, and home-loans, were withheld from neighborhoods of ethnic and low-income minorities classified as "hazardous" to investment. Working in parallel with highway construction, the exclusionary zoning and transportation policies in Los Angeles during this era segregated its geography by race and class boundaries.

9 Avila, *The Folklore of the Freeway*, 2014.

Grappling with the compounded spatial and social effects of the city's 2nd iteration, we are witnessing Los Angeles today develop in its 3rd iteration. This "Third L.A.", described by Hawthorne as taking place starting from the turn of the 21st century, is one that is trying to reinvent itself, grappling with prior decades of transportation decisions while facing new challenges in its evolving future. It is a city that is taking real and measurable steps to move beyond post-World War II urban growth models, including mainstays like the private automobile, freeways, and single-family zoning that defined its previous era. It is a city that "no longer dreams of infinite westward expansion" as Hawthorne describes, "and of growing its way out of every problem by expanding outward at its peripheries to gobble up new 'empty'

1939 Redlined map of Los Angeles issued by the Home Owners' Loan Corporation.

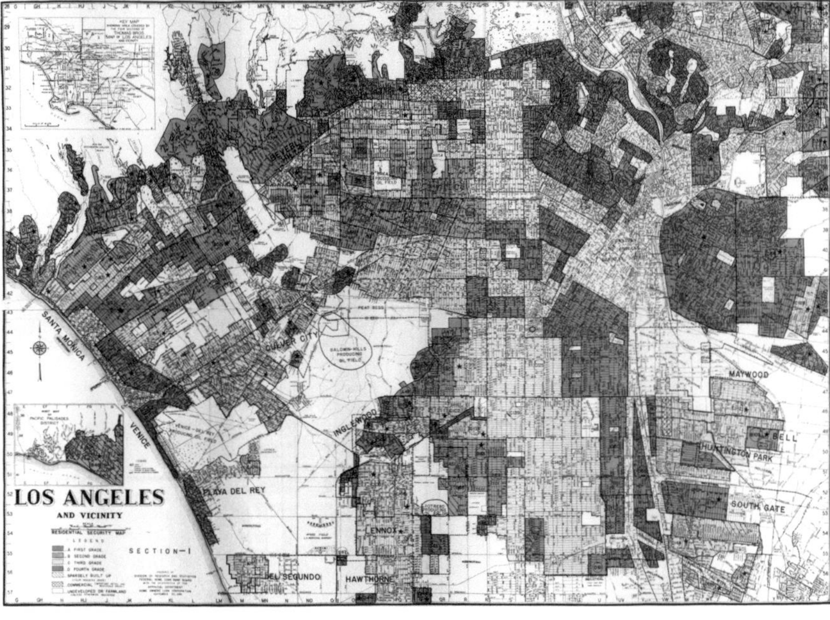

10 Artbound, "Third L.A. with Architectural Critic Christopher Hawthorne," *KCET*, 2016.

land through lower-density subdivisions."[10] Instead, the city is folding back upon itself, looking to develop more intensely inward through higher-densities, multi-family housing, mixed land uses, and shared public spaces. It is a city that is confronting increasing economic inequality and residential unaffordability, grappling with all the negative externalities of car culture while simultaneously contending with fast-landing new mobility technologies.

It is also a city that is now looking to offer increasingly diverse ways for its residents to move around, to free itself from the shackles of 20th century auto-oriented infrastructure and urban form. Decades of growing traffic congestion, environmental degradation, increasing populations, and escalating gas prices have led to a growing cultural appetite and public demand for the return of mass transit in the city. This unprecedented demand has led to various policy and funding measures (including Measures R and M) that have passed in landslide public votes, granting billions to the city's transportation agencies to expand and construct a 21st century transit system. In-and-of-itself, this is already a significant achievement, as prior decades of Motordom-fueled media and culture wars had solidified the embrace of the automobile and the necessity to accommodate it in the Angeleno way-of-life. But a return to implementing and expanding public transit in a city so previously dependent on the automobile is not such a simple task. Additionally, the proliferation of ride-hailing apps and new first/last mile mobility companies through bike and scooter sharing have already impacted the city's transportation choices in unintended ways. The arrival of the driverless vehicle will certainly add to this complicated mobility mix. Let's examine this "Third L.A." and its complex mobility contestations in more detail next.

View of downtown Los Angeles, showing an increasingly multi-modal city and street right-of-way.

Today's "Third L.A." is witnessing a major political and cultural shift towards expanding mass transit in the city. Interest in rebuilding its mass transit actually began in the early 1970s, when the city began dedicating efforts to opening up dedicated busways along the I-10 in 1973. However, it was only until 1976 that the state formed the L.A. County Transportation Commission (LACTC) to coordinate the planning of the countywide transportation systems. Prior to LACTC being established, the Southern California Rapid Transit District (RTD) oversaw various bus services, consolidating previously privately run bus companies under a public umbrella as well as piecemeal rail projects to mixed success. It was only until much later in 1993, that the city's modern day transit authority L.A. County Metropolitan Transportation Authority (LACMTA, branded and referred to as Metro) was established, merging the LACTC and RTD rival agencies into a single regulatory body.

Mass transit during these initial years expanded incrementally and piecemeal, resulting from the competing operations of these two agencies before Metro was formed, as well as from the political challenges of the many counties and constituents involved and the mixed public support transit received. Despite these challenges, by the 1990s, the city had completed the Blue Line, its first modern-day light-rail corridor. Eventually through measures like Proposition A, the city's Purple, Blue, Green, and Gold lines—designated at the time by color—as well as the Silver and Orange bus rapid transit lines followed and were established bit by bit. Some were a success, like the Blue line that underwent further expansion and platform lengthening, while others like the Red line saw mixed use in its early days. Many of these lines functioned in this era as regional commuter lines, connecting the city's downtown to its outer counties in Long Beach and Pasadena.

However, it was not until Measure R was approved and enacted in 2008 that larger comprehensive plans for mass transit were put into place. Measure R folded in the city's previous transit efforts while putting forth $40 billion in funding for the expansion of new Metro lines, bus operation improvements, and other transportation capital projects.[11] This rail revival renaissance received an even bigger boon in 2016 when 71% of Angeleno voters approved Measure M.[12] This second measure doubled the city-wide half-cent sales tax increase that funded Measure R, with an additional no-sunset, half-cent sales tax increase that also made Measure R permanent and in perpetuity with no expiration date to the sales tax revenue. Earmarked with a horizon year of 2057, Measure M granted Metro a further $120 billion, or an estimated $860 million annually, in decades of funding to expand upon Measure R. Measure M added new transit projects, further expanded transit lines in the metropolitan region, subsidized transit fares, and expedited other transportation capital projects previously approved under the original Measure R. These massive investments into transit and transportation improvements for the city carry the potential to fundamentally transform not only its mobility but the urban growth and densification of the Angeleno metropolitan region in extraordinary ways.

11 "Measure R," LA Metro, 2023.

12 "Measure M," LA Metro, 2023.

1992 Metro Rail map, 30-year projected buildout plan.

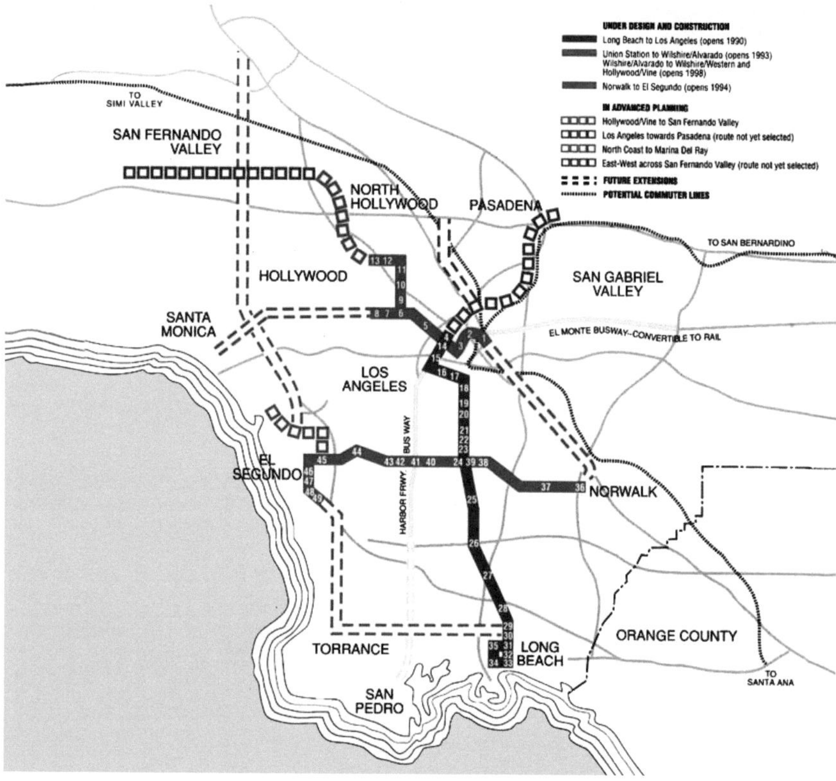

2024 Current Metro Rail and BRT service map.

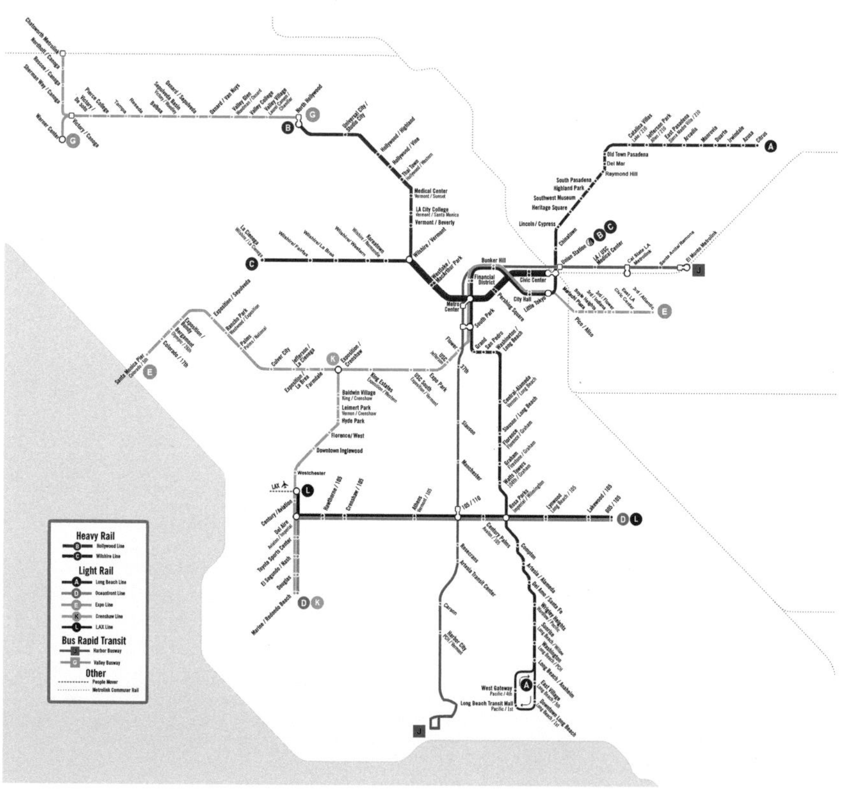

2057 Projected Measure M Metro Rail and BRT buildout.

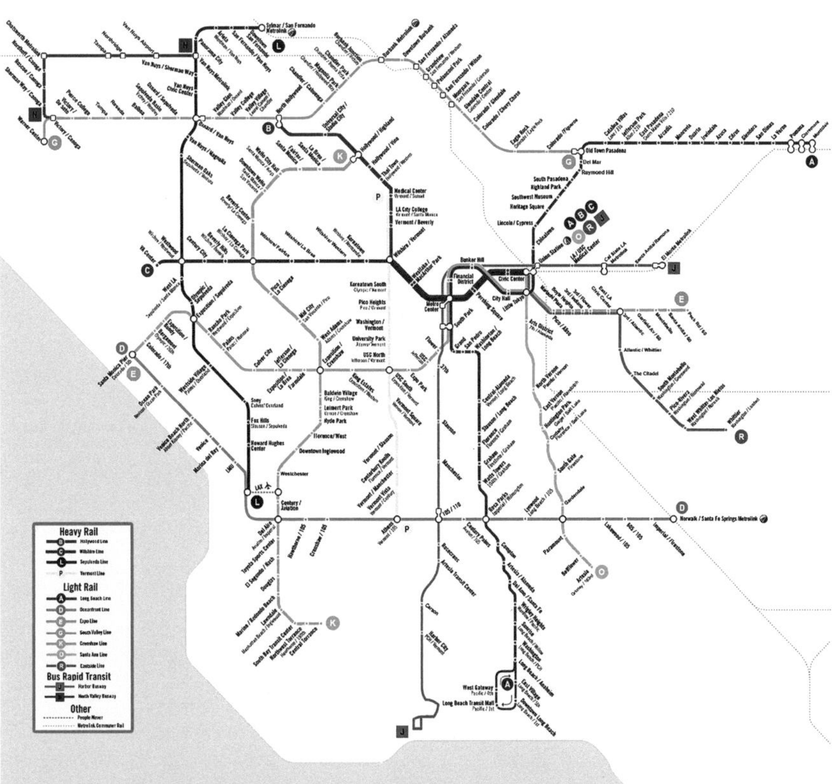

2090 Speculative (concept) future Metro Rail and BRT expansion.

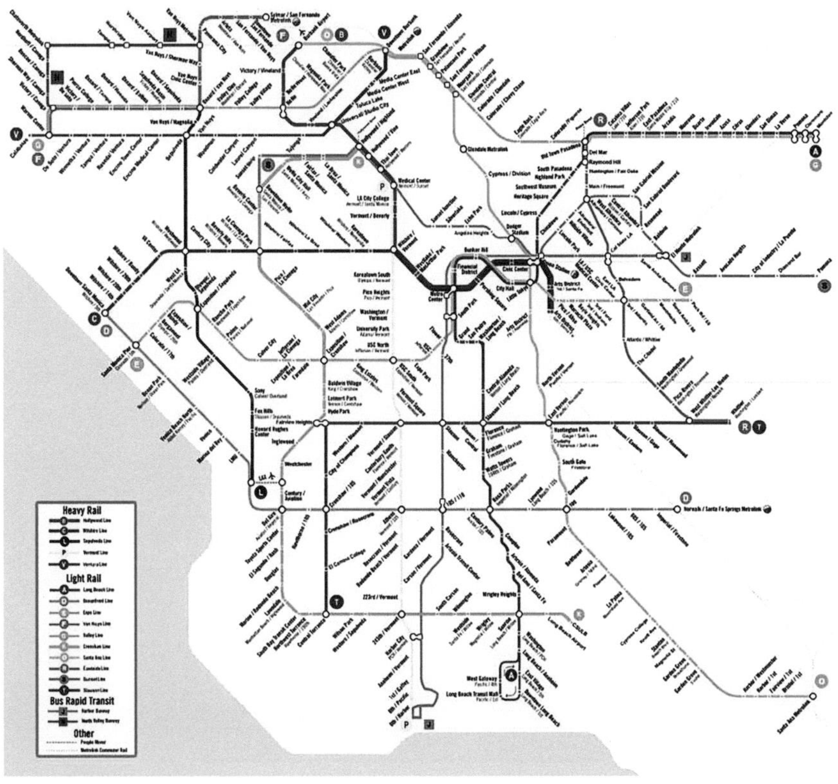

However, significant challenges remain in the way of Los Angeles becoming a transit-oriented city once again. This is a transformation that will be and has been contentious, as changes to the city's infrastructure and mobility cultures meets resistance by Motordom and those whose love of automobiles is intractable. This is a city that in the years since implementing Measures R and M, have had to contend with a continual battle over it's street right-of-ways. It is a city where recent bus and rail expansions, many of them at surface-grade, have had to grapple with sharing street lanes with the automobile. For example, the recent expansion of the majority at-grade Metro Exposition Line (the E Line) that runs all the way westward from East L.A. to Santa Monica, does not receive signal priority when approaching certain major intersections. This means that when traffic lights are red, one will

Metro Exposition Line light rail train waiting at a red light intersection in Santa Monica.

Congested HOV lane during rush hour traffic on an Angeleno freeway.

often see a multi-cabin subway carriage capable of carrying hundreds of passengers, waiting its turn to cross an intersection while single-occupancy vehicles zoom across. This is certainly an unusual sight for those from transit-rich cities accustomed to high-speed rail lines that receive hierarchical preference in the mobility pecking order. Other anecdotes include the fact that many of the high-occupancy vehicle or carpool lanes—meant to incentivize multi-occupancy vehicle use in the city's freeway network by providing dedicated fast-speed, low-traffic lanes—are often congested during rush hours. The few bus rapid transit lines that are allowed to run on Angeleno freeways must often contend with this congestion, and one often witnesses both express and local buses stuck in automobile traffic. Dedicated bus-only priority lanes on local streets are far and few, not to mention the lack of a robust and continuous network of bike lanes, bikeways, and even pedestrian sidewalks in some cases around the city.

Other challenges pertain to the unique geographic structure and sprawl of the city's metro region, as well as cultural and perceptual challenges to incentivizing modal shifts towards mass transit. When examining L.A. Metro's current rail transit network, one can clearly see that it is mono-centric in configuration, focused around its historic downtown core with the majority of its lines sprawling outward from this center towards the outer reaches of the city's peripheries. These mono-centric rail lines have been overlaid on the unrelenting Jeffersonian gridiron structure that underlies the city's street network.[13] This gridiron is primarily serviced by a network of local buses that crisscross the region in continuous linear routes. This means that often, transfers between lines are frequent when traversing from destination to destination. With the city's Measure M planned rail line expansions, one can see that future rail lines have been designed to expand from a mono-centric orientation towards a more democratic grid that follows the underlying structure of its street network. However, as we discussed in the previous section of this chapter, Los Angeles is a distinctly poly-centric city, featuring upwards of 17 different urban centers around its metropolitan area. The evolution of these poly-nucleated centers do not lie neatly on the structural grid of the city—a grid which has historically in Los Angeles been unable to sustain concentrated urban densities at a measure commensurate with its more transit-oriented counterparts. Therefore, it is fair to question whether Measure M's expansion of Metro's light rail network into this existing gridiron paradigm will provide comprehensive transit service that works to support the poly-centric urban densities unique to the city of Los Angeles and its diverse employment centers. Are there alternative transportation paradigms that can negotiate the city's poly-nucleated centers, gridiron structure, and existing mono-centric network while incorporating automation in transit? This alternative structural paradigm will be explored in the following chapter.

Furthermore, while Metro's Measure M plans are highly focused in both messaging and budgeting on expanding the city's rail and subway system, the city's bus network remains in reality the current backbone of Angeleno transit in both use and reach. According to the latest statistics, of the subset of Angelenos that use the city's public transit system, only 12% use the city's light rail system while

13 The Jeffersonian grid is a land surveying system based on an unrelenting grid-acre system that is laid out over a territory independent of topography or terrain, with its conceptual origins traced back to Thomas Jefferson's Land Ordinance of 1785.

another 10% use the city's heavy rail network.[14] However, the overwhelming majority of Angeleno transit users, 78% to be exact, take the city's comprehensive bus system instead. At present, there are just over a hundred rail stations servicing Metro's six total light and heavy rail lines reaching 109 miles of service. Meanwhile, there are nearly 12,000 bus stops and 120 bus routes servicing almost 1,500 square miles, far exceeding the geographic reach and capacity of the city's nascent rail system.[15]

Despite these disparate proportions, the majority of Measure M's funding goes towards its rail network. Prioritizing rail investment fundamentally comes at an opportunity cost of servicing, expanding, and improving the city's far more used and in-demand bus network. It is a bus network that could be argued is in dire need of service, operational, and infrastructural upgrades. Latest estimates put the average wait time at a bus station at 17-minute headways, with 50% of riders often waiting 20 minutes or longer.[16,17] Long wait times are further compounded by the fact that almost one-third (27%) of riders transfer at least twice during a single trip, meaning they have to wait for three different buses or trains in order to reach their final destination. These delays are attributed to traffic congestion and scheduling procedures, as the majority of the city's local bus network do not have dedicated bus lanes and must contend with unpredictable and ever-present Angeleno car congestion. Transit in L.A. takes nearly double the time (on average of 54 minutes) for commutes, versus commutes using a personal car (on average of 33 minutes) which is one of the biggest discrepancies of all American cities.[18] In addition, passengers must wait at spartan bus stops that often involve just a single uncovered bench and signpost exposed to the elements including desert heat and sun. Of its 12,000 stops, only a quarter (26%) provide any form of shade.[19]

Bus transit also faces significant social and perceptual barriers. Frequent complaints of the city's bus service include cleanliness and safety concerns. Moreover, of the 78% of transit riders who use the

14 "Annual Number of Passenger Boardings of LACMTA," *Statista*, 2023.

15 "Metro Facts at a Glance," L.A. Metro, 2023.

16 A headway is the amount of time between transit vehicle arrivals at a stop.

17 "Public Transit Statistics by Country and City," *Moovit Insights*, 2022.

18 "America's best (and worst) Commuter Cities," *GeoTab*, 2019.

19 Brozen et al., "Are L.A. Bus Riders Protected?" UCLA Lewis Center for Regional Policy Studies, 2023.

Typical Metro bus stop bench (uncovered) and signpost.

20 "Results of our 2022 Customer Experience Survey," L.A. Metro, 2022.

21 "The High Cost of Transportation," Institute for Transportation & Development Policy, 2019.

22 "Annual Number of Passenger Boardings of LACMTA," *Statista*, 2023.

23 Manville and Cummins, "Why Do Voters Support Public Transportation?" *Journal of Transportation*, 2015.

city's bus service, 88% are people of color.[20] 58% of L.A. bus users live below the poverty line. Driving is by definition a privilege that only those who can afford to own and operate a car have access to. In transit-rich cities, public transportation offers a primary and more affordable method of moving around, and thus becomes a system where populations of different walks of life come together. Enrique Peñalosa, former mayor of the city of Bogota, once said that "an advanced city is not a place where the poor move about in cars, rather it's where the rich use public transportation."[21] Photographer Andre Wagner famously described New York City's subway as a great equalizer in which a wide range of class and social statuses use the system ubiquitously. But in L.A., a transit system that in other cities acts as an equalizer only further reinforces income inequality, marginalizing already struggling communities. Unintentionally or not, Metro's prioritization of rail projects in Measure M funding disproportionately overlooks the fact that its largest user market (populations of lower socioeconomic status) take bus transit. Prioritizing bus transit is fundamentally a question of equity in the city.

All of these compounded issues—the long duration of crosstown travel, wait time lengths and transfers in low-quality station stops, first/last mile problems on pedestrian-unfriendly built environments, and the social barriers to use—not only demonstrate the need for investments and improvements but also the significant barriers that must be tackled in the city's bid to increase transit use. Funding the building and expansion of transit is one thing, getting people to use it is another. Despite the massive investments that Measures R and M have poured into the city since 2008, transit ridership remains extraordinarily and disproportionately low. Recent estimates from the latest 2020 census estimate that only 8.9% of the entire population of the city uses public transit.[22] Moreover, annual ridership has steadily declined in the past decade, nosediving in 2020 and 2021 due to the Covid-19 pandemic. The only increase in transit ridership in the past decade was from 2021 to 2022, which saw a rebound from the pandemic, but 2022 figures still remain far significantly lower (nearly 33% less) than pre-pandemic years. A study by the Institute of Transportation Studies at UCLA determined that while voters overwhelming voted in favor of Measure M and want more mass transit options, the same voters don't necessarily use those options even when provided.[23] This is because few Angelenos view transit as an amenity that directly benefits them. As theorized by this study, Angelenos voted for Measure M as an expression of their political beliefs and in support of the broader good—that *someone else* will use this public service and improve congestion, *just not me*. People who voted for transit because they believe it reduces congestion are voting in favor of it because they want driving to be easier. However, transit works best in places where driving is harder. The same study found that Measure M proponents weren't nearly as supportive for the type of land use, urban form, and structural changes that could actually make driving less appealing in the city through policy and spatial changes like paid parking, highway tolls, congestion pricing, increased housing density, and streets dedicated to bus and bike lanes. Building good service is one thing, breaking old habits and cultural attachments is another—one in which Book 2 (*The Experience*) propositionally tackles.

While a modal shift to mass transit in L.A. may be challenging, it is not insurmountable. As Chapter 3 has discussed at length, driverless vehicle technology can be used to support these efforts. In fact, they *must* be used by emboldened Angeleno public transportation agencies or else they risk proliferating into the private market through individualized transport that will only exacerbate these challenges and further reinforce the car-oriented structure and built environment of the city.

While a few AV regulatory standards have been put forth at the state level, and initial research reports have been conducted at the city's Department of Transportation (LADOT), no AV funding or implementation plan has been put forth through Measure M.[24] Current plans remain silent about the potential role of automation in bus travel, even though bus transit in particular has the exciting and distinct potential to be revolutionized. In 2018, Metro initiated a study called the NextGen bus plan that was developed to understand and update the bus network of the city, a system that hadn't seen a significant change in 25 years. This NextGen bus plan was just recently approved in 2020, and Metro has begun to roll out incremental improvements to the city's bus network post-disruptions brought on by the pandemic.[25] This is exactly where driverless technology can be deployed most impactfully and to transformational effect. Automating transit could significantly improve service and efficiency, as well as network reach and frequency. Additionally, it could spark cities to not only reprioritize but reconceptualize how local and rapid bus networks work in tandem with the planned rail expansion of Measure M. Multiple scales of local bus transit, including the city's neighborhood DASH service could be folded into this new network hierarchy.[26] If the city's automated transit network were to shift towards a poly-nucleated structure, as the next chapter will propose, a new spatial paradigm for movement, densification, and development in the "Third L.A." could emerge. Automated bus rapid transit along with rail could form the regional backbone between the city's poly-centric centers, while a range of smaller AVs could service each of those poly-nucleated hubs within a concentrated local area.

These particular changes for an alternative automated transit paradigm are illustrated geographically in the following chapter. The history and contemporary context that frames it demonstrates that the city is at a ripe moment for receiving and incorporating it. In particular, L.A. has already established contemporary precedence when it comes to regulating and incorporating new fast-landing mobility technologies on its streets and built environments.

When on-demand, ride-hailing apps and the transportation network companies (TNCs) pioneering them landed and proliferated in the city in the early 2010s, Los Angeles' transportation agencies were unprepared and caught off-guard, as almost all cities in the U.S. were at that time. As a result, traffic congestion in the city ballooned as their popularity grew exponentially. By the time regulations on TNCs were eventually put into place at the state and city levels in 2013, their use and subsequent effects were already ubiquitous. Furthermore, these mobility technology companies have a history of ignoring and evading local regulations. With their policy of permissionless innovation, they are criticized for their intentional strategy of commencing operations in a city without

24 Hand, *Urban Mobility in a Digital Age*, L.A. Department of Transportation, 2016.

25 "NextGen Bus Plan," L.A. Metro, 2020.

26 DASH, short for Downtown Area Short Hop, is a free-of-charge neighborhood-scale local bus shuttle service operated by LADOT outside of Metro, designed to serve travel within certain neighborhoods on internal fixed routes. Since its launch in downtown L.A., it has expanded to 27 other neighborhoods in the city.

seeking permission, as well actively lobbying politicians to change regulations in favor of their business models. In addition to their documented contributions to traffic congestion, which we have previously discussed, ride-hailing vehicles also increased the contestation over curbside use and space in the city. A recent study found that on the frequented Santa Monica Boulevard in West Hollywood, only 5% of its curb space had been reserved for passenger loading. This led to ride-share vehicles that were double-parked in active traffic lanes to load or unload passengers (41% of ride-share vehicles double-park a cumulative average of 37 minutes per hour) adding to already congested city streets.[27] Additionally, no mobility data-sharing standards were required by city agencies, making it difficult to regulate let alone even understand the impact that TNCs were having on urban mobility systems. This fight over the right to data is one that continues to play out today, but one that private mobility companies have typically won.

27 Sanders, *Renewing the Dream*, 2023.

Recognizing this contentious battle that continues to play out post-TNC proliferation, Los Angeles transportation agencies were better prepared when the recent wave of on-demand, first/last mile micro-mobility solutions began proliferating in the city in the late 2010s and early 2020s. When on-demand e-scooters and e-bikes through companies like Bird, Lime, Spin, Veo and many others deployed in the city, the L.A. Department of Transportation (DoT) launched the Mobility Data Specification (MDS) standard and required all new mobility technology companies to comply.[28] Pioneered by DoT General Manager Seleta Reynolds, the MDS standard requires all ride-shared bike and e-scooter companies to connect their services to a city-run application programming interface (API) that keeps track of where pick-ups and drop-offs take place on different mobility modes. This forward-thinking regulatory standard—one of the first and few cities in the U.S. that has implemented it—allows L.A.'s DoT to electronically manage various private transportation services in real time. By doing so, it can geo-fence certain parts of the city off as "no scooter zones,"

28 "What is MDS," LADoT, 2023.

Misparked on-demand e-scooters as urban hazards on sidewalks and street right-of-ways in cities without micro-mobility regulatory standards.

Metro bike-share dock hub at rail station.

Metro rail line with new transit-oriented housing construction behind.

Tranzito-Vector bus station shelter with shade, seating, lighting, integrated display, and charging outlets.

including popular pedestrian sidewalks or parks. It can even electronically reduce the speeds of e-scooters and e-bikes when entering such zones wirelessly. MDS can also restrict users from dropping off scooters or bikes in high-use sidewalks or urban areas where parking those micro-mobility vehicles is prohibited. MDS also allows regulators to send live instructions to mobility consumers and companies to inform them of dynamic street closures, traffic updates, and other useful information that can smooth transportation flows in the city. These standards have allowed the city to generally avoid incidents of parked bikes and scooters taking up huge swaths of sidewalks that leave little room for pedestrians in high-demand areas of the city. These documented curb space issues produced by micro-mobility free-for-all parking has plagued other cities without such standards. The MDS standard allows DoT to serve in effect, as both a physical and digital custodian of the public realm, protecting the safety, use, demand, and organization of the city's public right-of-way.

The implementation of MDS in Los Angeles is an example of a city that stands out in its capability to confront and tackle new mobility technologies, and the shifts they bring, with multi-pronged planning approaches. More so than in a number of other large American cities, Los Angeles has strong and capable Department of Transportation, a well-funded Metropolitan Transit Authority, as well as a forward-thinking City Planning Department that work reasonably well in synchrony. These public institutions have already begun leading the charge to adapt the city to shifting cultural and transportation preferences. In addition to MDS, DoT and Metro run a publicly-owned bike-share program called MetroBikes, which has allowed them to trip-chain bike-share use-fees to public transit fares. Taking a Metro bike to a transit station and switching modes to bus or rail doesn't incur a second fare charge for the rider.[29] Along with regulatory oversight over private micro-mobility apps, DoT and Metro have also been piloting innovative first/last mile service options around rail stations in partnership with private TNC companies like Via. Just like Metro bikes, ride-hailing a Via car that drops a passenger off at a Metro station doesn't incur extra ticketing costs for switching to transit. Along with affordable housing construction requirements stipulated by Measure JJJ, which led directly to the creation of the Transit-Oriented Communities Affordable Housing Incentive program, a push towards developing greater allowable densities, affordable housing, and amenity clusters have begun to emerge around transit hubs in the city.[30,31] Other public-private partnerships include a recent 2022 initiative called the Street Transit Amenities Program that approved a city partnership with Tranzito-Vector—a private company that was given the green light to install, upgrade, and maintain 3,000 new bus stop shelters in exchange for the rights to digitally advertise on them.[32]

These and other mobility innovations that are already under way in Los Angeles have created unique institutional capabilities at the city level to tackle future mobility challenges with collective public interest at the forefront—positioning the city well to grapple with the arrival of driverless technologies. These various mobility innovations and partnerships also signify an increasingly blurred boundary between those responsible for providing transportation

29 "Metro BikeShare," LA Metro BikeShare, 2023.

30 Measure JJJ is a 2016 voter-approved measure in L.A. that requires developers requesting certain entitlements for residential projects to provide either affordable housing units or pay an in-lieu fee.

31 Chiland, "Measure JJJ triggers New Incentives to Encourage Affordable Housing near Transit," *Curbed Los Angeles*, 2017.

32 Tu, "L.A. tries to address its flailing Bus Shelter Program with a New Contract," *Dot.LA*, 2022.

services in the city. While the ebb and flow between public transit and private automobile use in the city's history was clear and immutable, today this demarcation is muddled and intertwined. This blurred boundary could in fact be considered an exciting opportunity in Los Angeles' quest to provide more accessible and better-serviced transportation for its residents. It is one that takes advantage of both market innovations and regulatory standards to do so for all of its citizens. Along with integrated new technological systems and digitally enabled devices, this interweaving of public and private services could further bolster the city's transportation paradigm shift in its third era.

While in very nascent stages, the city has also begun exploring some of the critical transportation policies that Chapter 5 has delineated. Their implementation is crucial to incentivizing Angeleno modal shifts out of personal automobile use. Recently in 2023, Metro has begun considering the implementation of congestion pricing in the city including on a 16-mile stretch of the I-10 freeway between downtown L.A. and Santa Monica.[33] If approved, the pilot program would be launched right before the 2028 Olympics scheduled to take place in Los Angeles.

33 Fonseca, "L.A. could try Congestion Pricing. Will drivers go for it?" *The Los Angeles Times*, 2023.

The city has also piloted other infrastructural changes that have begun to shift the cultural associations that Angelenos have with their automobiles. CicLAvia, a non-profit car-free streets initiative that temporarily closes miles-long stretches of streets to motor vehicles by turning them into uninterrupted bikeways accessible to the public, was introduced in the city in 2010.[34] Based on Bogota's Ciclovia, the event now runs six times a year on new and repeating routes that stretch all over the metropolitan reaches of the city. The event organizers work in close coordination with the city's transit agencies, and funding to CicLAvia is even provided by Metro. Other street-closure initiatives have been introduced since then, including the PlayStreets initiative that temporarily closes a neighborhood block's streets, transforming them into a place where residents of all ages can reclaim their streets for recreational and public uses.[35] Even more recently in 2023, the city closed a six-mile stretch of the Arroyo Seco Highway between Pasadena and downtown Los Angeles to vehicular traffic for four hours on a weekend morning. Organized as the 626 Golden Streets ArroyoFest, the partnership between city agencies and neighborhood organizers opened up the freeway to thousands of pedestrians, runners, cyclists, and other non-motorists to great fanfare, press, and cultural excitement.[36]

34 "What is CicLAvia?" CicLAVia, 2023.

35 "What is the Play-Streets program?" Los Angeles PlayStreets, 2023.

36 "Home/About 626 Golden Streets," 626 Golden Streets, 2023.

By themselves, these seemingly small steps can't relieve dependency on automobile use. However, they are an inexpensive first step that can help a city's residents question the assumptions that perpetuate automobile dependency. In the longer term, if permanent implementation of these initiatives fueled by automated transit can be realized together with expanded transit service, transportation policies, and reduced vehicular demand, then paradigmatic spatial and structural transformations for the city are not only possible but achievable.

While the city pilots these incremental gains and initial reclamations, in the background, self-driving technology is beginning to arrive in the city. In fall of 2023, the autonomous driving technology company Waymo began deploying pilot services in six

CicLAVia street closure event in East Los Angeles.

PlayStreet street reclamation event in Boyle Heights.

ArroyoFest freeway closure event on the Arroyo Seco Parkway.

37 Carpenter, "Waymo will offer Robotaxi rides in L.A." *Spectrum News*, 2023.

Los Angeles neighborhoods.[37] If left unregulated, private AV use threatens to combat the recent mobility gains achieved by the city and even erase the efforts put forth by Measure M. The urgency for city-wide agencies to get ahead of AVs through regulatory and infrastructural standards is real. The urgency to incorporate driverless technologies to assist its multi-modal transit-focused efforts is even more so critical.

All of these contemporary examples illustrate a city that is ready to move past the automobile in its future iteration. It is a city in which its political and social culture has embraced new mobility technologies in recent years. It is a city that has shown the appetite to invest in and prioritize multi-modal shifts, in recognition of the decades-long effects of accommodating to and prioritizing the automobile in its urban form and structural evolution. It is a city that could yet again reclaim its status as a leader in mass transit, but only if behavioral and experiential attitudes change with it. It is a city that holds vast latent spatial opportunities for densification and reconfiguration—including its 25.4 square miles of surface parking, much of which could be released with automating vehicles and transit mode shifts. This book hopes to build upon these recent gains and exciting opportunities, offering an alternative path forward that incorporates AVs into a new transportation paradigm. Understanding the context for which its transportation choices have shaped Los Angeles in its previous two eras, as well as how its urban structure and form have evolved in tandem with such changes, what possible futures and urban visions can be put forth for a city with such impulse, appetite, and readiness for change?

Urban Spatial Implications by Scale

As the first half of this book has illuminated, driverless transport could catalyze transformational spatial and physical changes in our built environments. A variety of key urban typologies in our cities' histories have evolved in response to the automobile and the land dedicated to parking and transporting them. These spatial relationships will be distinctly affected when that vehicle is automated, at multiple scales and through new service models. What exciting design opportunities could emerge when these fundamental spatial relationships change? How will current car-centric building typologies evolve and what new ones could develop in their place? How can block patterns and their associated zoning and land uses change in more mixed, denser, and sustainable ways? What alternative design opportunities exist for our mobility infrastructure that has currently relinquished its majority right-of-way to the individual automobile? How could such paradigm shifts transform the greater structure of the city and how we collectively move around in it?

These questions are what the second half of this book explores, through visual illustrated drawings. The intent of these drawings is to allow readers to imagine, and therefore inhabit, these potential futures. Drawings become as much a physical and digital tool as it is a social one, provoking questions about how these futures could be realized as much as it also forecasts certain possibilities. They not only represent and visualize but also carry with them the potential to actualize. The power of drawing is its capacity to become the blueprint for that future to become realized. In particular, they invite readers to consider the experiences and encounters these urban typologies will impact, both through axonometric aerial views as well as eye-level perspectives—appearing also in the narrative stories that accompany this text in Book 2 (*The Experience*).

These explorations are also grounded in real geographic places, positioning themselves in the city of Los Angeles as a testbed. This tangible locale allows them to be specific to the urban fabric and cultural contexts that they have emerged from. They are rooted in real places—applicable to the urban systems and mobility infrastructures of the city—and yet at the same time able to be applied to a wide range of new contexts and cities exhibiting similar urban traits. These *typological* traits leave open alternative possibilities rather than predetermining them through absolute fixed form. Explorations compare an existing typological condition today, against a possible future scenario unlocked by automation—using Measure M's benchmark year of 2057 as a goal—so that one can understand the latent potential all of them hold. They take advantage of the opportunities that driverless technology catalyzes but are not necessitated by them, leaving open the possibility that these transformations could be achieved regardless. AVs, however, could serve to accelerate and spark these spatial transformations if implemented correctly.

These design implications range from several scales in the following chapters. *XL: Transportation Network Changes* explores how conceptual changes to transportation network and mobility hierarchies can incorporate and leverage driverless technology to form a framework for the densification and development pattern of cities. *L: Mobility Infrastructure Restructuring* explores how our city streets and highways at a variety of scales can leverage driverless technology to reprioritize and reclaim public right-of-ways for alternative uses.

M: Block-Scale Land Use and Density Transformations explores new land use and zoning opportunities unlocked by the potential of land lots released from parking. *S: Adaptive Conversion of Car-Centric Building Types* explores how architectural building types that have evolved around the automobile can evolve when that spatial relationship is disrupted as vehicles become driverless. Collated together, these multiple scales form a diverse toolkit/toolbox of design strategies to guide designers and planners as AVs becomes adopted around the world. They are enabled by the key transportation policies previously discussed and certainly require the right economic and development incentives to be produced. However, if done so, they have the potential to collectively set a radical blueprint for a paradigm shift, unlocking multiple scales of rethinking and therefore redesigning a city's future built environments.

Mobility Assumptions

Understanding the uncertainty that is yet unknown around how AVs will be developed, deployed, and used means that a variety of competing futures and their permutations may yet emerge. As this book puts forth a vision for one of these futures—a best-case scenario for cities to aim for—it is thus important to explicitly outline the important assumptions underlying that vision. These indicators frame the design explorations in the following chapters and the experiential future illustrated in Book 2. This future year 2057 is based on the assumption that there is approximately:

- **70% Overall AV Absorption**
  An overwhelming majority of mobility consumers have embraced driverless technologies. AVs have proliferated, absorbed, and been accepted into everyday use, with only a minority subset of the population still owning and operating non-autonomous vehicles.

- **90% Fleet-based AV Operation and Management**
  Of this majority of AV users, driverless mobility is primarily consumed as a service model. AVs are owned and operated through fleet-based models either directly by public transportation agencies or through private transportation owner-operators in partnership with public oversight. The overwhelming majority of driverless vehicles are not individually owned and those that are require Certificate of Entitlements that limit their number and use.

- **100% Automated Public Transit Network**
  The entire public transportation network of the city has become automated. Driverless technology has been comprehensively rolled out to update and revolutionize public transit operations in a range of new vehicle sizes and hierarchies. On-demand para-transit, mini-shuttles, local bus, rapid bus, and rail operate driverlessly, which allows for more flexible, responsive, and frequent service operations and network routes.

- **100% Regulatory Oversight by Public Sector Agencies**
  Public transportation agencies, including Department of Transportations as well as Metropolitan Transit Agencies, oversee the launch of automated transportation at all service levels and scales. Public sector agencies work closely in hand with private mobility operators to regulate pricing, fares, subsidies, and network reach, route, and service. Transportation agencies also work closely with their regulatory departmental partners like City Planning and Housing to coordinate the urban structure, development, and densification of the city in tandem with mobility. All mobility data obtained by private transportation operators are shared with public agencies to allow them to effectively and holistically manage digital and physical urban space.

XL: Transportation Network Changes

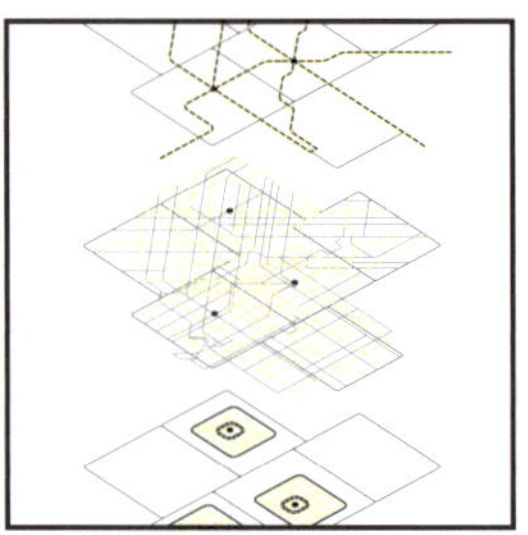
Mobility Hierarchies
Page 185

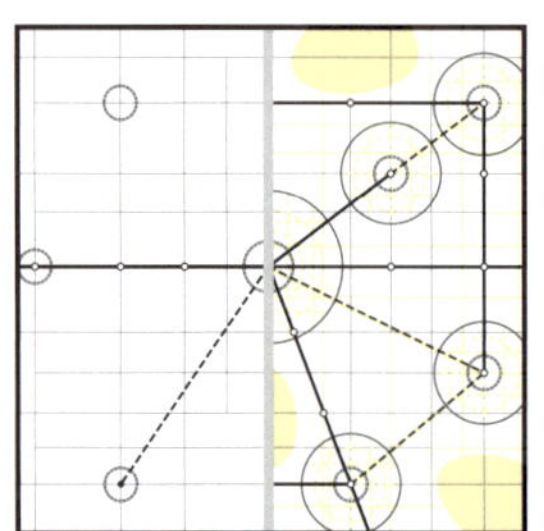
Mobility Framework
Evolution
Pages 186–187

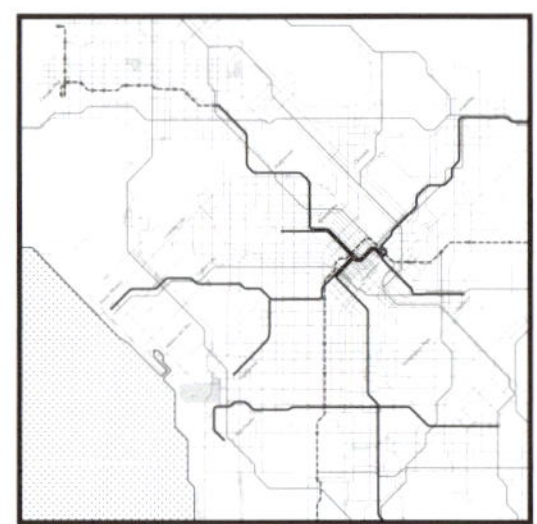
Mono-Nucleated
Grid Model 2024
Page 188

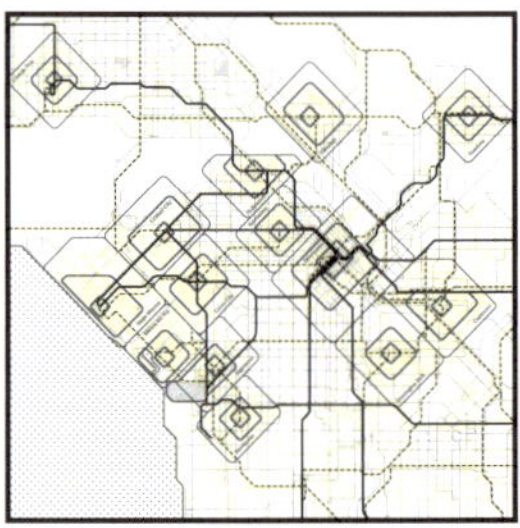
Poly-Nucleated
Spoke Model 2057
Pages 189–193

L: Mobility Infrastructure Restructuring

Highway, At-Grade
(Removal)
Page 200

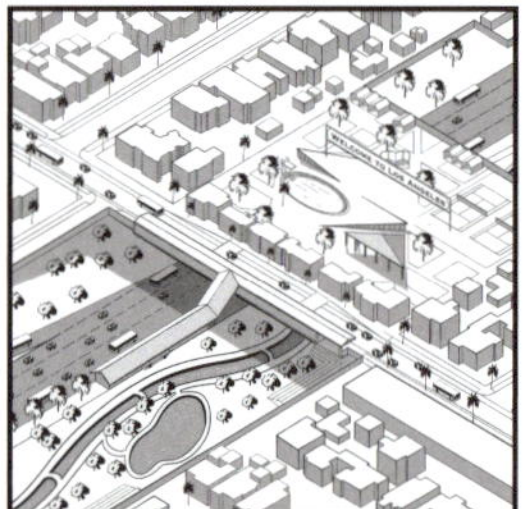
Highway, Sunken
(Decking and Conversion)
Page 206

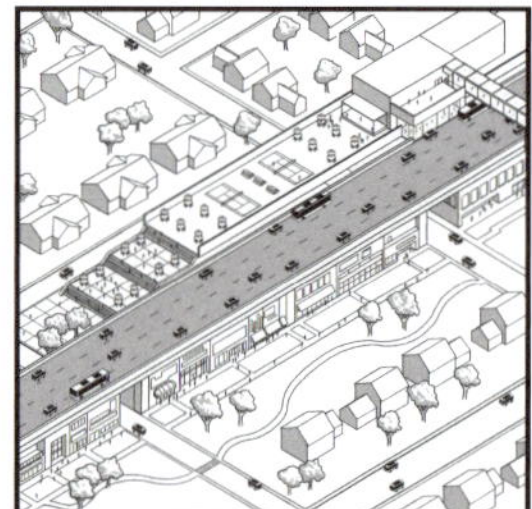
Highway, Elevated
(Conversion)
Page 212

Major Boulevard
Page 220

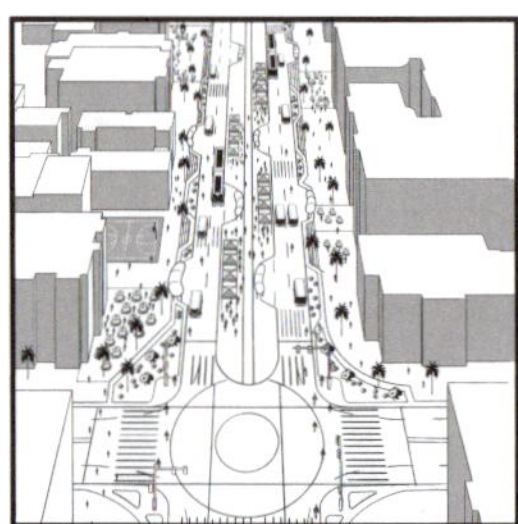
Commercial Street
Page 226

Community Avenue
Page 232

Dynamic Lane-Managed
Street
Page 238

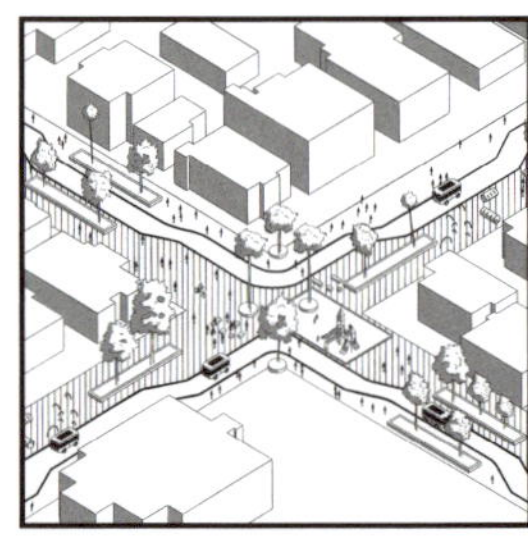
Residential Intersection
Page 246

M: Block-Scale Land Use and Density Transformations

S: Adaptive Conversion of Car-Centric Building Types

7 XL: Transportation Network Changes

7 XL: Transportation Network Changes

Automating our transportation is a seemingly small-scale technological change that will have transformative, large-scale impacts on the way we move around the city. In particular, these technological changes *cannot* be considered in a vacuum, affecting only individualized transport or the personal automobile. Rather, they must be understood in relationship to our comprehensive mobility systems and network layers in order to forecast their impacts and take advantage of their opportunities. This requires a multi-scalar, system-level framework for projecting a vision for how we could move across the spatial extents of the city. The mobility modes and travel choices that we make as individual passengers ripple across the aggregated geography of a city's transportation networks, just as much as the city-wide transportation services that public and private mobility providers offer have individually scaled impacts on these modal choices. These metropolitan-scale network implications range from how we move in the first/last miles, engaging micro-mobility options, all the way to long distance commutes using rapid transit rail or bus.

When considering these multi-scalar impacts, several critical questions arise. What advantages do automating these mobility layers bring to the overall network? How can cities reconceptualize mobility options into scaled and tiered hierarchies? Can AVs address the weaknesses and limitations of current transportation geographies? Can they fill the first/last mile gap challenge in transit? Can they support rather than compete with transit operations as a whole? How will AVs impact accessibility and access to other mobility modes, including walking and biking? Can our networks and services be more responsive and flexible to variable demand? Cumulatively, what blueprint for future growth and densification does this network paradigm shift spark?

The arrival of driverless vehicles sparks an opportunity to redesign our transportation networks with answers to these questions in mind. Transportation modeling and planning is a highly complex and often theoretical endeavor, one in which this chapter does not purport to provide expertise in. However, what it does do is suggest a conceptual framework for rethinking these complicated layers in diagrammatic terms. Today, fledgling transit systems are typically organized through a network of lines radiating outwards from the central downtown of a city. Typically, this spatial planning model reinforces the mono-centric nature of a city's historical development, which displays a high population density and business activity in a center core with decreasing density as one moves farther from it. The development of Los Angeles's transit system also follows this structure, as we've discussed in the previous chapter. However, the city exhibits a far less strongly defined downtown core than cities of comparable size, instead having developed around multiple poly-nucleated urban centers of competing economic activity. There are up to 17 diverse job centers in L.A., which means that traversing between and accessing them requires complex, multidirectional journeys. However, due to the mono-centric nature of its current rail network, journeys between poly-nucleated centers often involve inefficient and indirect travel to its traditional downtown, as well as transfers between lines and often between modes. This mono-centric rail structure is overlaid onto the underlying gridiron of the city, the latter of which enables the point-to-point direct route connection that automobiles provide. What this means is that the larger transit structure of the city, which follows a hub-and-spoke model, is often at times independent of the point-to-point travel demands occurring between multiple centers serviced by its urban gridiron.[1]

Can this democratic grid network of the city, and the multiple urban centers that don't lie neatly on it, be incorporated into a transportation model that better services L.A.'s distinct poly-centric characteristic while building upon its current transit network and future expansion? This alternatively envisioned transit model, explored in the following diagrams, suggests such a structural blueprint. It transitions the mono-nucleated grid model of the city's current network into a poly-nucleated spoke model that more comprehensively facilitates driverless mobility between the many poly-centric urban centers of the city.

Within this overarching framework, a hierarchy of new transportation layers, including expanded express buses and first/last mile shuttles, are introduced into the city's existing transit network that currently are comprised of only rail and local buses. Hierarchically, Metro rail and bus rapid transit work together to form the backbone of the poly-nucleated spoke system of the network. Rail lines would be expanded to service long-distance, high-capacity trunk lines that not only connect Los Angeles' downtown to its poly-nucleated centers, but also between each of these urban centers directly. These direct connections are also serviced by expanded express bus rapid transit corridors, which could be established as precursor lines prior to the implementation of rail infrastructure. Bus rapid transit (BRT) is a far more cost-effective way to implement long-distance transit service than rail, which require cities to build out new infrastructure, establish new right-of-ways, or tunnel underground—all which can be expensive. BRT corridors serviced by dedicated priority lanes and transit signal priority can be established first, testing variable demand, and once successful, could eventually be replaced by the fixed infrastructure of rail. BRT corridors could also be established on the city's extensive highway network, taking advantage of the existing high-speed, long-distance connectivity they provide. At the next scale of the network, the local bus grid of the city is expanded in quantity of lines and in frequency of service through the fiscal cost-saving opportunities that automation brings. A higher density of autonomous local buses could then crisscross the gridiron network of the city, servicing medium-to-local distance trips while supporting travel between multiple urban centers.

Within each of these poly-nucleated centers, three co-centric zones are established: the existing district core (Zone A), an expanded first/last mile area (Zone B), and the district's transport zone boundary (Zone C). Zone A's core is centered around major rail or BRT transit hubs and is serviced by new micro-mobility options like scooter and bike-sharing through pedestrian-oriented streets. This zone can be thought of as akin to traditional transit-oriented development areas. Overlaid on this core, Zone B expands the first/last mile service range of this transit-oriented area through new automated transit shuttles carrying three to five passengers. These local district shuttles operate on fixed routes to collect and feed passengers in this zone to long-distance BRT and rail lines, expanding the geographic

1 In the context of transportation, a hub-and-spoke model is a distribution network that organizes routes as a series of spokes to connect to a central hub. Resembling a bicycle wheel in diagram, the hub sits at a central location (typically representing a city's downtown) with lines radiating outward along spokes.

range of the first mile, up to several miles for denser, transit-serviced development. Finally, Zone C, which divides poly-nucleated centers from each other, delineates the transport zone boundary that incurs a transport zone fee on any individual private vehicles crossing from zone to zone. This zone incentivizes private AV use for short distance trips within a poly-nucleated urban center, rather than through longer-distance commutes which compete with the efficiency of rail and BRT. Any areas that fall outside of and between Zone Cs are serviced by flexible-route, on-demand mini-shuttles carrying one to three passengers. These driverless mini-shuttles provide services, akin to ride-hailing, to areas of the city outside of the poly-nucleated cores. These are areas of the city that do not typically have enough population density and ridership to support fixed-route transit.

The framework for this future mobility network does not negate the individual vehicle, autonomous or not. What it does do, however, is expand the reach and modes of mobility options to *counter* the car-dependency that many cities, and in particular Los Angeles, suffer from. A variety of trip types and scales service the diverse needs and distances required by the urban dweller. Instead of having to choose between undependable local bus transit requiring multiple transfers, rail transit that underserves the first/last mile gap, or the cost-prohibitive car that faces traffic-ridden streets and highways, there is now a holistic mobility hierarchy that covers a wider extent of the city's geography. What this framework also does, is better incentively tie a passenger's trip length to vehicle size, reducing the spatial costs that single-occupancy individual vehicles cumulatively impose on valuable urban space especially when traversing long distances. This framework also suggests how point-to-point ride-hailing services can be better incorporated into the network so as to support lower-density areas of the city rather than competing with densely serviced zones. Together, the network is supported by transportation policies like transport zone fees, congestion pricing, and dedicated bus-only lanes, all of which make using automated transit more convenient, efficient, faster, and affordable then a private automobile or AV.

This model for Los Angeles' NextGen transit system, and in particular its buses and shuttles at the aforementioned scales, can be revolutionized by driverless technologies. Automating the existing fleet while introducing new vehicle fleets could reduce operational costs, of which funding reallocation must go towards expanding more lines and implementing more frequent and reliable service schedules. The entire network would be regulated by a coordinated public transit authority, which would oversee service types ranging from flexible on-demand zones to more traditional fixed-route services. Similar to successful public-private models in other transit-rich cities, individual lines, service types, and zones could be operated by public agencies or bid out to private companies to encourage quality of service and competition to the benefit of mobility consumers. The city would work in partnership with private mobility operators to determine routes, schedules, fare pricing, fare subsidies, and ensure minimum quality of service. With these frameworks, the entire network could transition into an integrated hierarchy of mobility scales, and a paradigm shift for moving around the city would emerge.

The design of this transportation framework serves not only the city of Los Angles, of which it is applied to, but it also acts as a typological blueprint for other car-dependent cities looking to expand their transit systems. The diagrammatic nature of the drawings in this chapter open up possibilities for rethinking the geographic and spatial relationships of moving around a city and how they could synergistically work with urban development to constitute a greater whole. If implemented, Los Angeles could develop inwards around its multiple urban centers and the higher-density corridors that connect them, rather than continuing to expand in its outer peripheries through lower-density form and segregated land uses. Fundamentally, it not only sparks a driverless mobility paradigm shift, but it becomes an exciting structural template for the city's densification and growth in its future eras.

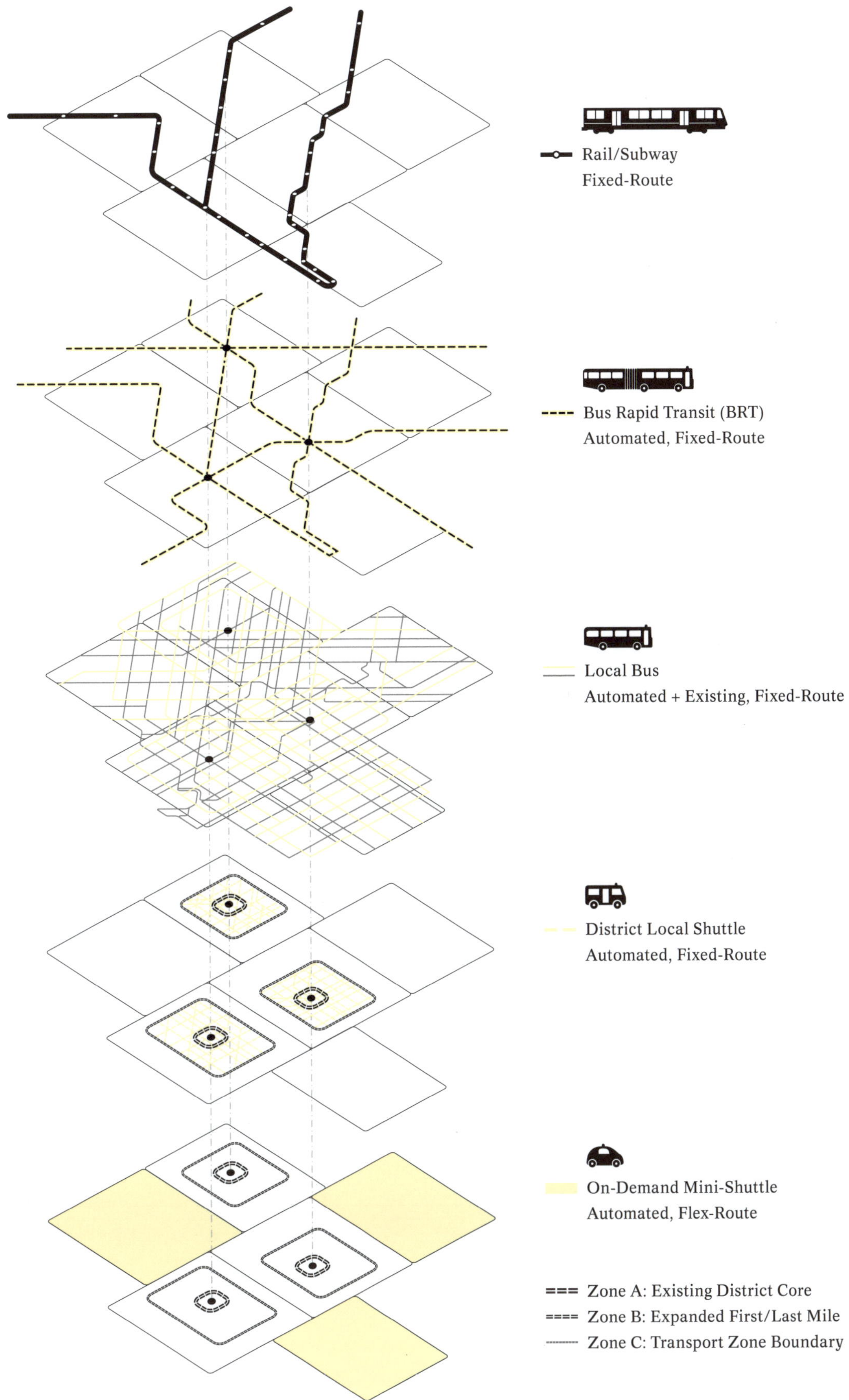
Rail/Subway
Fixed-Route
Bus Rapid Transit (BRT)
Automated, Fixed-Route
Local Bus
Automated + Existing, Fixed-Route
District Local Shuttle
Automated, Fixed-Route
On-Demand Mini-Shuttle
Automated, Flex-Route
Zone A: Existing District Core
Zone B: Expanded First/Last Mile
Zone C: Transport Zone Boundary

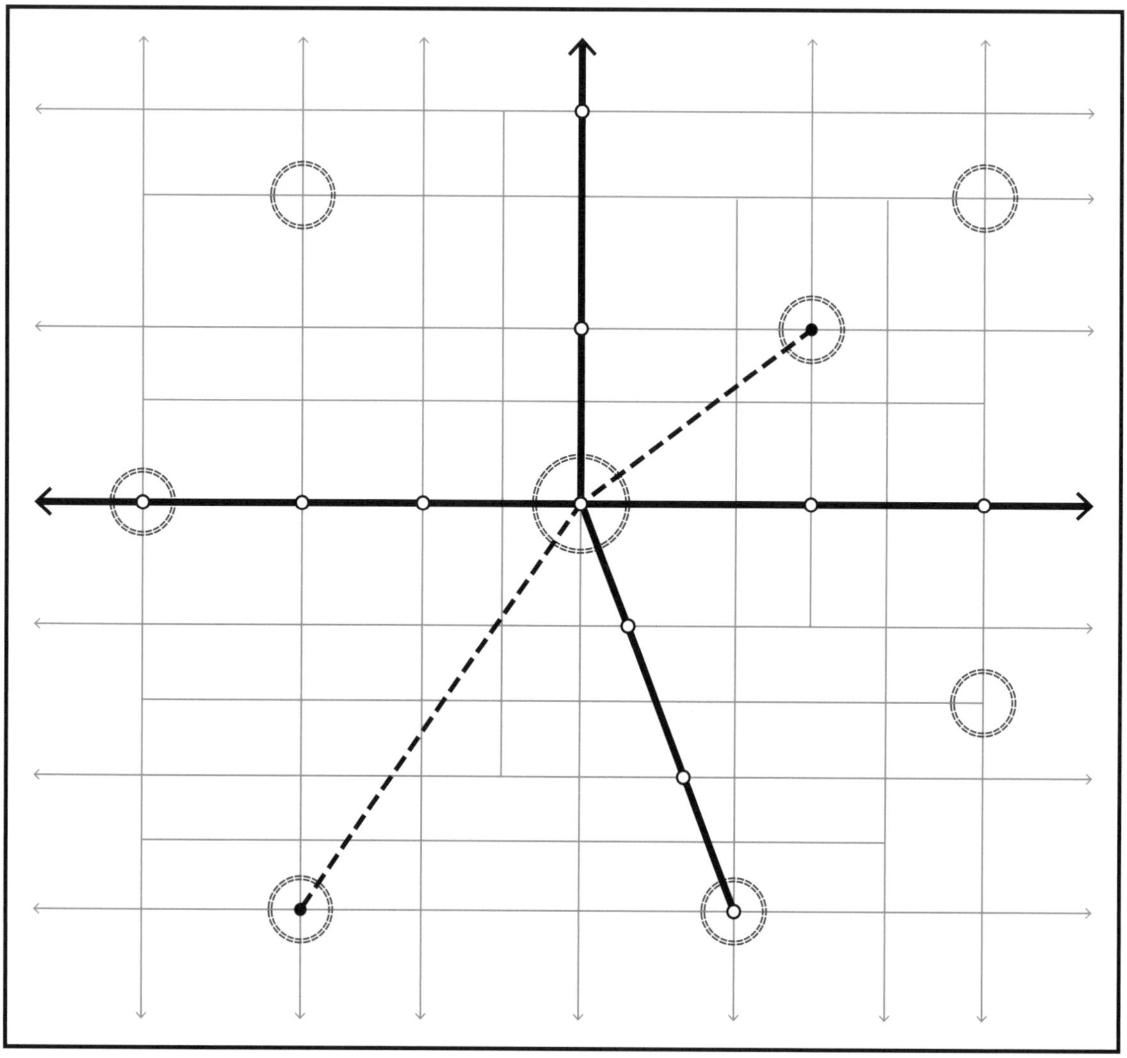

Rail/Subway
Fixed-Route

Bus Rapid Transit (BRT)
Existing, Fixed-Route

Local Bus
Existing, Fixed-Route

Zone A: Existing District Core

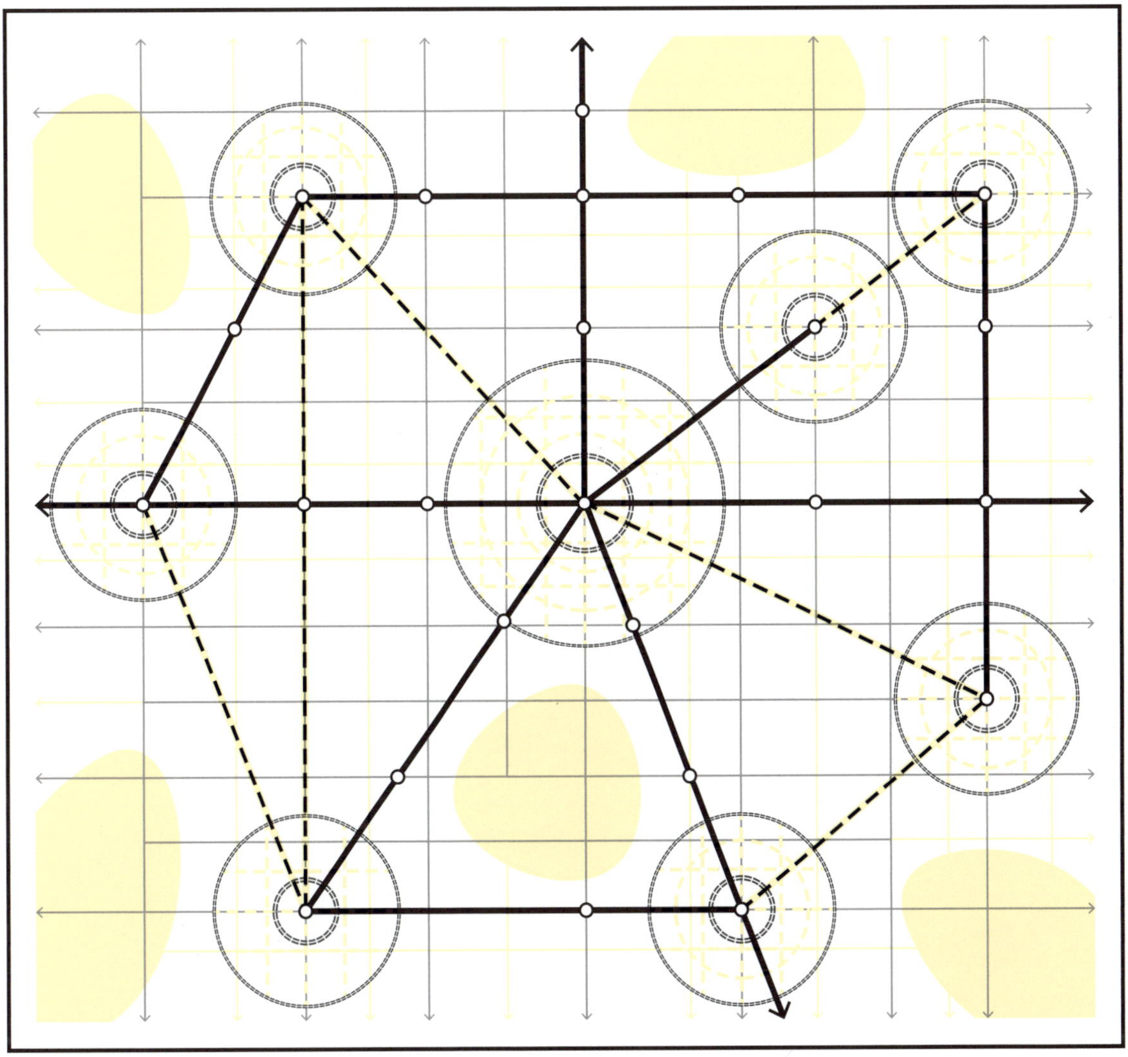

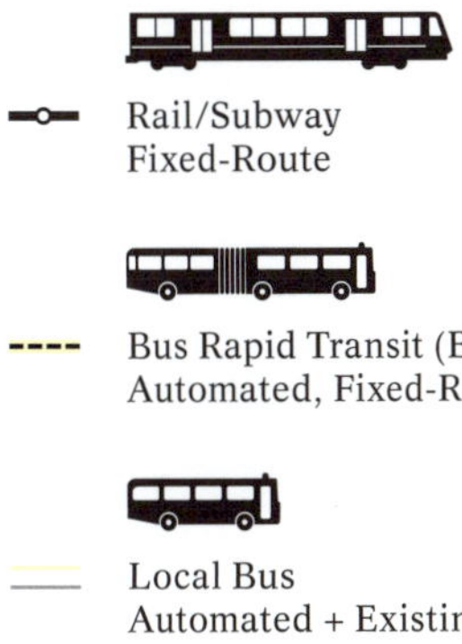

Rail/Subway
Fixed-Route

Bus Rapid Transit (BRT)
Automated, Fixed-Route

Local Bus
Automated + Existing, Fixed-Route

District Local Shuttle
Automated, Fixed-Route

On-Demand Mini-Shuttle
Automated, Flex-Route

Zone A: Existing District Core
Zone B: Expanded First/Last Mile
Zone C: Transport Zone Boundary

Mono-Nucleated Grid Model 2024

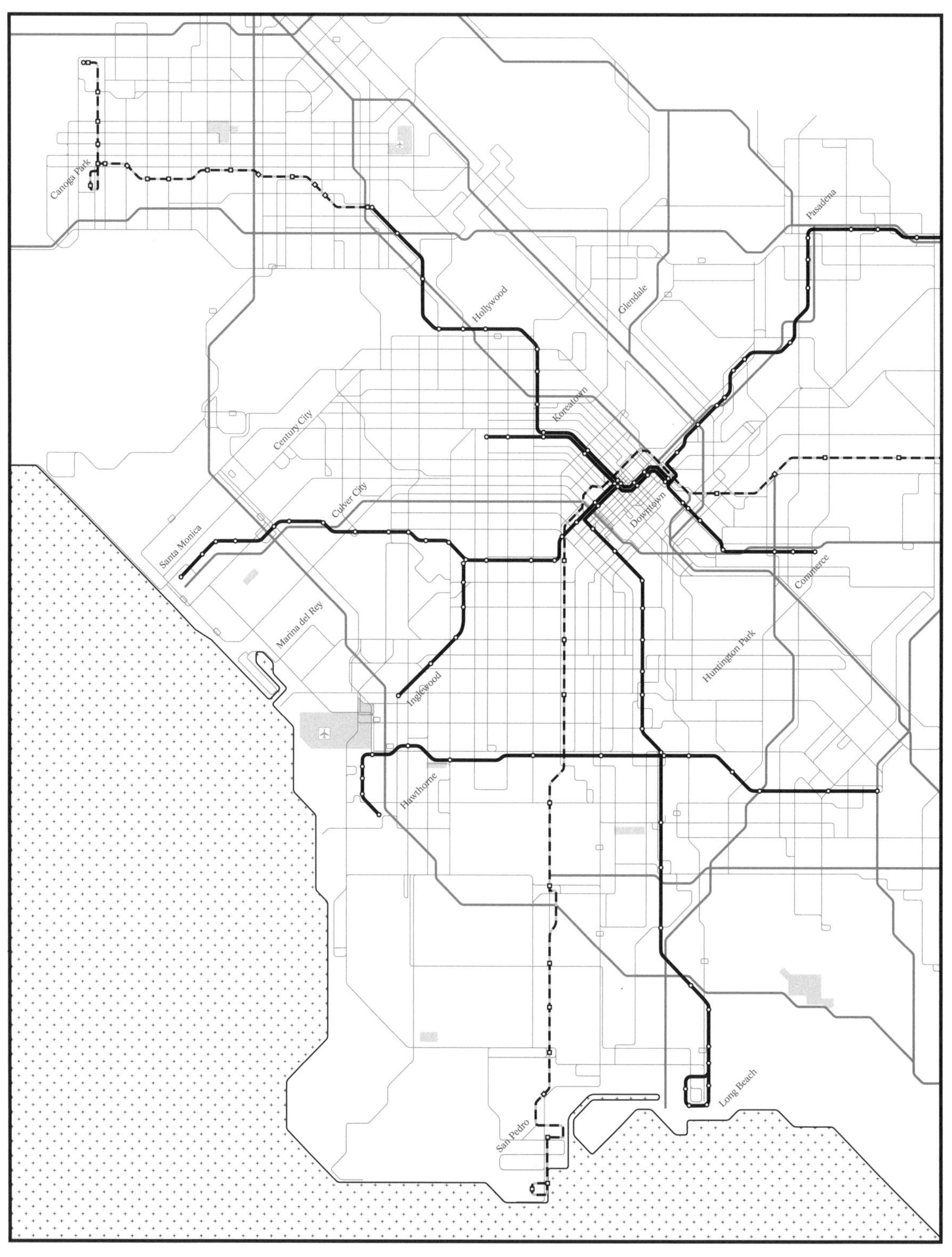

Rail/Subway
Bus Rapid Transit (BRT)
Local Bus
Highway

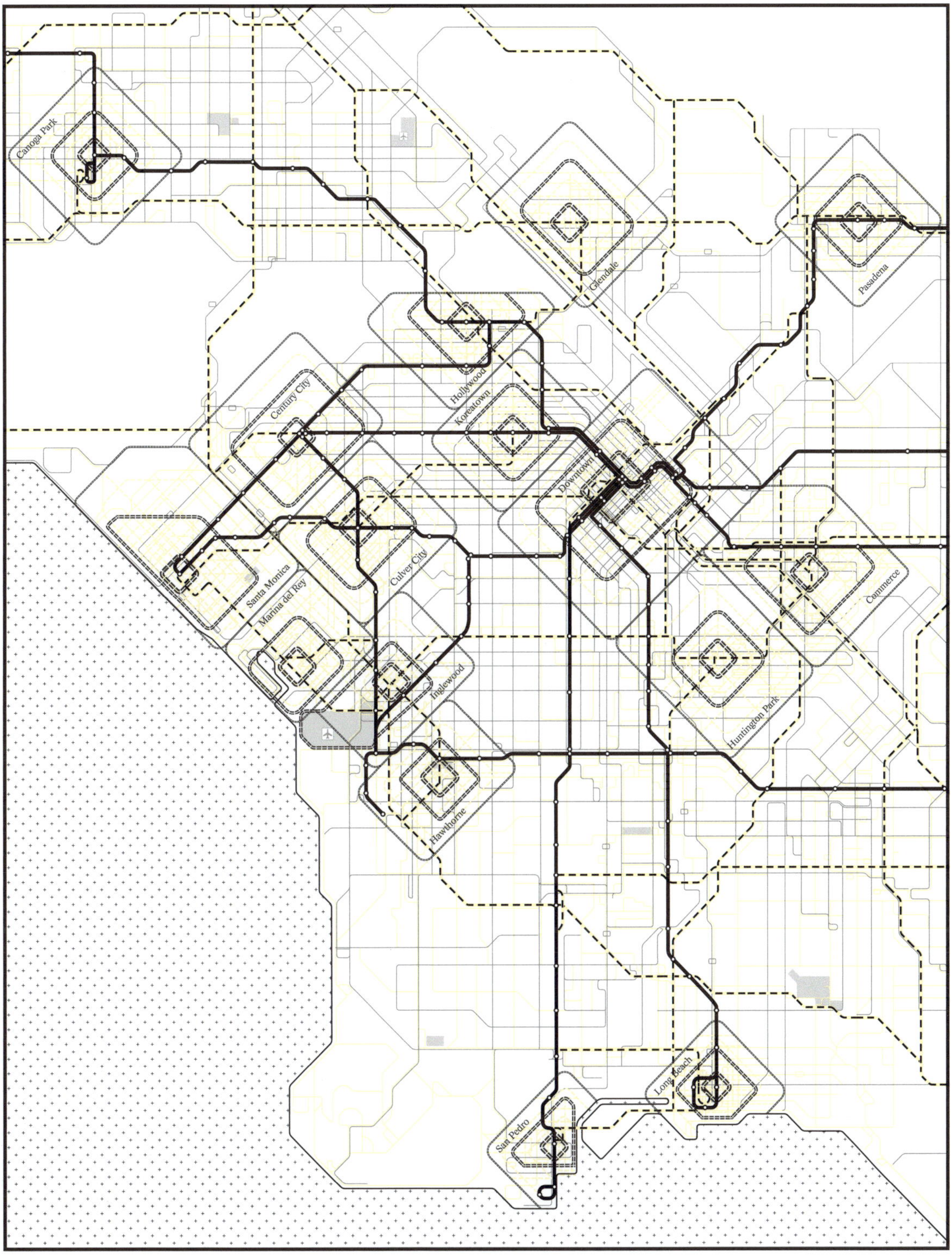

Canoga Park
Glendale
Pasadena
Hollywood
Koreatown
Century City
Downtown
Santa Monica
Marina del Rey
Culver City
Commerce
Inglewood
Huntington Park
Hawthorne
Long Beach
San Pedro

Rail/Subway
Bus Rapid Transit (BRT)
Local Bus
District Local Shuttle
Zone A: Existing District Core
Zone B: Expanded First/Last Mile
Zone C: Transport Zone Boundary

Rail/Subway
Fixed-Route

Bus Rapid Transit (BRT)
Automated, Fixed-Route

Local Bus
Automated + Existing, Fixed-Route

District Local Shuttle
Automated, Fixed-Route

On-Demand Mini-Shuttle
Automated, Flex-Route

Zone A: Existing District Core
Zone B: Expanded First/Last Mile
Zone C: Transport Zone Boundary

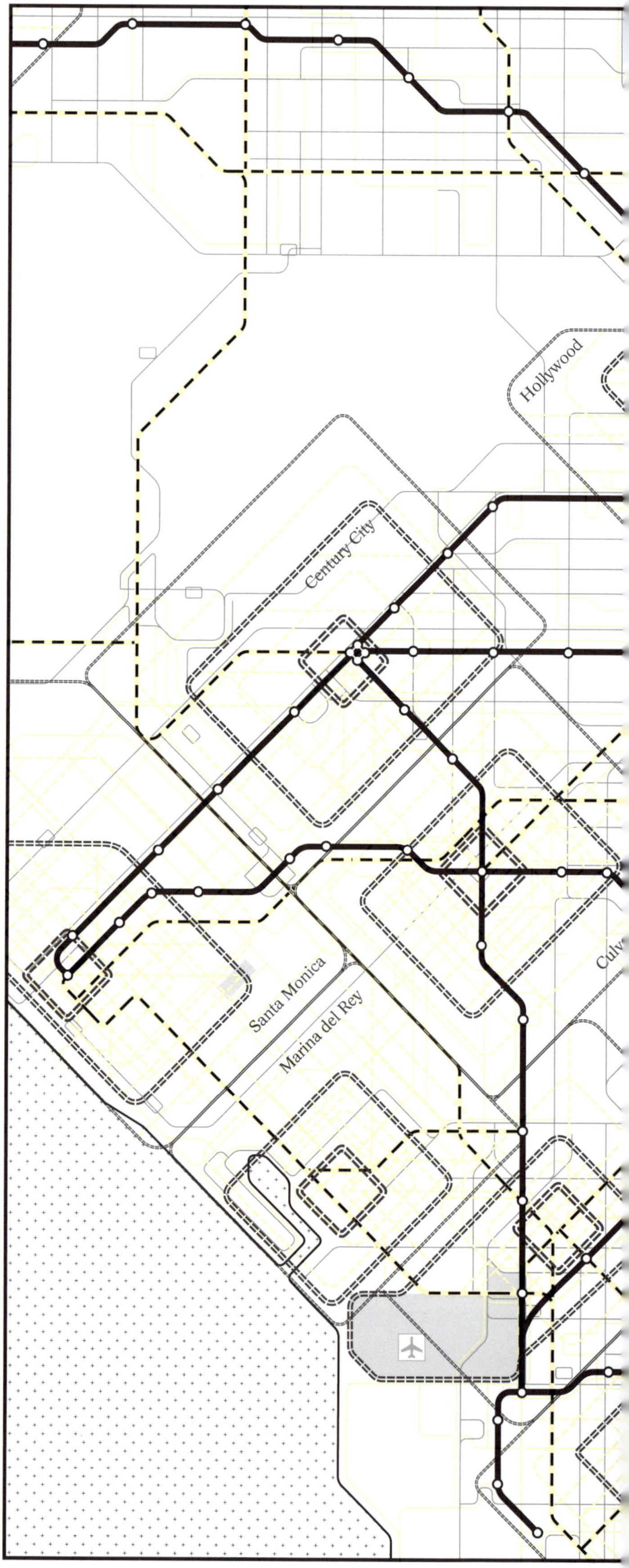

Hollywood
Century City
Santa Monica
Marina del Rey

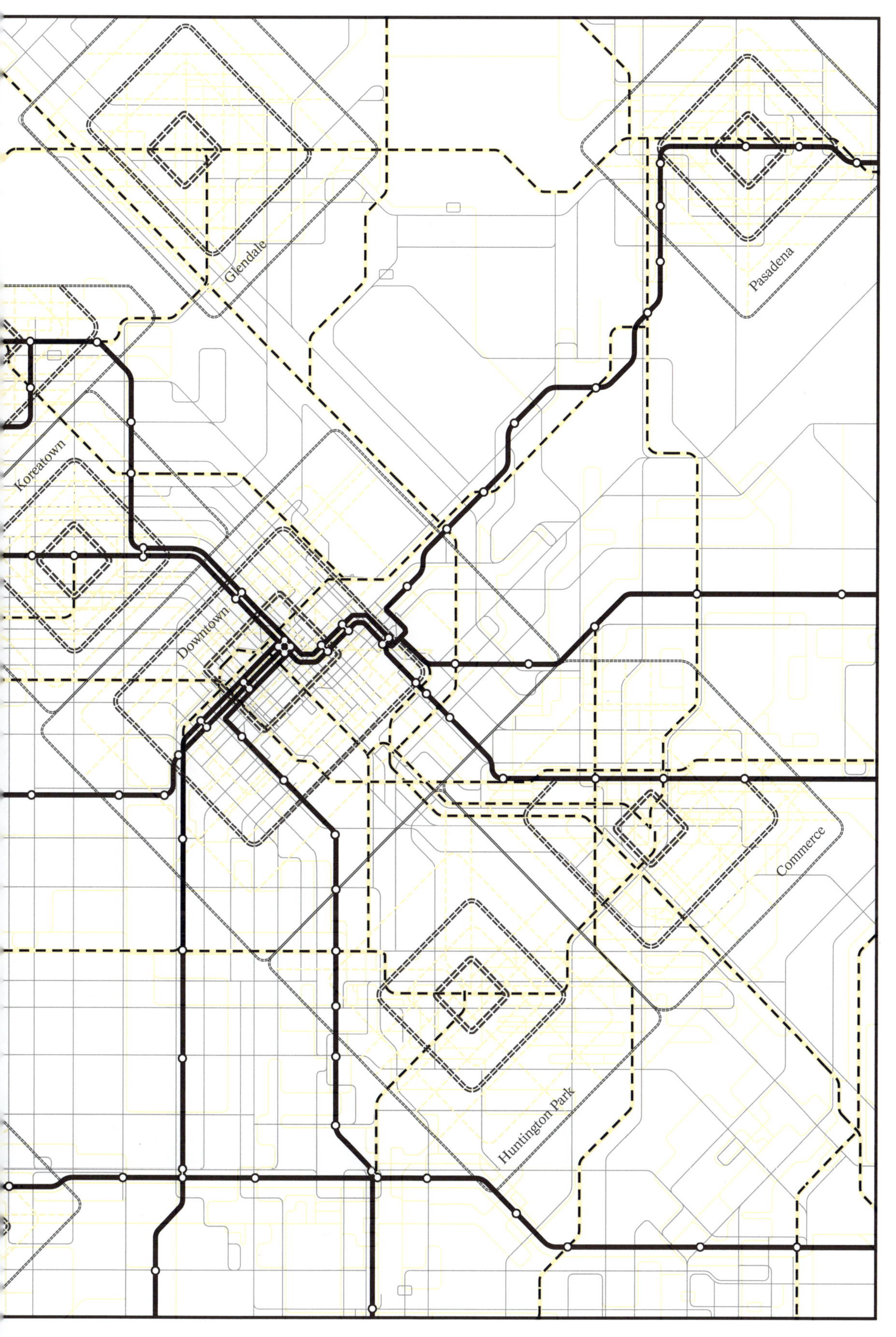
Glendale
Pasadena
Koreatown
Downtown
Commerce
Huntington Park

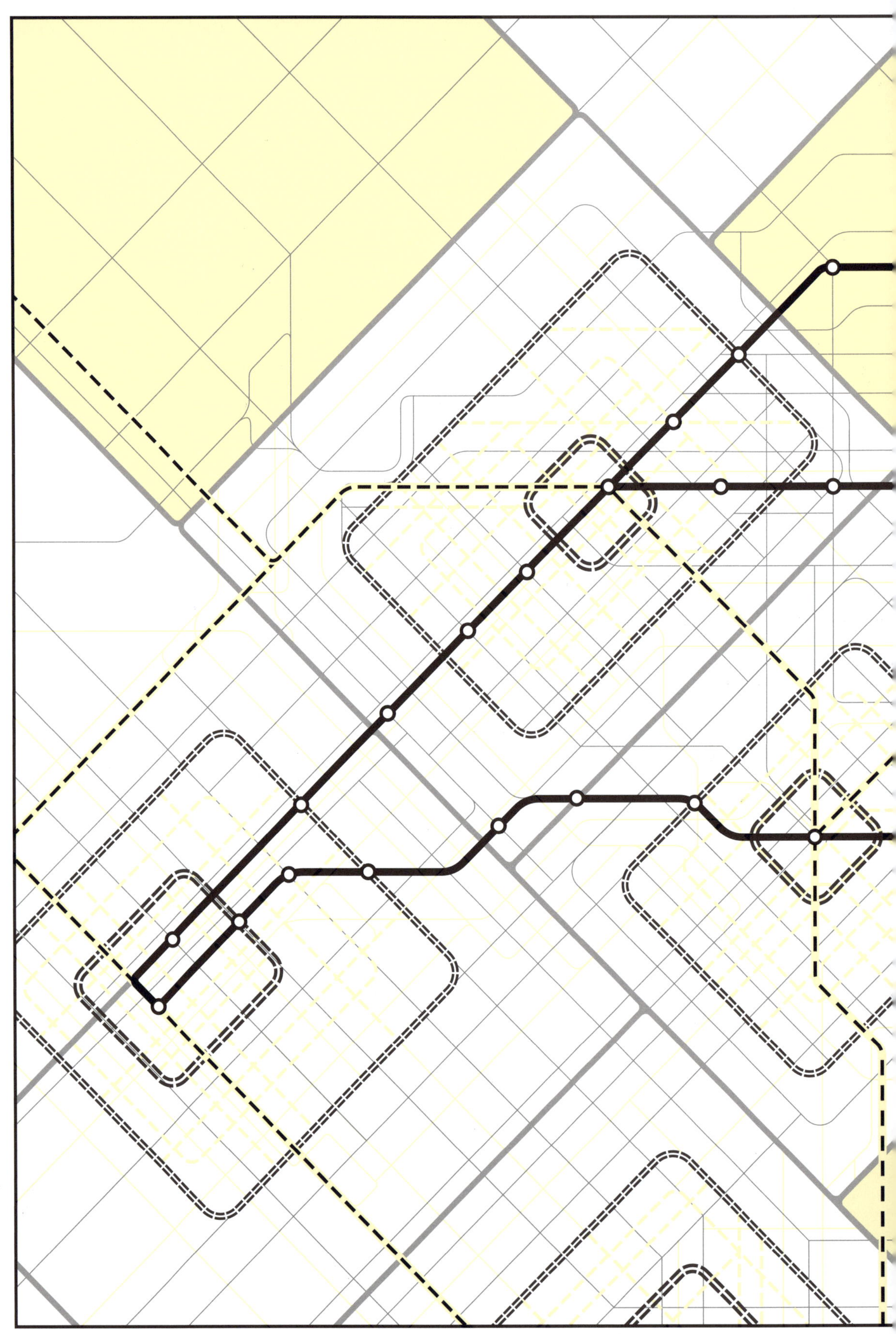

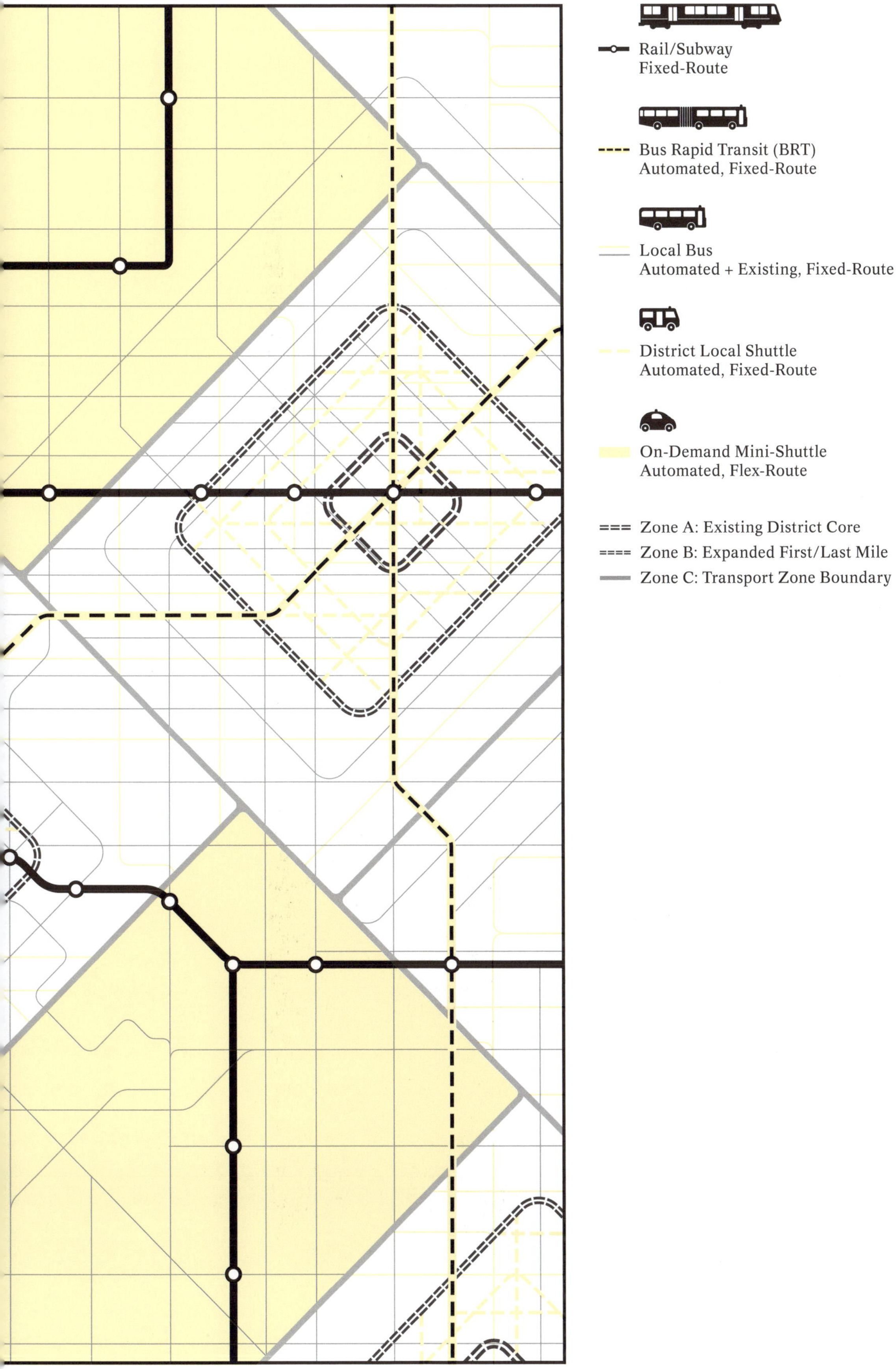

Rail/Subway
Fixed-Route

Bus Rapid Transit (BRT)
Automated, Fixed-Route

Local Bus
Automated + Existing, Fixed-Route

District Local Shuttle
Automated, Fixed-Route

On-Demand Mini-Shuttle
Automated, Flex-Route

Zone A: Existing District Core
Zone B: Expanded First/Last Mile
Zone C: Transport Zone Boundary

8 L: Mobility Infrastructure Restructuring

Roads take up an extraordinary proportion of urban space in cities. Averaging 55 feet wide, they occupy 18% of American cities' total average land area.[1] In some areas, this percentage is more than double; for example, 36% of Manhattan in New York City is dedicated to its streets.[2] Designing this infrastructure is a delicate and intricate dance between multiple demands and uses. These delicate balances have also been disrupted in recent years by new mobility technologies including not only AVs, but micro-mobility vehicles from e-bikes and e-scooters to autonomous robot package and food delivery services. While we typically think of our mobility infrastructure as the connective circulatory systems of cities—moving people from point-to-point—they also serve a multitude of other complex and vital uses. These range from sustaining commercial and economic activities, to acting as essential political and public spaces, to supporting critical environmental and ecological functions.

1 Millard-Ball, "The Width and Value of Residential Streets," *Journal of the American Planning Association*, 2021.

2 Mboup, *Streets as Public Spaces*, UN-Habitat, 2013.

Certainly, they must service vital transportation needs including moving urban dwellers from place to place at a range of speeds and vehicle types, to transporting packages and goods, to facilitating walking pedestrians. They have, at certain times in their recent history, been used to park and store these vehicles, through street parking lanes and under-highway parking lots. But streets also serve critical roles in supporting and giving access to the varied land uses that they connect, from commercial businesses to the workplaces and residences that depend on them. They are often used both formally and informally as commercial spaces, supporting street vendors or food-trucks. Streets are also, in their most fundamental form, a space of democratic access, a public space of the city that every citizen has a right to use. Jane Jacobs describes streets and their sidewalks as "the main public places of a city… its most vital organs. Sidewalk contacts are the small change from which a city's wealth of public life may grow."[3] They support acts of political demonstration and protest, in addition to daily recreational and social uses. Our mobility infrastructures also serve essential latent roles in the environmental systems of our cities. They have typically been completely concretized and paved over with impermeable surfaces, contributing to stormwater management and inland flooding issues that continually plague cities during major hydrological events. When they have been designed with the ecological health of the urban environment in mind, however, large-scale street plantings and coverings have been documented to reduce urban heat island effects and create more comfortable microclimates in cities. Introducing nature-based blue-green systems into our mobility infrastructures—like bioswale zones and rain gardens designed to capture and filter water runoff into permeable landscape channels—can also help reduce stormwater overflows in addition to fostering other non-human habitats and biodiversity.

3 Jacobs, *The Death and Life of Great American Cities*, 1961.

Despite these varied demands, our mobility infrastructures have in our recent history been primarily allocated to and designed for the spatial dimensions and operating maneuvers of automobiles. Automating these vehicles sparks an opportunity to not only rethink the structure, configuration, and infrastructure that facilitate this movement but also revisit the fundamental assumptions that underlie who and what uses our streets and highways should ultimately prioritize.

How much of our street right-of-ways should be dedicated to AVs and buses, pedestrians and cyclists, public spaces and urban landscaping? How do these percentage allocations and use proportions change at a variety of street-scales, from major boulevards to local residential street intersections? How can a shift towards multi-occupancy, space-efficient transit free up right-of-ways for alternative uses that prioritize the public good? How can a reduction in the vast amounts of space dedicated to street parking be reclaimed and for what uses? What alternative spatial futures exist for limited-access highways to transform their use into multi-performative economic, ecological, and community-building opportunities? How can we address the negative externalities that highways wrought, while leveraging their benefits for long-distance travel and their typologically varied spatial conditions (ranging from sunken to elevated)? Can the fixed and unchanging design of our physical infrastructures be rethought of as more flexible and mutable in use to respond to their varying temporal demands? How can automation work with other technologies to transform fixed infrastructures into more dynamically reconfiguring ones? How can our streets and highways integrate nature-based blue-green systems and even be converted into greenways and landscape corridors? And lastly, what are the implications of this cumulative restructuring on urban fabric, form, and future development at large?

These are challenging, contested, and in-demand infrastructures that hold latent and transformational potential for reorganizing urban space. Sparked by driverless technologies, if we reprioritize the way we design and think about them, they could not only alter our movement but also our fundamental experiential relationship with this essential public space of our cities.

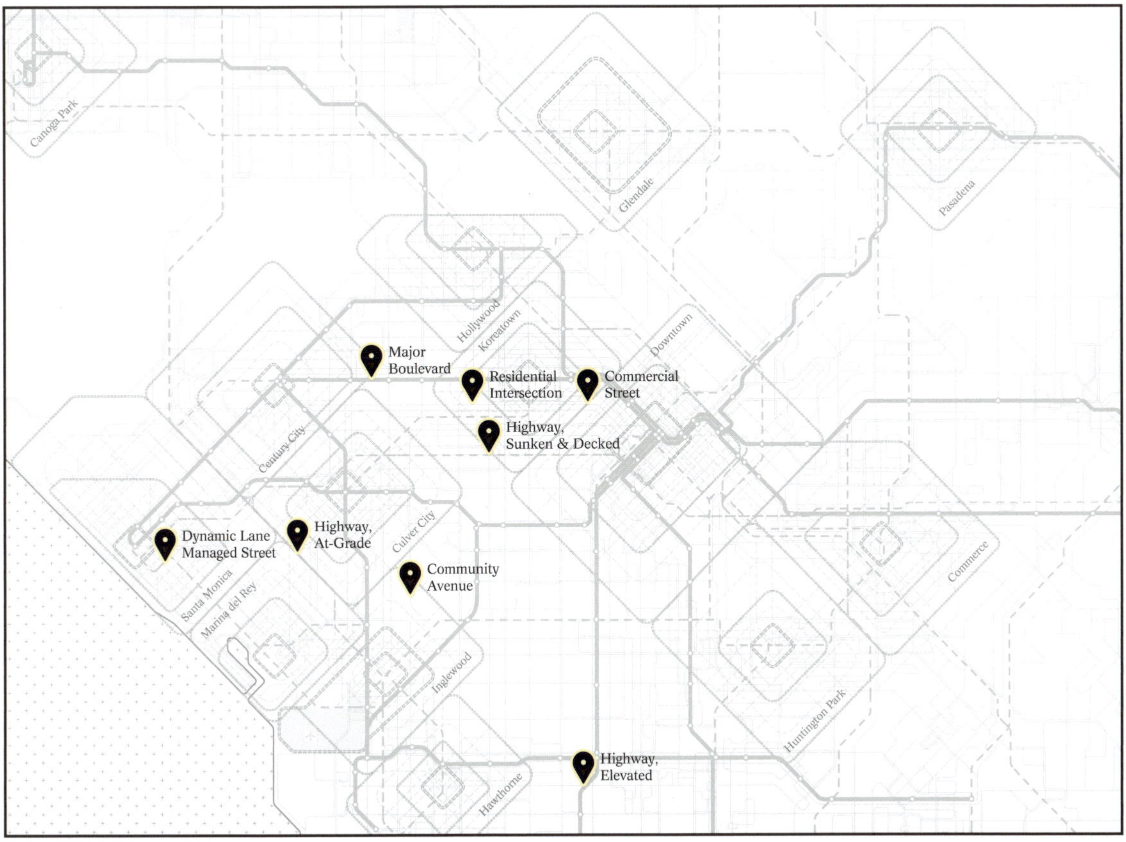

Canoga Park
Glendale
Pasadena
Hollywood
Koreatown
Downtown
Major Boulevard
Residential Intersection
Commercial Street
Highway, Sunken & Decked
Century City
Dynamic Lane Managed Street
Highway, At-Grade
Culver City
Community Avenue
Santa Monica
Marina del Rey
Commerce
Inglewood
Huntington Park
Highway, Elevated
Hawthorne

Highway, At-Grade (Removal)
I-405 Freeway

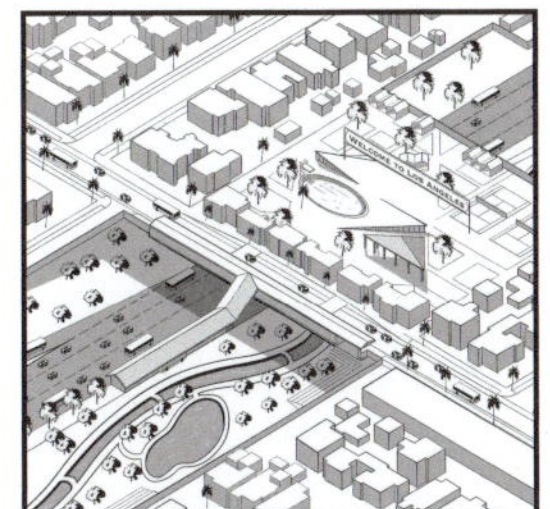

Highway, Sunken (Decking and Conversion)
I-10 Freeway

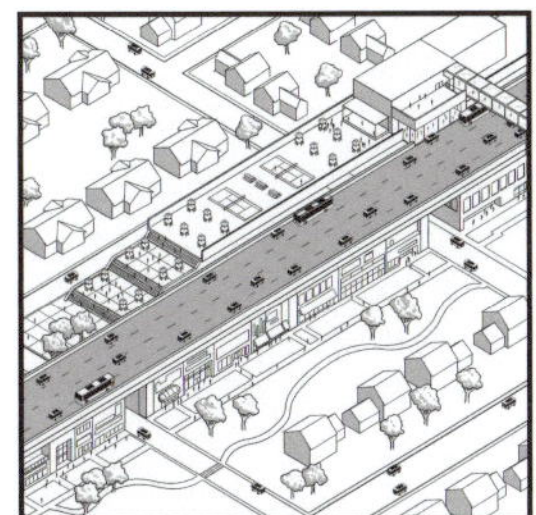

Highway, Elevated (Conversion)
I-110 Freeway

Major Boulevard
Wilshire Boulevard

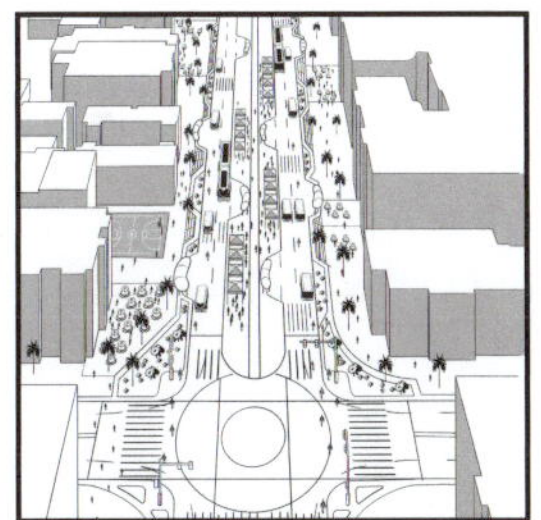

Commercial Street
Alvarado Street

Community Avenue
Crenshaw Boulevard

Dynamic Lane-Managed Street
3rd Street Promenade Extension

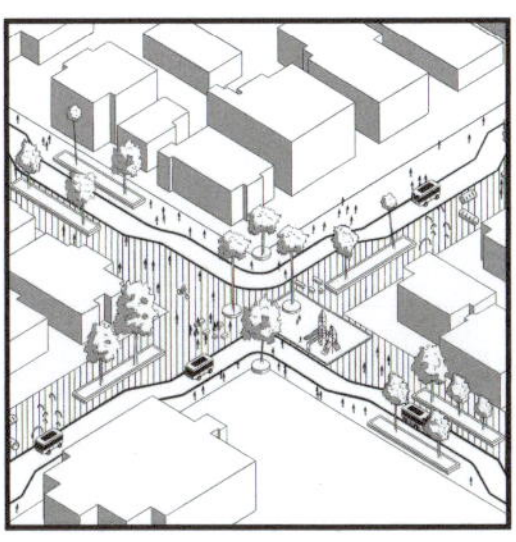

Residential Intersection
Larchmont

Highway, At-Grade (Removal)
I-405 Freeway

One of Los Angeles' most infamous and busiest freeways is "the 405." The freeway's constant congestion problems have led to jokes that it was named as such because drivers have to spend four or five hours on it in traffic. The 405 is in such high demand because it is the only major north-south freeway and commuter corridor for the densely populated neighborhoods connecting the San Fernando Valley all the way to Long Beach and Irvine. It is one of the widest freeways in the city, with up to 14 lanes in portions of its length. Along this vast width and length, it is typically elevated from the ground through earthen berms or a field of columns which sever the adjacent urban fabric. In order to reach either side, one must traverse wide and dark tunnels underneath. In recent years, the 405 has even been astonishingly widened through the city's I-405 Improvement Project, which added four more lanes (two general traffic, two toll lanes) for 16 miles at its southernmost stretch, resulting in induced automobile use, pollution, and various other externalities on already burdened communities fronting the 405.

Instead of widening, this exploration illustrates a radically diametric scenario, one in which the freeway is removed in its entirety and reconstructed with an urban greenway. Instead of facilitating pollution-emitting cars, a riparian waterway and landscaped linear park is introduced into the heart of the city, providing a variety of environmental carbon-capturing and hydrological benefits. Urban fabric on either side of the former highway is reconnected through pedestrian walkways and streets. This transformation would work in tandem with the construction of a new Metro rail line along the greenway's edge, along with automated express bus and local transit to absorb the transportation demands of the corridor. Adjacent urban form that had previously turned its back to the highway would instead by fronted by a new urban living room. This would spark wider economic development and densification benefits for the previously affected neighborhoods along its length.

While perhaps seemingly extreme to those who fear the displacement of 405 traffic, successful closure of the 405 is not without precedent. In 2011, the city closed a 10 mile stretch of the freeway in order to demolish the Mulholland Drive Bridge over a weekend period. Dubbed "carmageddon" and "carpocalypse," local media widely publicized the closure trumpeting warnings of potentially dramatic consequences for the 300,000+ motorists who traversed that stretch of the 405 daily. Despite estimations that traffic gridlock disasters would abound, the city surprisingly saw the opposite happen. There was a documented 61-73% decrease in traffic volume leading towards that closure.[4] Traffic levels were also dramatically reduced along detour routes on nearby highways and local street bypass routes. It was also documented that the city's Metro system saw an approximate 50% jump in ridership over that weekend. While temporary freeway closure over a short period is a far cry from permanent removal, these short-lived effects demonstrate how long-term reduced demand, through paradigmatic shifts in mobility behaviors, could set the stage for provocative spatial transformations of our urban infrastructure.

4 Taylor and Wachs, "Lessons of the 1st Carmageddon in L.A.," RAND Corporation, 2012.

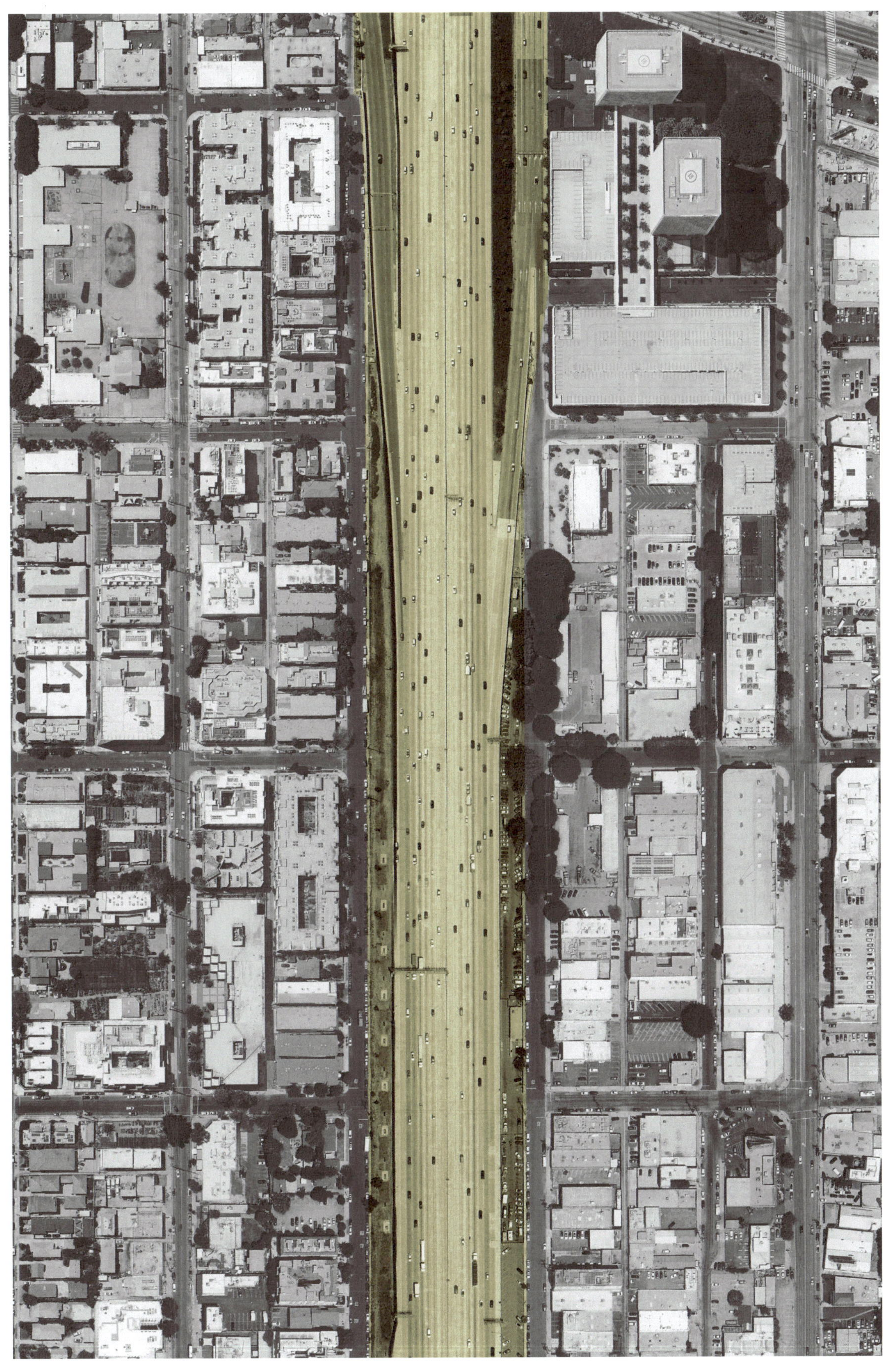

Highway Elevation

Highway Perspective

Highway, At-Grade (Removal)
I-405 Freeway

2024

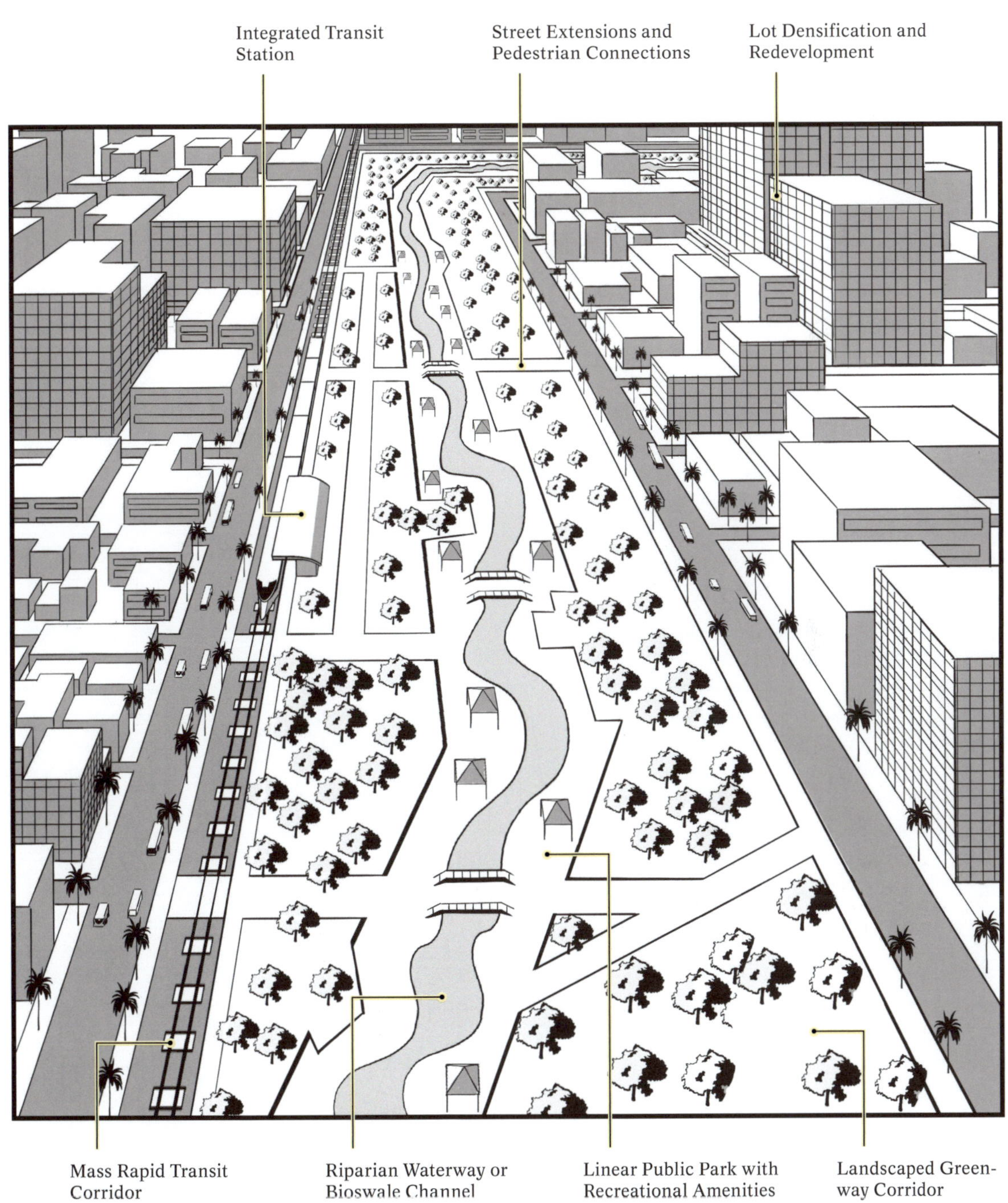
Integrated Transit Station
Street Extensions and Pedestrian Connections
Lot Densification and Redevelopment
Mass Rapid Transit Corridor
Riparian Waterway or Bioswale Channel
Linear Public Park with Recreational Amenities
Landscaped Greenway Corridor

Highway, Sunken (Decking and Conversion)
I-10 Freeway

Colloquially referred to as the Santa Monica and San Bernardino Freeways in Los Angeles, "the 10" is an interstate highway running the length of the transcontinental U.S. that terminates at the Pacific coast in the metropolitan region. In Los Angeles, the freeway serves as one of the few high-speed, east-west corridors that connects the most western edge of the city to its most eastern metropolitan reaches. For Angelenos whose daily commute involves any sort of east-west long-distance mobility, the freeway is nearly unavoidable. Despite facilitating high-traffic volumes, most of the freeway remains automobile-only in use, with just its eastern-most portion allowing bus rapid transit lines (Metro's J Line) on a dedicated lane that shares use with high-occupancy vehicles. The 10 varies in width between eight and ten lanes, even expanding to 14 lanes wide in segments reaching the I-110 Harbor Freeway interchange. The sectional condition of the 10 is quite varied, ranging from elevated (raised on columns or berms), all the way to trenched (sunken into the ground). All of these typological conditions sever the urban fabric on either side of the highway in spatial, social, and economic ways. This severance was so severe that in some cases it led to the complete erasure of communities like Sugar Hill, which was a historically vibrant and affluent Black neighborhood that declined and disappeared after the 10 was constructed through its center in the 1960s.

In this illustrated exploration, the sunken spatial conditions of the 10 become exciting potentials for decking and conversion to non-automobile uses. With a sunken typological condition, adjacent urban fabric can be physically reconnected at-grade through spanned decks. Decks could introduce new public recreational spaces, community amenities, pavilions, and outdoor markets in previously disconnected neighborhoods in dire need of these investments. Streets that had been previously separated could be reconnected though retail corridors. Furthermore, the use of the highway's eastern portion as a transit corridor is expanded through multiple BRT and HOV lanes on half of the highway's right-of-way. With a dramatic shift towards efficient mass transit, other lanes could be transformed into a landscaped waterway for hydrological uses taking advantage of the highway's trenched spatial condition to naturally collect water. Other landscapes could emerge in this space, from productive urban farming to recreational urban parks. An intermodal transit hub integrated into this reconfigured highway could facilitate level change access between deck and street at-grade with the new landscaped corridor below. This typological exploration suggests how hybrid design strategies could take advantage of a highway's longitudinal connectivity but stitch separated urban fabric strategically back together.

Highway Elevation

Highway Perspective

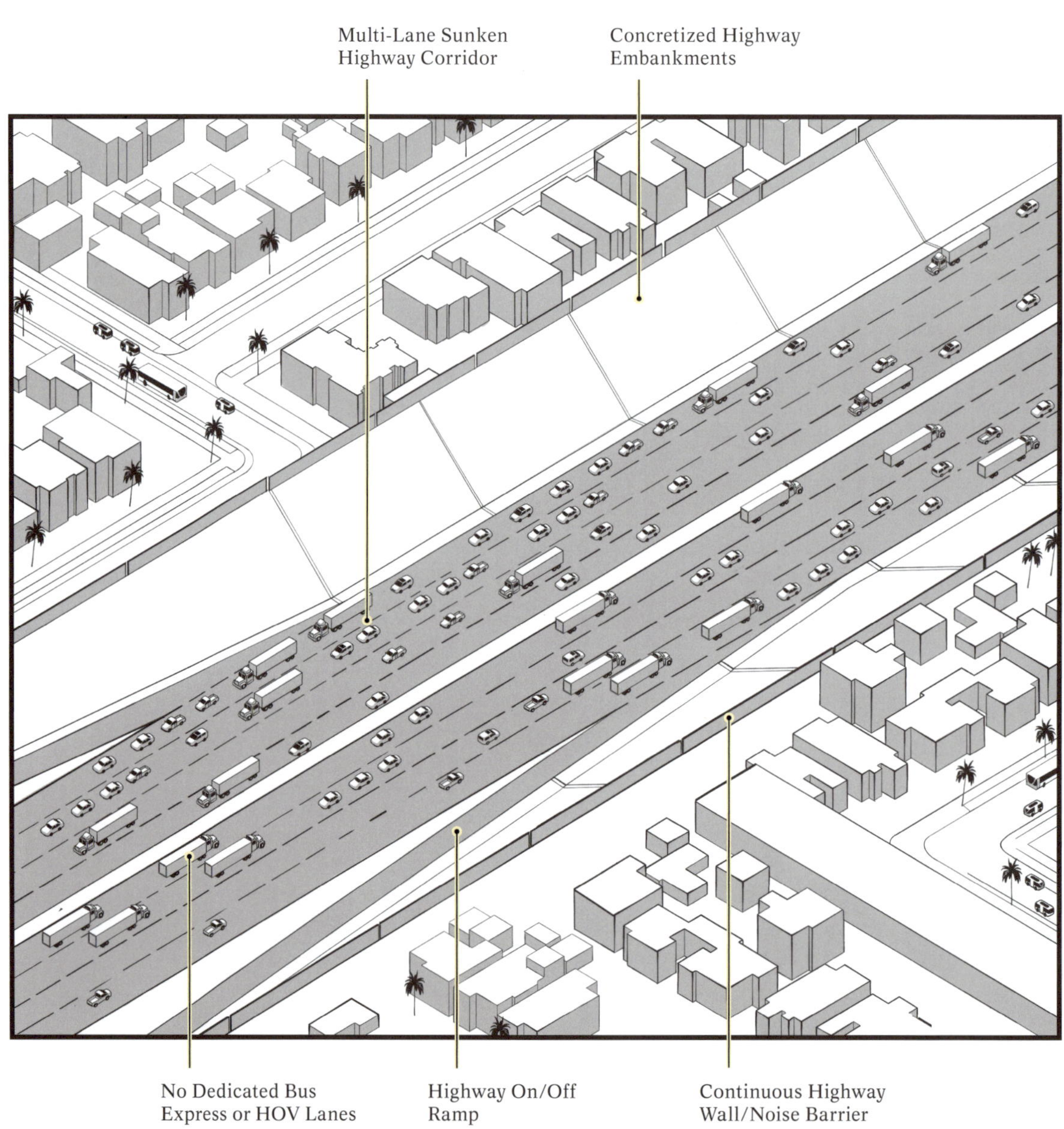
Multi-Lane Sunken
Highway Corridor
Concretized Highway
Embankments
No Dedicated Bus
Express or HOV Lanes
Highway On/Off
Ramp
Continuous Highway
Wall/Noise Barrier

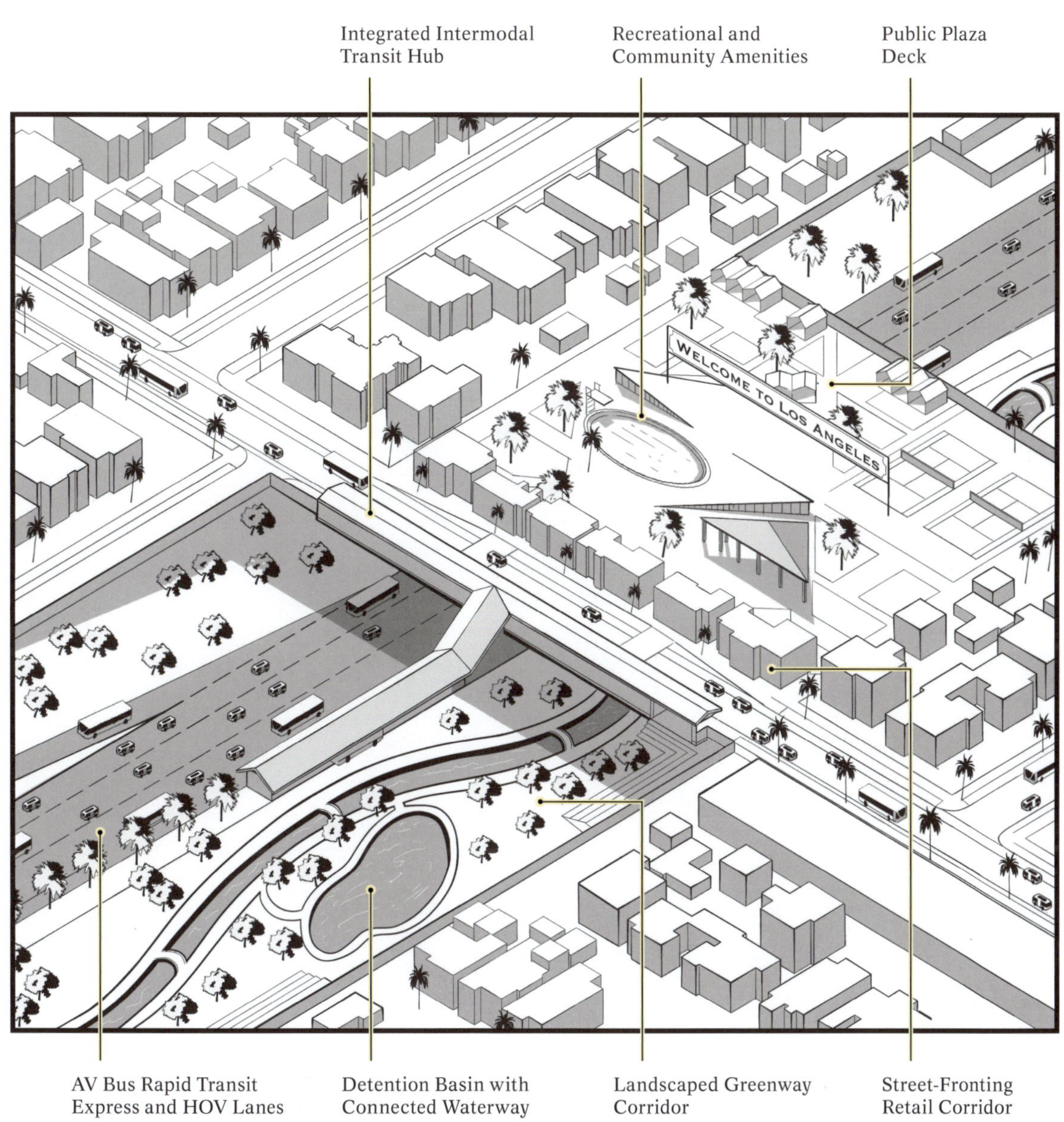
Integrated Intermodal Transit Hub
Recreational and Community Amenities
Public Plaza Deck
WELCOME TO LOS ANGELES
AV Bus Rapid Transit Express and HOV Lanes
Detention Basin with Connected Waterway
Landscaped Greenway Corridor
Street-Fronting Retail Corridor

Highway, Elevated (Conversion)
I-110 Freeway

"The 110" or the Harbor Freeway, is a major north-south highway that runs the entire metropolitan length of the city, connecting Los Angeles' southern-most point to its northeast. The 110 is also an infrastructural divider that prevents downtown Los Angeles, which it runs through, from naturally expanding westward. It is a major commuter corridor for residents of Los Angeles' South Central, where much of the city's Hispanic and Latino populations reside. Primarily elevated on columns or bermed along its length, the 110 ranges from around eight to fourteen lanes in width. In instances, the highway also holds expansive parking lots underneath it due to the lack of light and the high levels of noise and emission pollution that prevent other land uses from occupying that space. A unique feature of the 110 is that it is one of the few Angeleno highways that actually services express bus rapid transit in 22 miles along the majority of its length. Designated the Harbor Transitway, this shared-used BRT corridor runs Metro's J Line all the way from the I-10 freeway down to San Pedro, featuring multiple dedicated lanes for express buses and high-occupancy vehicles. In order to service these express buses, eight transit stations are integrated into the infrastructural design of the highway at several points. Located in the center median zone between multiple lanes, however, these stations exist as isolated platforms in a sea of lanes, disconnected from the adjacent urban fabric. Not only do these island stations expose passengers to two-way, high-speed, highway traffic, pedestrians also face challenges in accessing them. These multi-level stations often require overhead, precariously-spanning pedestrian bridges or long walks down underpass tunnels to reach them, meaning that many residents are often not aware of their existence and availability.

In a driverless future, these stations could be redesigned into integrated intermodal hubs that transfer long-distance highway express transit to at-grade, local automated bus, mini-shuttle, and micro-mobility options servicing the adjacent neighborhood. They could also be integrated with station amenities and services that anchor the transformation of the highway infrastructure itself. Multiple elevated lanes of the highway could become adaptively converted to recreational public spaces and community amenities with a raised view of the city. Portions of that elevated infrastructure could be replaced with terraced steps and seating to transition public access vertically. Parking lots or under-highway dead zones could be reclaimed for retail land uses that activate new public spaces and pedestrian malls in the buffer spaces running alongside. This transformation could only be made possible by a paradigm shift toward electric vehicles and mass transit use on remaining highway lanes, which could reduce many of the latter's associated negative externalities. If done so though, this infrastructure could become excitingly rethought through an adaptive reuse lens, sparking the development of a transit-oriented commercial corridor, public spaces, and mixed land uses. This would turn the highway network of Los Angeles from one that neighborhoods have turned their back on, into a blueprint for future densification.

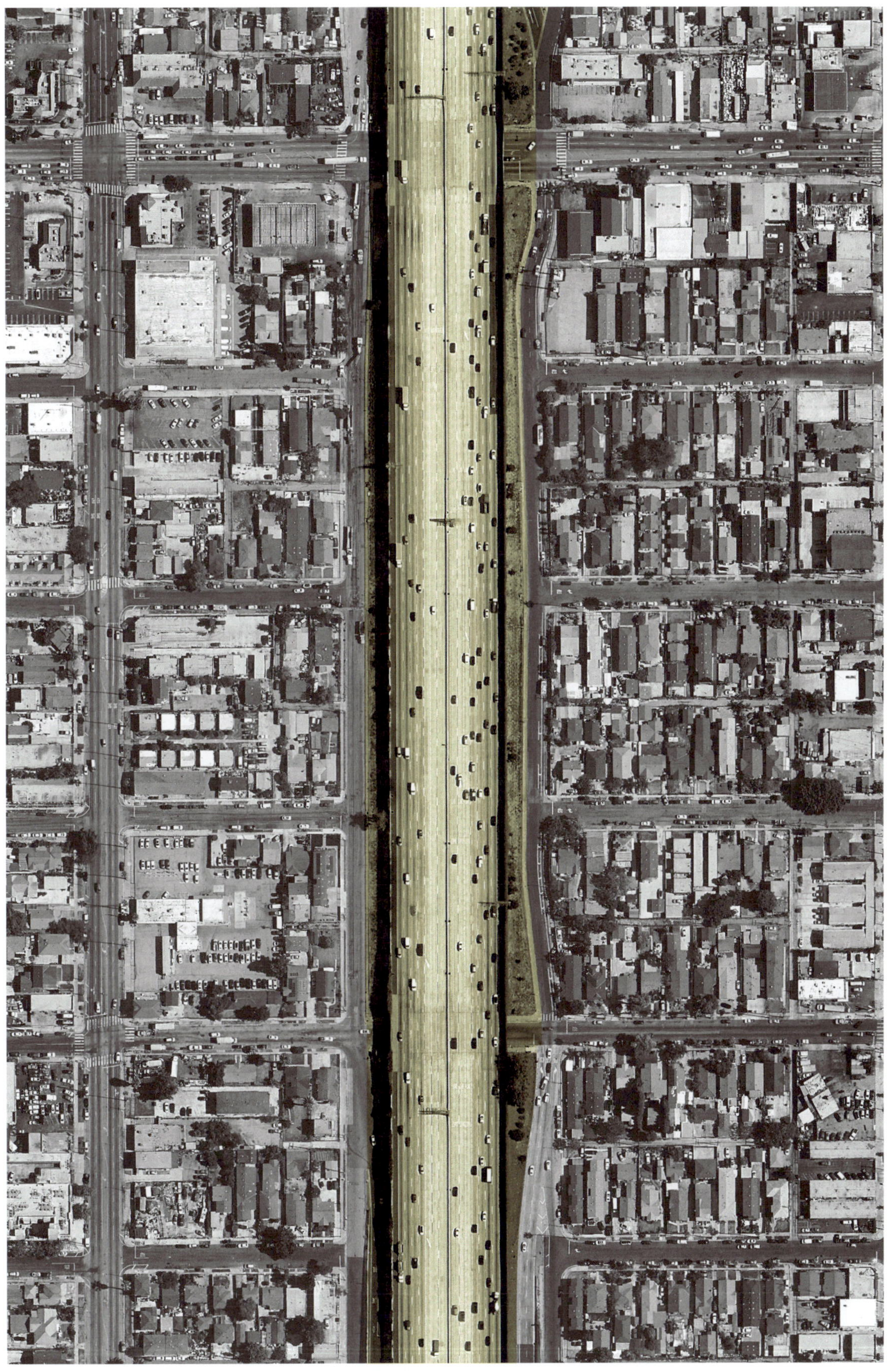

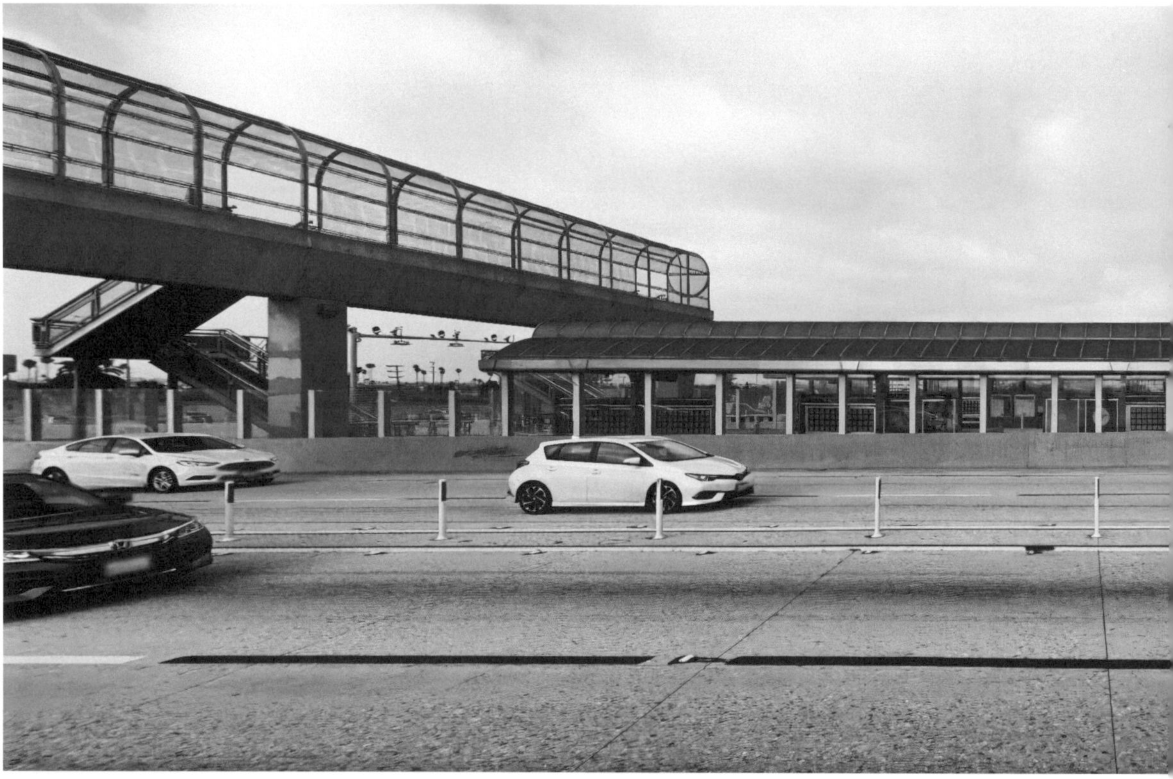

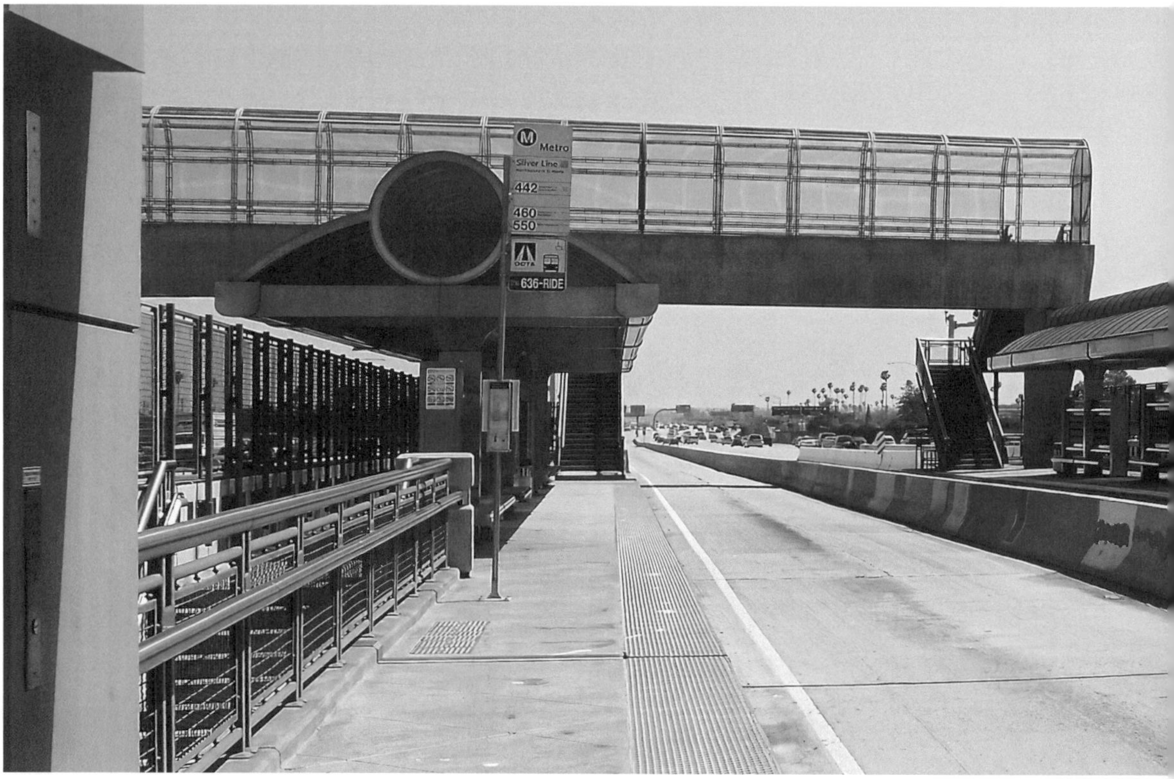
M
Metro
Silver Line
442
460
550
OCTA
636-RIDE

Highway Station
Elevation

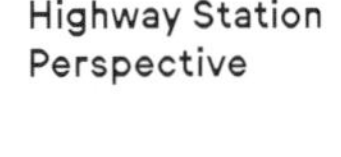
Highway Station
Perspective

Highway, Elevated (Conversion)
I-110 Freeway

2024

Multi-Lane Elevated Highway Corridor

Monolithic Highway Infrastructure

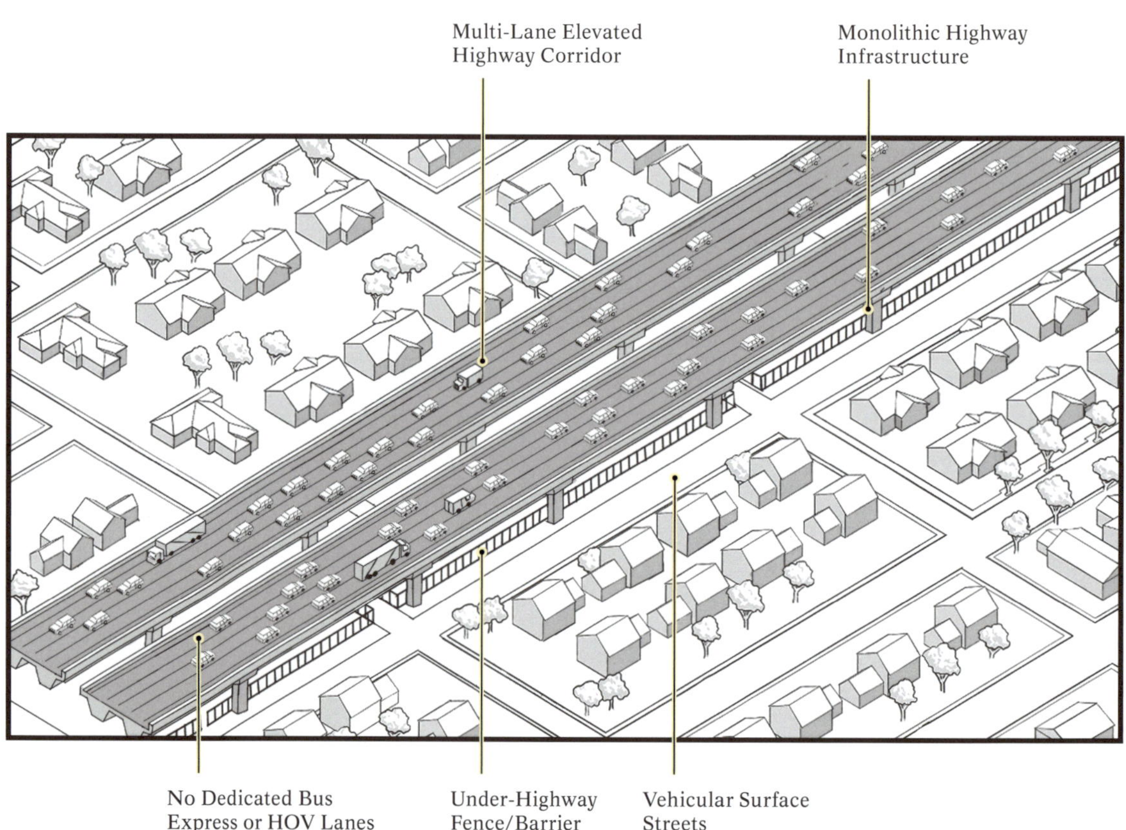

No Dedicated Bus Express or HOV Lanes

Under-Highway Fence/Barrier

Vehicular Surface Streets

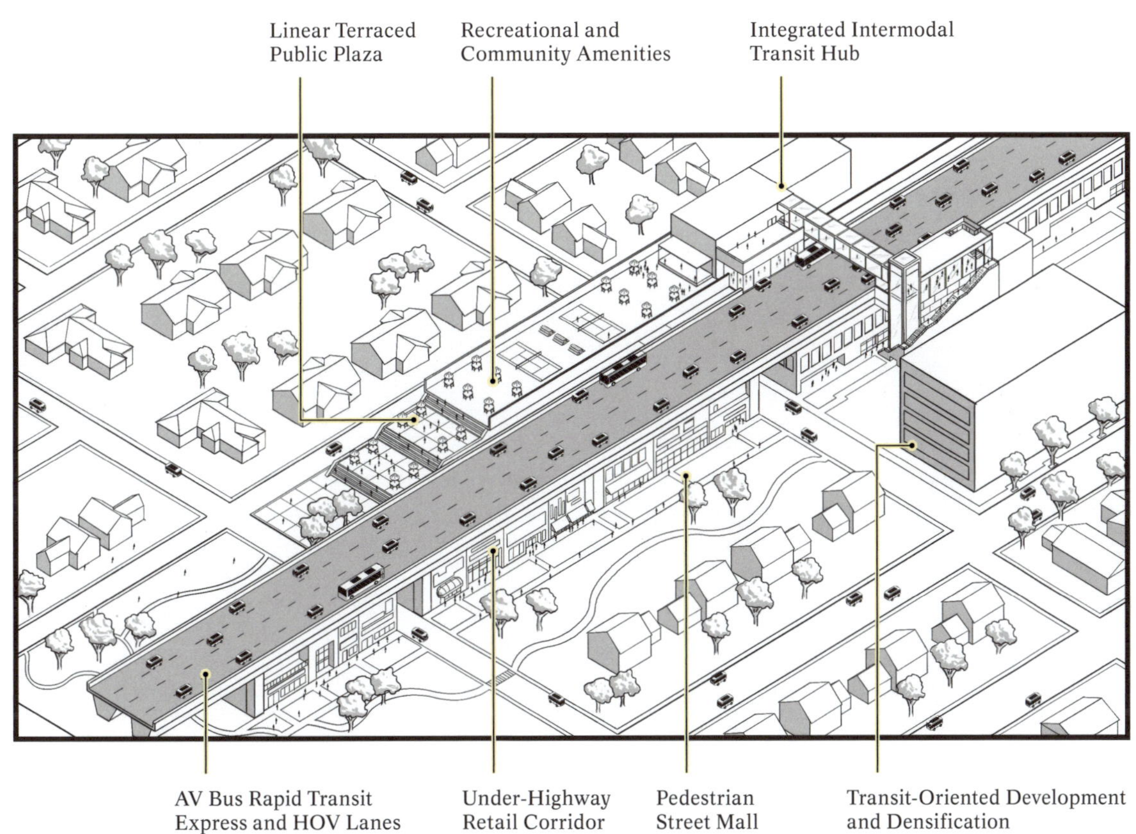
Linear Terraced
Public Plaza
Recreational and
Community Amenities
Integrated Intermodal
Transit Hub
AV Bus Rapid Transit
Express and HOV Lanes
Under-Highway
Retail Corridor
Pedestrian
Street Mall
Transit-Oriented Development
and Densification

Highway, Elevated (Conversion)
I-110 Freeway

2024

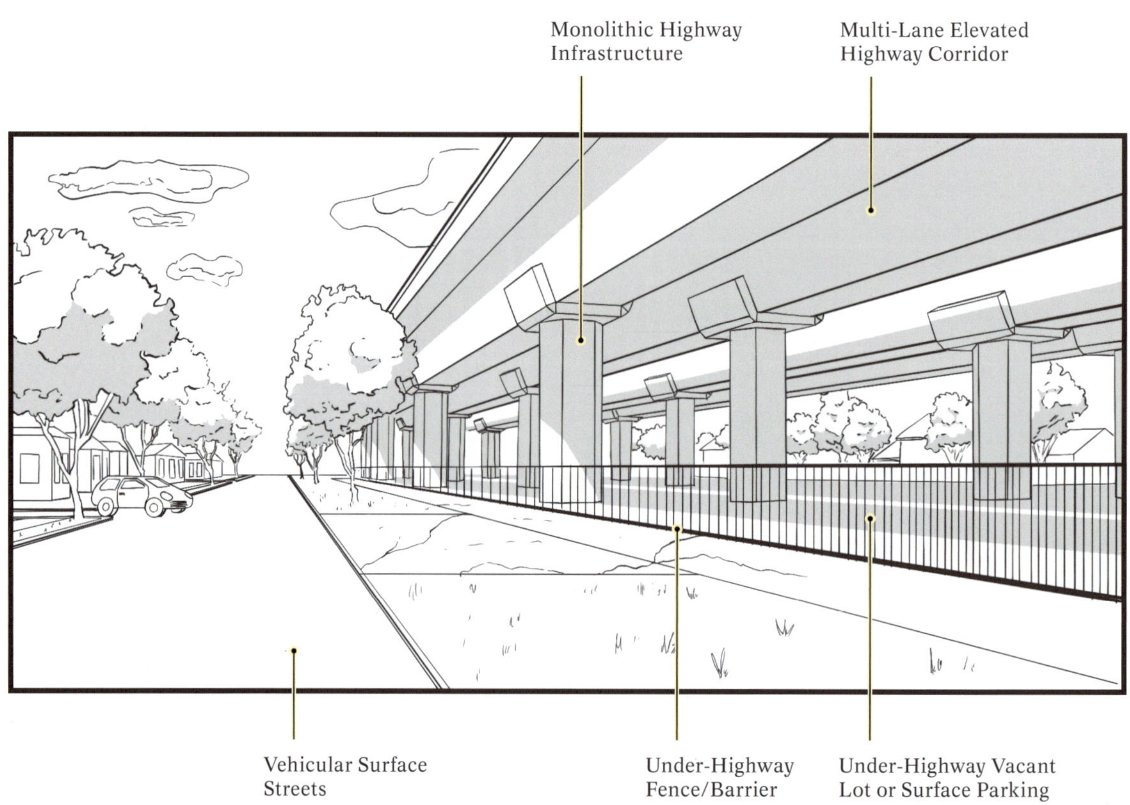

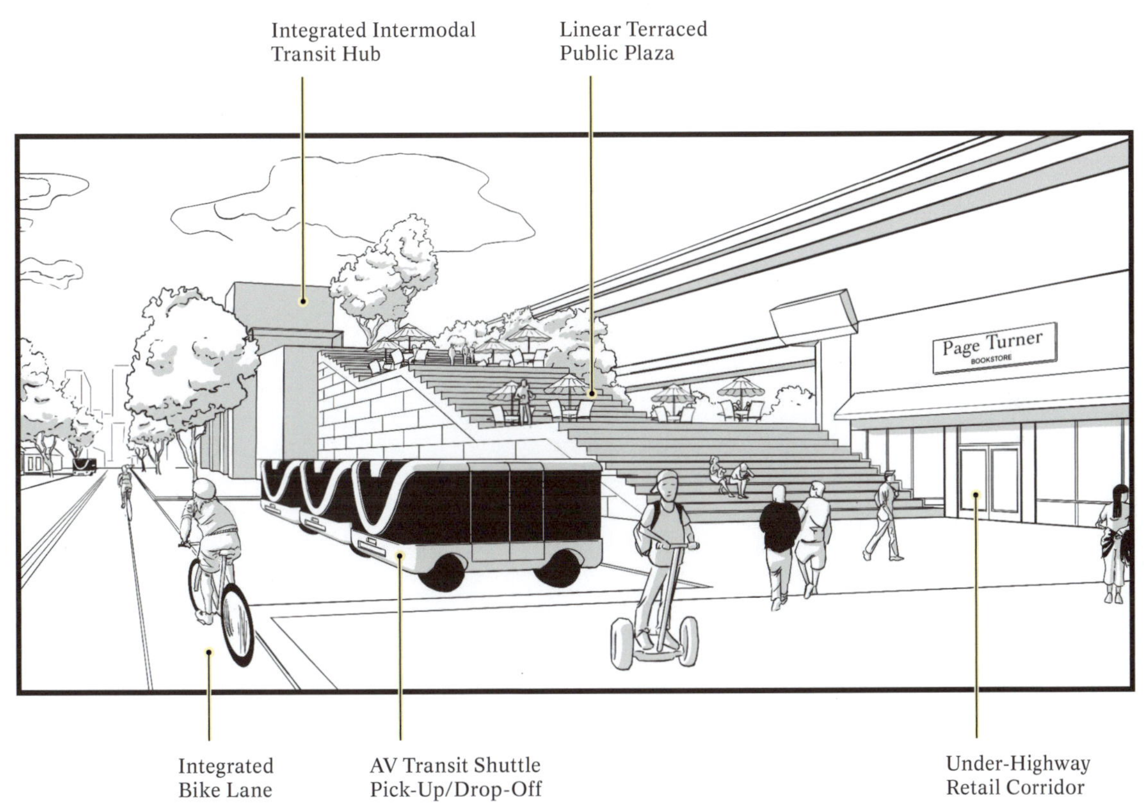
Integrated Intermodal
Transit Hub
Linear Terraced
Public Plaza
Page Turner
BOOKSTORE
Integrated
Bike Lane
AV Transit Shuttle
Pick-Up/Drop-Off
Under-Highway
Retail Corridor

Major Boulevard
Wilshire Boulevard

Although Los Angeles is not typically recognized for holding sustained and concentrated high-density urban environments, several of the urban corridors that cross the city do exhibit these traits. These boulevards serve not only as vital transportation corridors in the large swaths of urban fabric between the city's highway network, they have also become important corridors for economic and commercial land uses holding denser urban form. Many of these major arterials run parallel to each other east-west, including prominent thoroughfares likes Sunset and Santa Monica Boulevard. One of the most emblematic of these corridors is Wilshire Boulevard, which connects downtown Los Angeles all the way to its westside before terminating at the Pacific Coast. It is densely developed throughout most of this span, being the site of various contemporary high-rises, historical Art Deco buildings, and cultural institutions at various points of its length.

Despite its cultural and civic significance, its public right-of-way prioritizes the passage of single-occupancy vehicles above other forms of mobility. The whole boulevard is at least four lanes in width, with the majority of it being six to eight lanes of traffic flanked by street parking at the curb. Portions of Wilshire have a raised center median lane that separates two-way traffic. One of the widest and busiest portions of the boulevard is located in the Westwood neighborhood, where pedestrians of the business district must traverse ten lanes, including two left-turn pockets to cross to the other side. Even though local Metro buses run frequently along Wilshire, using it as a major local transit corridor, the boulevard does not have designated bus lanes to ensure predictable travel times nor are many of its bus stops shaded or sheltered. Aging sections of Wilshire are also notorious for their potholes. Metro's D and B rail lines run underground along limited portions of Wilshire, serving four Metro stations along its eastern length. One of the major earmarks of Measure M funding is the extension of the D line westward along Wilshire in multiple phases.

Before waiting for this future rail expansion to occur, however, automated bus express transit could be implemented as a precursor line along the length of Wilshire. This shift towards expanded express transit could run in dedicated central lanes of the boulevard right-of-way, free from slower-speed outer-lane traffic. Underutilized center lane medians could be reclaimed for covered BRT station platforms that facilitate shorter crossings for pedestrians across the wide boulevard. New expanded transit stations could be integrated into the street and sidewalk right-of-way with station amenities like bike-share docks and shaded canopies. Street and lot parking could be reclaimed for bike lanes, expanded sidewalks, street furniture, and other public spaces. Coupled together, this could transform a restricted high-density commercial corridor that still relies on automobile access, to an expanded transit-oriented corridor spurring new density and development along its length while stretching multiple blocks in width. These adaptations could ensure that future rail expansions below-grade would have the density above-grade to support their use in a future poly-nucleated transit network.

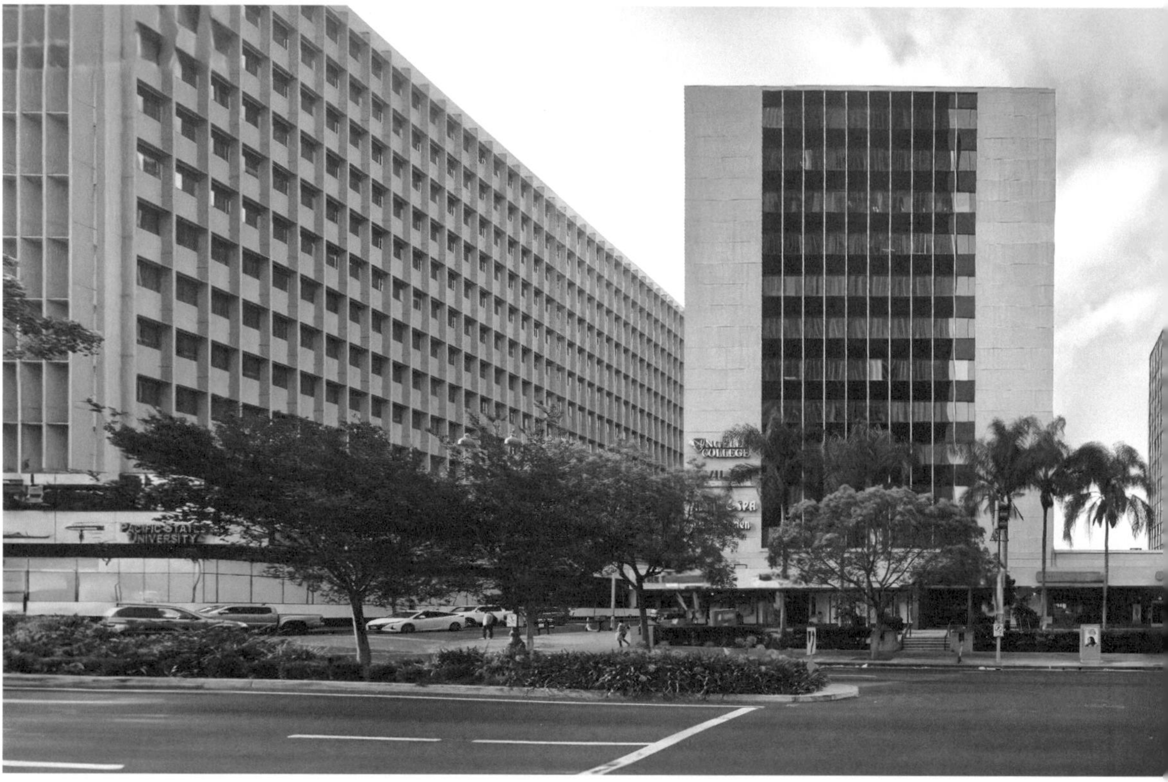
PACIFIC STATE
UNIVERSITY

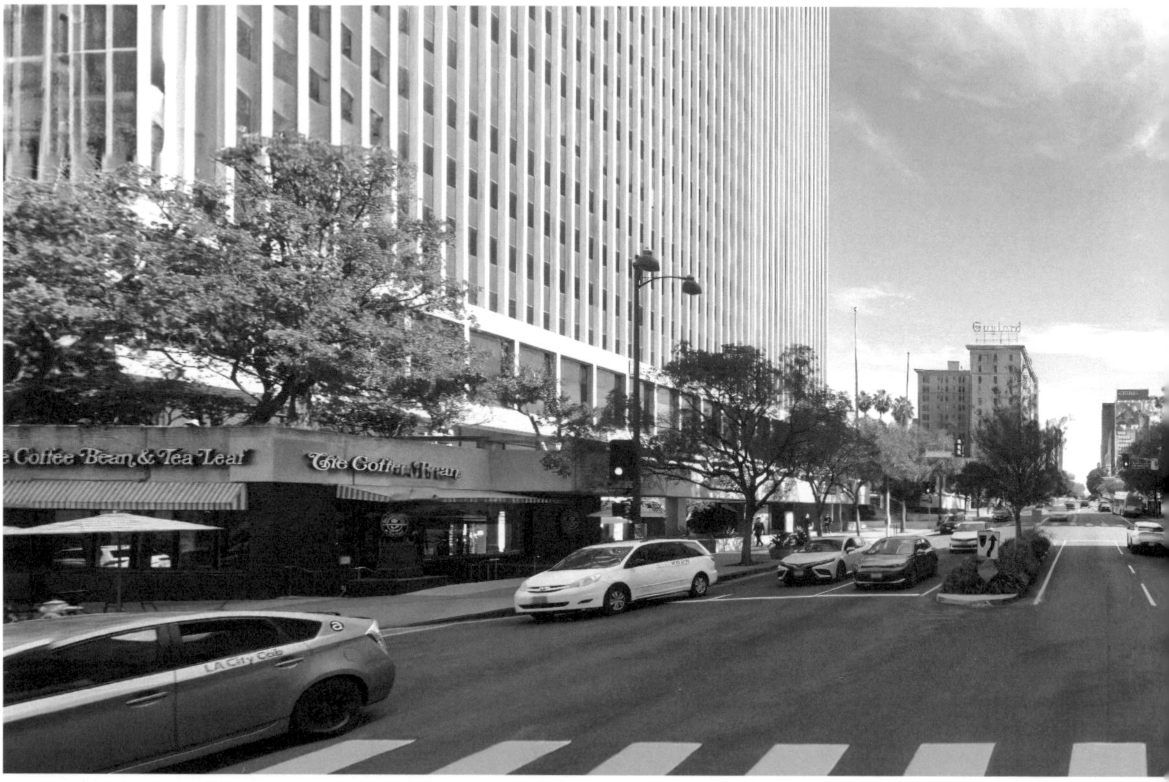
e Coffee Bean & Tea Leaf
LA City Cab

Street Elevation

Street Perspective

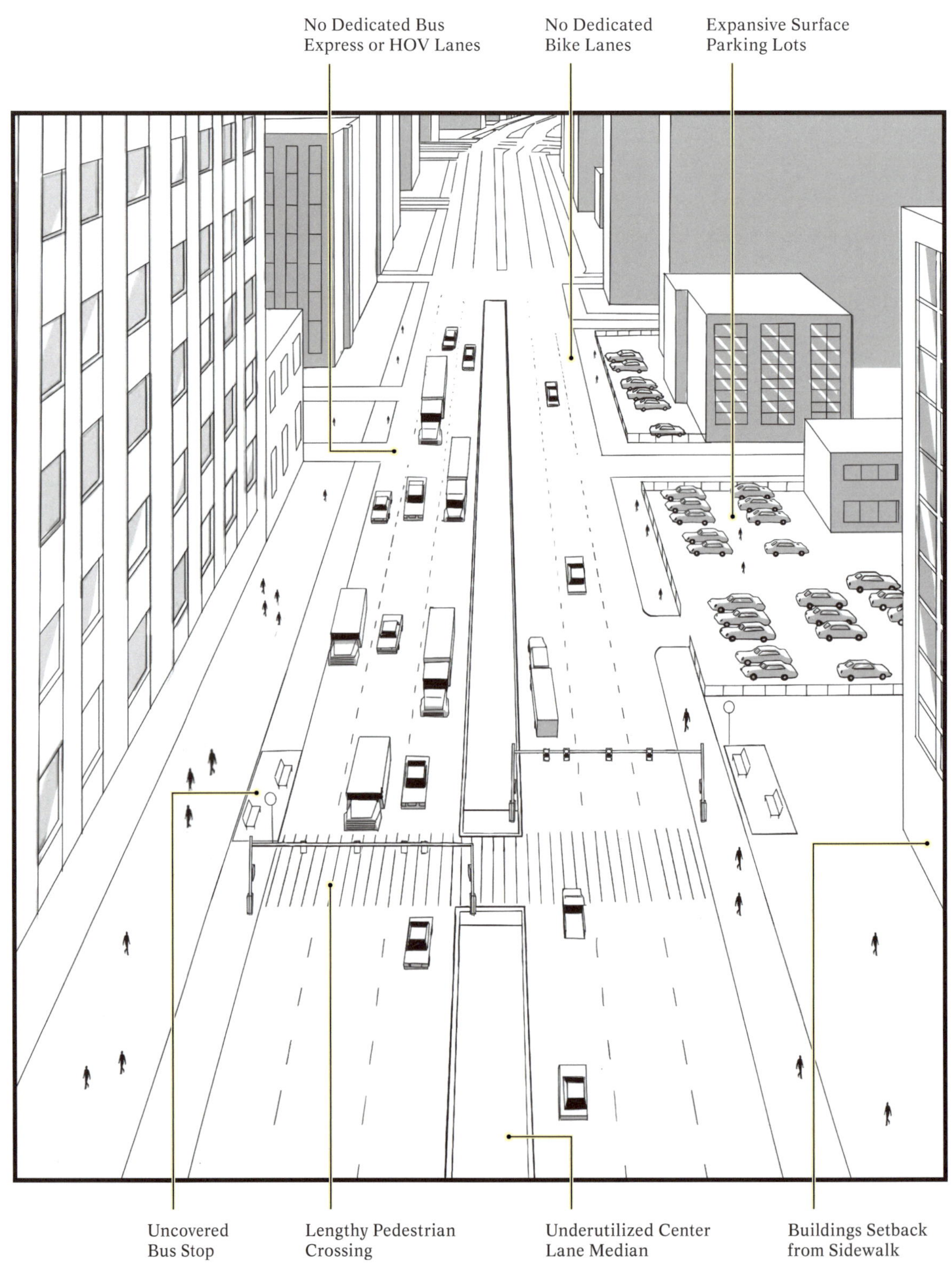
No Dedicated Bus Express or HOV Lanes
No Dedicated Bike Lanes
Expansive Surface Parking Lots
Uncovered Bus Stop
Lengthy Pedestrian Crossing
Underutilized Center Lane Median
Buildings Setback from Sidewalk

Street Amenities, Urban Furniture
Integrated Bike Lanes
Expanded Sidewalk and Public Realm
Integrated Bus Stop with Station Amenities
AV Bus Rapid Transit Express Lane
Covered Medians for Safe Pedestrian Crossings
Urban Landscaping, Bioswale Zones

Commercial Street
Alvarado Street

Embedded within Los Angeles' gridiron network is a series of commercial and retail street corridors that frame the residential fabric of the city. Often running north-south from the foothills of the Santa Monica mountains that overlook the city in the north, all the way down south to the Palos Verdes Hills near San Pedro and Long Beach, these avenues provide the perpendicular mobility connectivity that funnel automobiles to the nearest highway. They also hold a lot of the commercial land uses of the city, with retail businesses, strip malls, and other commercial uses that support its residential fabric. Because of the auto-dependent evolution of Los Angeles' land use and density patterns, businesses tend to set up shop in low-density building forms that extend along these arterial streets, rather than clustering around major centers or transit hubs. As access to these buildings is dependent on automobiles, these one- to two-story retail buildings are interspersed with surface lots and street parking often set back from the street wall. This has produced a stark, unfriendly, and at times perilous street environment for pedestrians and bicyclists, compelling individuals on foot to walk long distances between businesses. Local Metro buses frequent these corridors, using them as the foundation of the city's local transit network but without any lane accommodations or sheltered stops to encourage their use and prioritize their speed.

One of these commercial thoroughfares is Alvarado Street, which runs north-south from the termination of the Glendale Freeway until Hoover Street near the 10 freeway. In a driverless future, the Metro Local Line 2 and Metro Shuttle 603, which currently operate on the street, are automated and given lane priority. The center of this commercial street is transformed from an underutilized two-way turning lane, to a pedestrian mall and bikeway. New public spaces for street commerce and outdoor markets could support existing commercial businesses with expanded economic opportunities and patrons traversing by foot or by transit. This transformed public realm would be accessible by frequent pedestrian crossings, street improvements, and AV pick-up/drop-off curbside zones. A street with commercial activities that previously fostered only automobile traffic and access could be reorganized into one that supports transit-activated pedestrian traffic, new densities, and far more vibrant uses.

ENCIA DE VIAJES
OMPRE SU BOLETO!
BARGAIN
DICA
OLYMPIC
Hola!
POLLO
BEAUTY SUPPLY

LATIN AMERICAN
DENTAL CLINIC
* HIERBAS *
(213) 483-4330
DENTISTA
BOTICA
La Original
DEL PUEBLO
PRESCRIPTIONS
VITAMINS
COSMETICS
ABIERTO LOS 7 DIAS DE 9 AM A 9 PM
OK
Ria

Street Elevation

Street Perspective

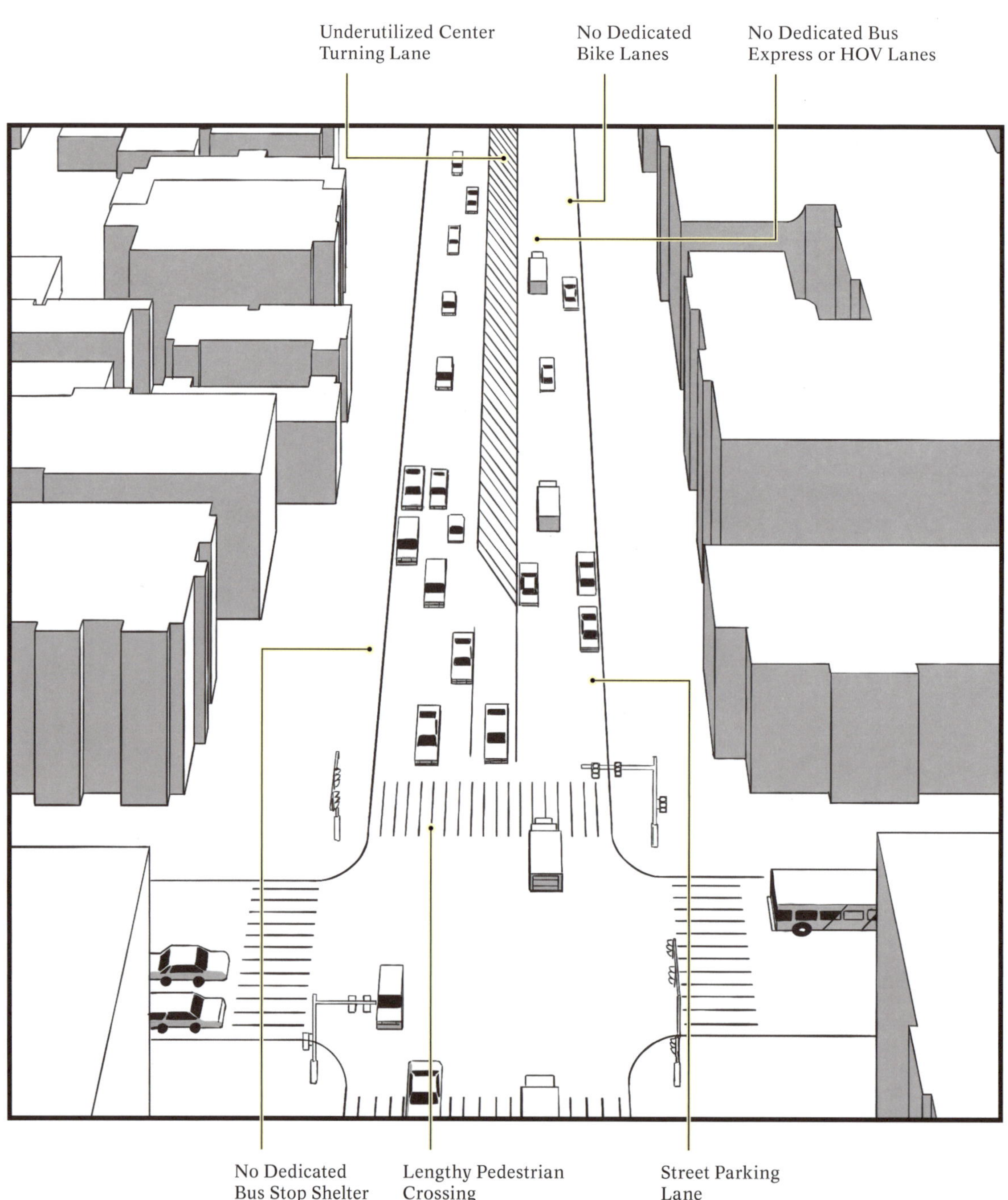
Underutilized Center Turning Lane
No Dedicated Bike Lanes
No Dedicated Bus Express or HOV Lanes
No Dedicated Bus Stop Shelter
Lengthy Pedestrian Crossing
Street Parking Lane

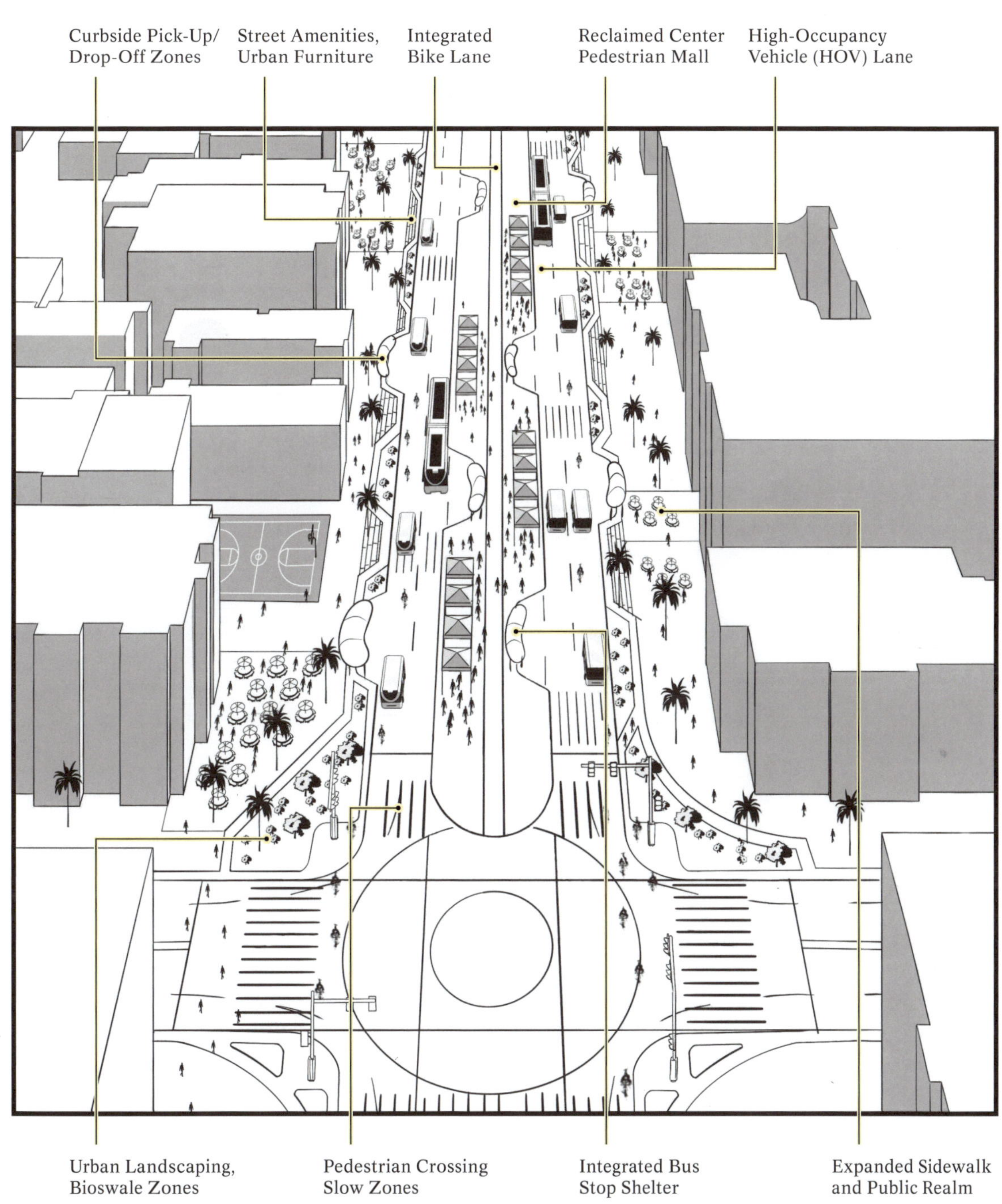
Curbside Pick-Up/
Drop-Off Zones
Street Amenities,
Urban Furniture
Integrated
Bike Lane
Reclaimed Center
Pedestrian Mall
High-Occupancy
Vehicle (HOV) Lane
Urban Landscaping,
Bioswale Zones
Pedestrian Crossing
Slow Zones
Integrated Bus
Stop Shelter
Expanded Sidewalk
and Public Realm

Community Avenue
Crenshaw Boulevard

Another example of an important commercial thoroughfare in Los Angeles is Crenshaw Boulevard, which runs along a 23-mile route in the west-central part of the city. Crenshaw serves as an important mobility route for an area that does not have a highway slicing through it. Its central thoroughfare runs through the Crenshaw neighborhood, Leimert Park, and Hyde Park in southwest Los Angeles, neighborhoods known as the "heart of African American commerce." These areas are home to large Black populations that have witnessed decades of disinvestment, resulting in urban environments that have seen trees removed (from the construction of the Metro K Line), and lack of quality public spaces and urban amenities that more affluent areas routinely receive. These are areas that could benefit greatly from street right-of-way reclamation projects as well as transit investments that support a demographic population of the city that relies heavily on it for mobility. The corridor is serviced by a variety of transit lines including the recently planned Metro K line extension funded by Measure M, as well as various local bus lines.

While in the previous typology the commercial street was converted into a pedestrian-oriented retail corridor, this illustration explores a scenario in which the majority right-of-way of the street is reclaimed as a cultural plaza instead. It builds upon the city's recent Destination Crenshaw revitalization project, a publicly funded plan that adds four acres of park space, street furniture, public art installations, and murals to a 1.3 mile segment of the boulevard. In this illustration, this transformation is imagined to further reclaim street lanes themselves, providing even more space for all of these planned uses. Together, these changes could transform the central boulevard into an open-air museum dedicated to preserving the history and culture of its local residents. Street right-of-way reclamations would be supported by mass transit mobility provided by the Metro K Line and its new expansion southward down the boulevard below-grade. Along with new dedicated express bus lanes at-grade, increased service and connectivity would be brought to the area to support the existing neighborhood. Typologically, this exploration reconceptualizes the street as not only a place to pass through but equally a place to occupy, support, and celebrate vital community functions and cultures.

DETOUR
AHEAD

Benjamin
Moore
Paintcraft
Supply & Hardware
Paintcraft
HARDWARE
ACE
Hardware

Street Elevation

Street Perspective

Community Avenue
Crenshaw Boulevard

2024

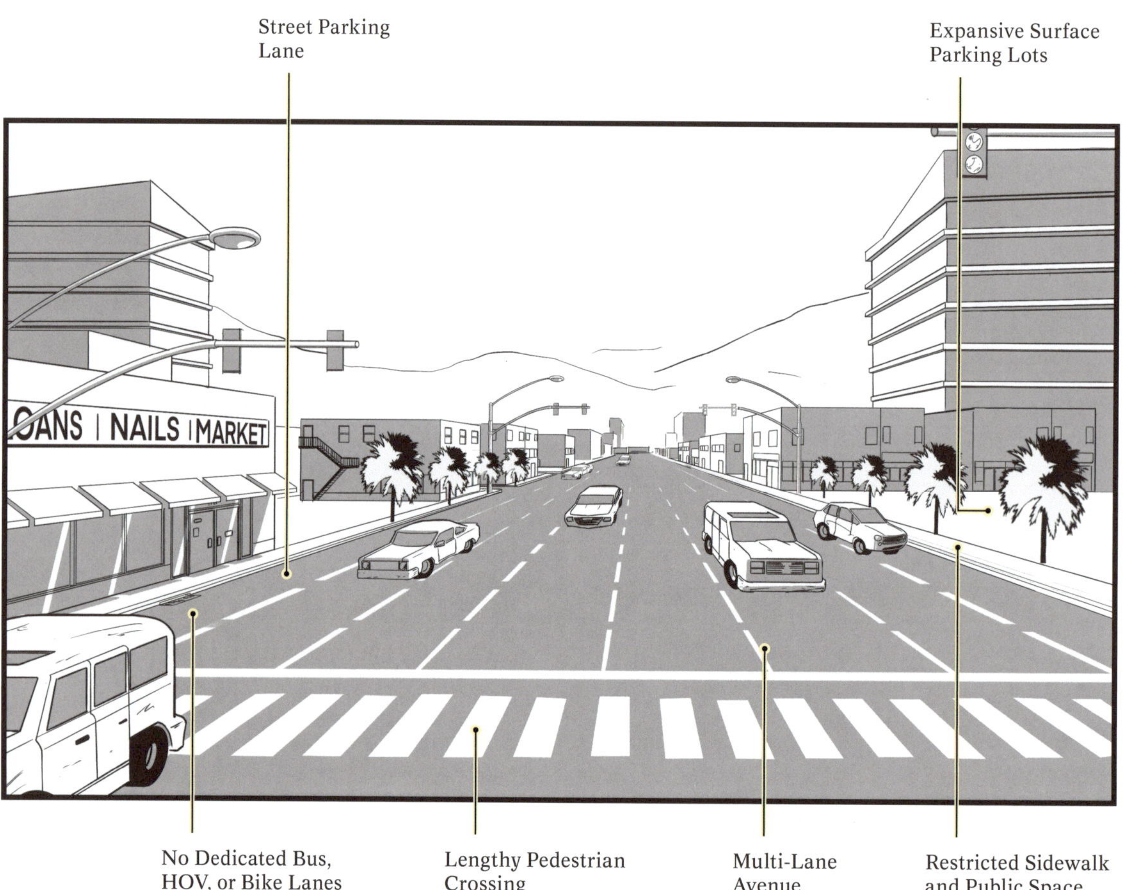

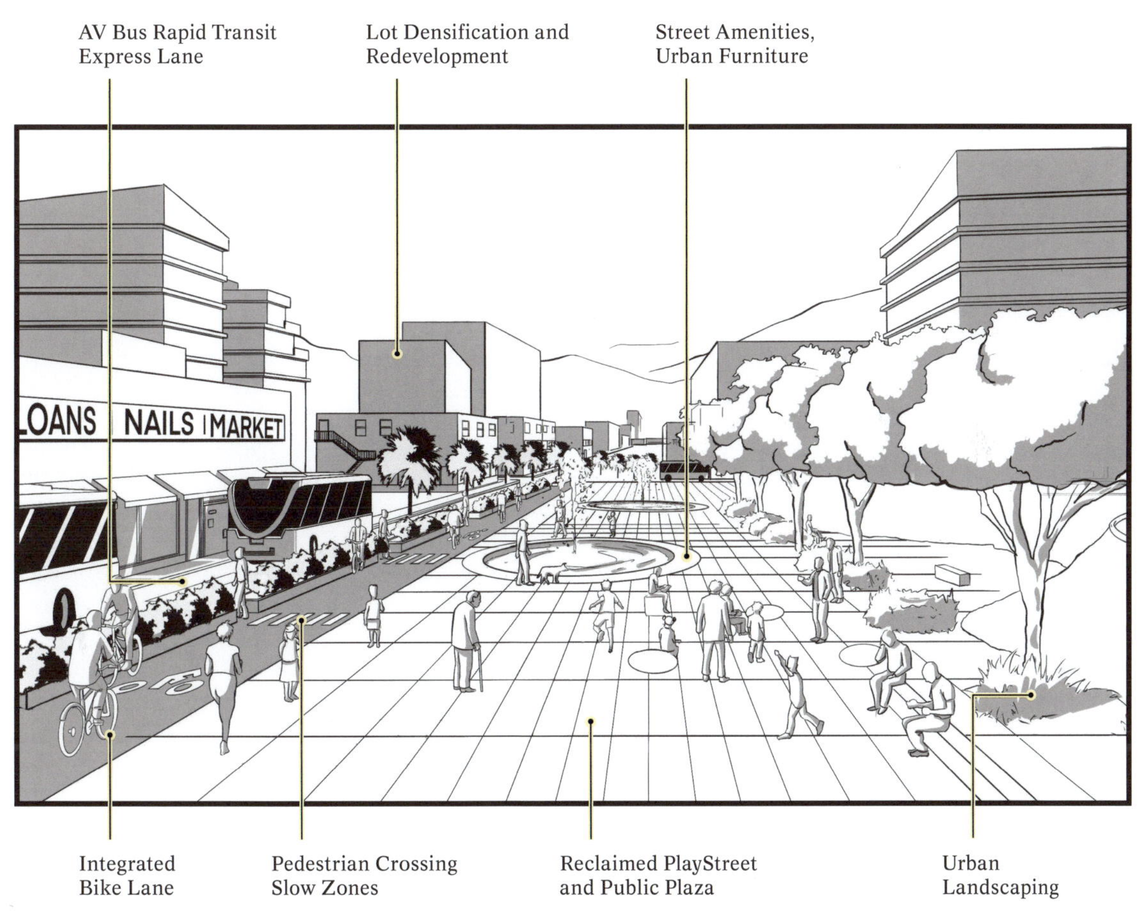
AV Bus Rapid Transit
Express Lane
Lot Densification and
Redevelopment
Street Amenities,
Urban Furniture
LOANS
NAILS
MARKET
Integrated
Bike Lane
Pedestrian Crossing
Slow Zones
Reclaimed PlayStreet
and Public Plaza
Urban
Landscaping

Dynamic Lane-Managed Street 3rd Street Promenade Extension

As discussed in earlier chapters, dynamic lane management describes a system of hard and soft infrastructural changes to a street right-of-way that involves reallocating street lanes in response to changes in demand during different times of day. Dynamic lane-managed streets allow what we typically think of as fixed infrastructure and street right-of-ways, to become more flexible spaces that could service a diversity of demands ranging from vehicular traffic to pedestrian use. Reversible lanes, lane closures, mechanical bollards, and electronic street signage are technologies that could communicate with AVs to determine priority use of a street right-of-way. In its current form, the technology has been used to only change lane directions or redirect traffic flow to optimize vehicular movement. In the future, this strategy is capitalized further to reclaim lane space for non-automobile uses in its entirety. Lunch hours could allow restaurants to expand their outdoor seating to accommodate an increase in patrons, with the entire street closed for pedestrian and public use. During rush hour traffic, lanes could be diverted for mass transit, cycling, and vehicular use, while off-peak hours requiring urban freight deliveries to businesses could have dedicated space allocations at those times.

Illustrated in this section, dynamic lane-managed streets are implemented in downtown Santa Monica, as an extension of its highly successful 3rd Street Pedestrianization project. This three-block street in Santa Monica was converted in 1965 to an outdoor pedestrian mall in an attempt to draw residents back to downtown areas as they fled to the suburbs. While its success ebbed and flowed in the decades since its first conversion, the pedestrian mall has seen a boon in use after a renovation in the 1980s addressed many of the outdoor street's shortcomings. Designed initially with a central roadway providing automobile access down its center during off-peak hours, the renovation was so successful that its right-of-way was ultimately fully converted to pedestrian-only use with permanent bollards at its entrances. 3rd Street Promenade is a prime example of how pedestrian-oriented street reclamations can serve as commercial and economic drivers for not only the businesses fronting it but also revitalize areas and blocks around it. In this exploration, dynamic lane management expands the length of the pedestrian mall, or is implemented on parallel streets. Typologically, it could serve to address the concerns of motorists demanding continued automobile access at certain times of days, while demonstrating how streets could also be fully reclaimed in other temporal instances. In futures where that demand has been fully absorbed by mass transit use, the street could be fully reclaimed for pedestrians only, as in the case of 3rd Street. This evolution could have larger paradigmatic implications on the flexible and dynamic use of *all* street infrastructure and right-of-ways in the overall network of the city.

3rd Street Promenade Perspective

Typical Street Perspective

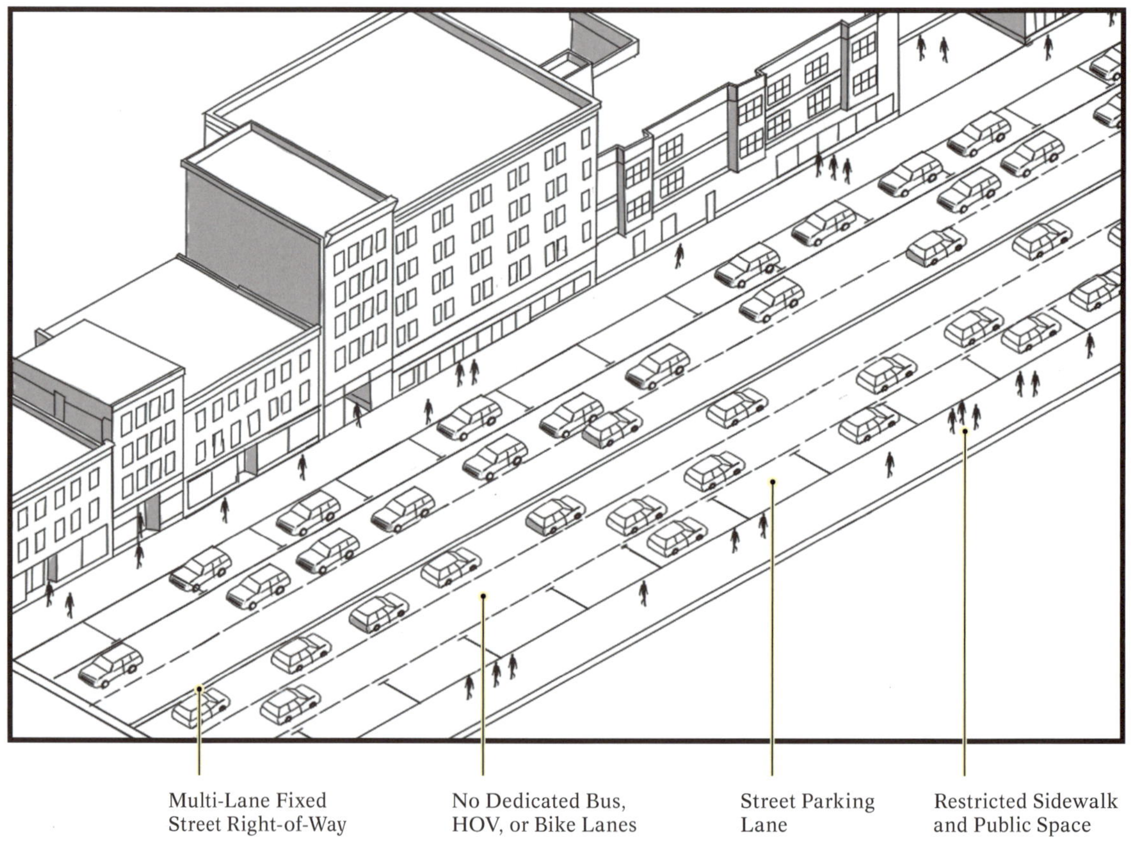
Multi-Lane Fixed
Street Right-of-Way
No Dedicated Bus,
HOV, or Bike Lanes
Street Parking
Lane
Restricted Sidewalk
and Public Space

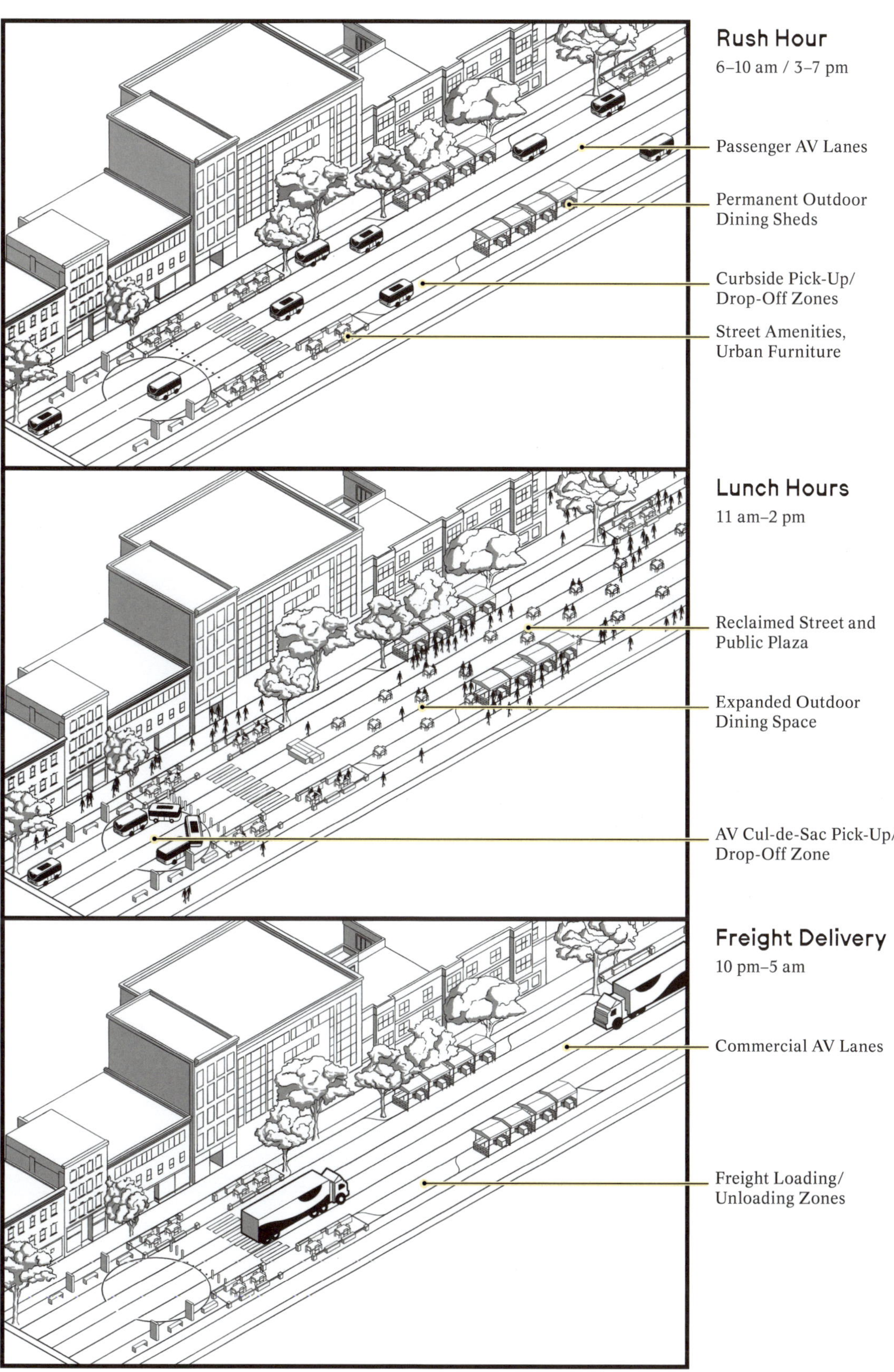
Rush Hour
6–10 am / 3–7 pm
Passenger AV Lanes
Permanent Outdoor Dining Sheds
Curbside Pick-Up/ Drop-Off Zones
Street Amenities, Urban Furniture
Lunch Hours
11 am–2 pm
Reclaimed Street and Public Plaza
Expanded Outdoor Dining Space
AV Cul-de-Sac Pick-Up/ Drop-Off Zone
Freight Delivery
10 pm–5 am
Commercial AV Lanes
Freight Loading/ Unloading Zones

2024

Restricted Sidewalk and Public Space
No Dedicated Bus, HOV, or Bike Lanes
Multi-Lane Fixed Street Right-of-Way
Street Parking Lane

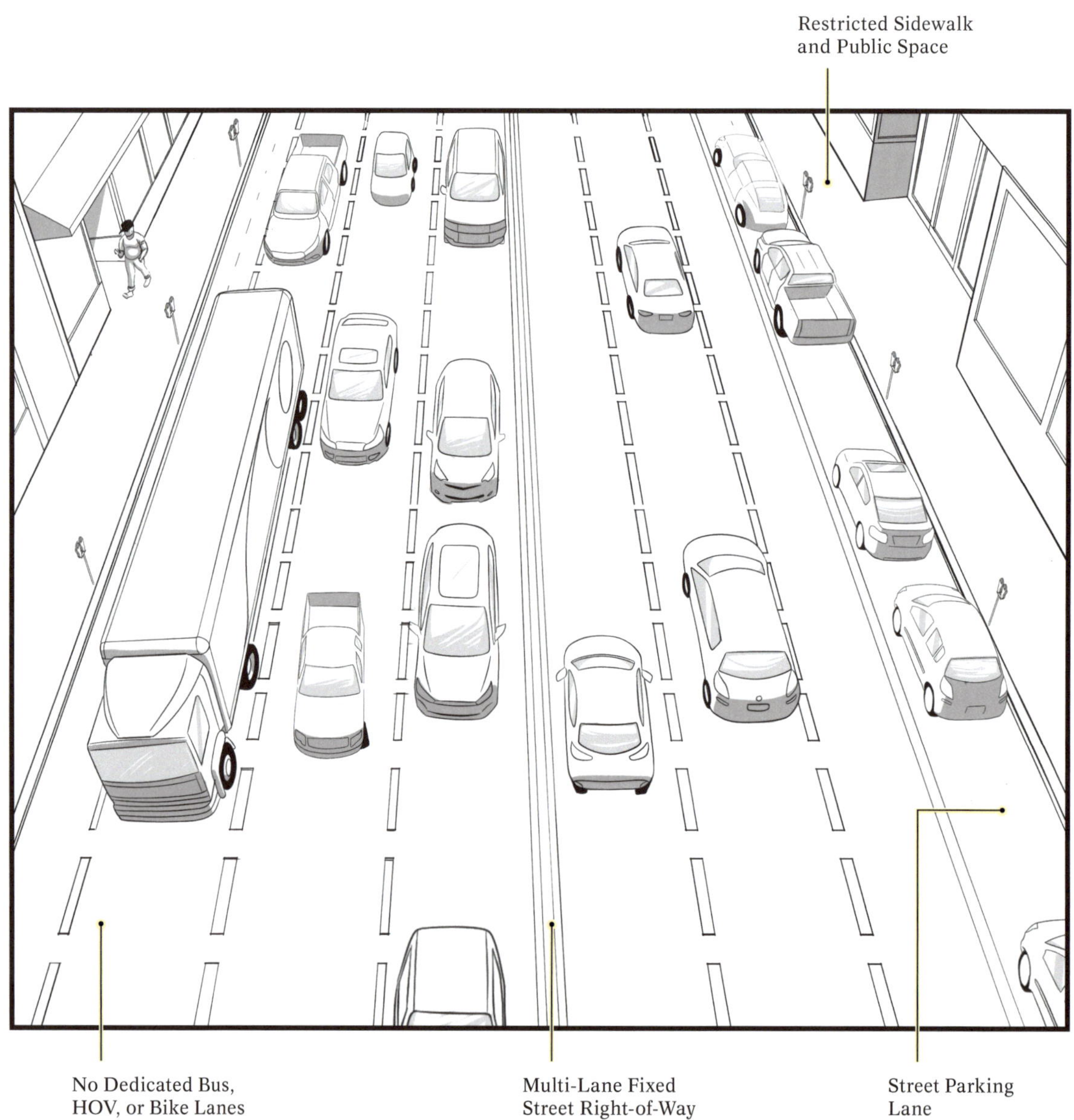

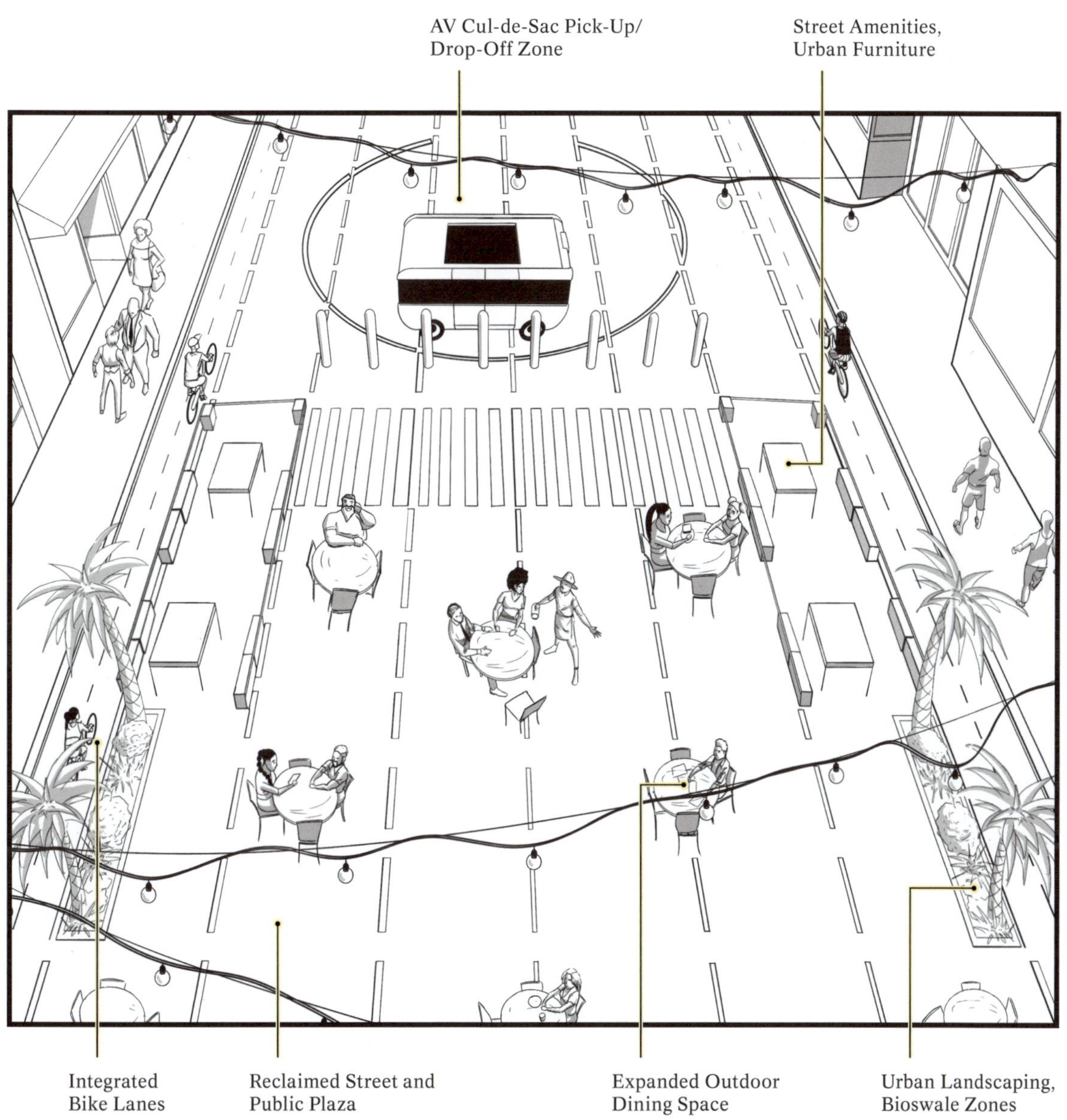
AV Cul-de-Sac Pick-Up/
Drop-Off Zone
Street Amenities,
Urban Furniture
Integrated
Bike Lanes
Reclaimed Street and
Public Plaza
Expanded Outdoor
Dining Space
Urban Landscaping,
Bioswale Zones

Residential Intersection
Larchmont

With over 22,000 miles of roads, the gridiron network of Los Angeles results in thousands of intersections. With almost all of these intersections consisting of vehicles crossing in two directions in two ways along with pedestrians and cyclists, it is no surprise that almost one-quarter of all fatal accidents and one-half of all crashes resulting in injury, happen at intersections.[5] In particular, many of these intersections fall in the vast metropolitan landscape of mid- and lower-density residential fabric. This network, while important in providing local connectivity to residents, is also fairly underutilized, especially in low-density residential areas of the city gridiron. In proportion to the low-traffic utilization by vehicles at these intersection right-of-ways, their space allocation far exceeds their demand.

5 "About Intersection Safety," Federal Highway Administration, 2023.

Historically, street intersection corners can act as important social spaces for communities, linking neighborhood blocks and also marking where people's paths naturally cross. When these spaces are built to welcome pedestrians instead of vehicles, they can not only become far safer but also more social and recreational for pedestrians, children, and families alike. Illustrated in this section, automation sparks an opportunity to reexamine the intersection of residential streets. Instead of two-way crossings, vehicular traffic is rerouted though one-way, right-turn only lanes in order to free up the center of these intersections for reclamation. Playstreets and plazas with urban landscaping, outdoor furniture, and street amenities could be designed into the intersection itself and supporting laneways. Vehicular lanes could jog back and forth when nearing intersections to naturally reduce vehicular speeds for safety. Bike lanes and pedestrian pathways could suddenly have continuous through-street access without fear of crossing vehicles, fostering pedestrian connectivity through multiple blocks in transformational ways. While not likely practical to be implemented at every single intersection, key intercommunity intersections could introduce these street typologies. This would not only transform the street itself but also redefine the structural relationship *between* blocks and the residential urban fabric of the city in exciting ways.

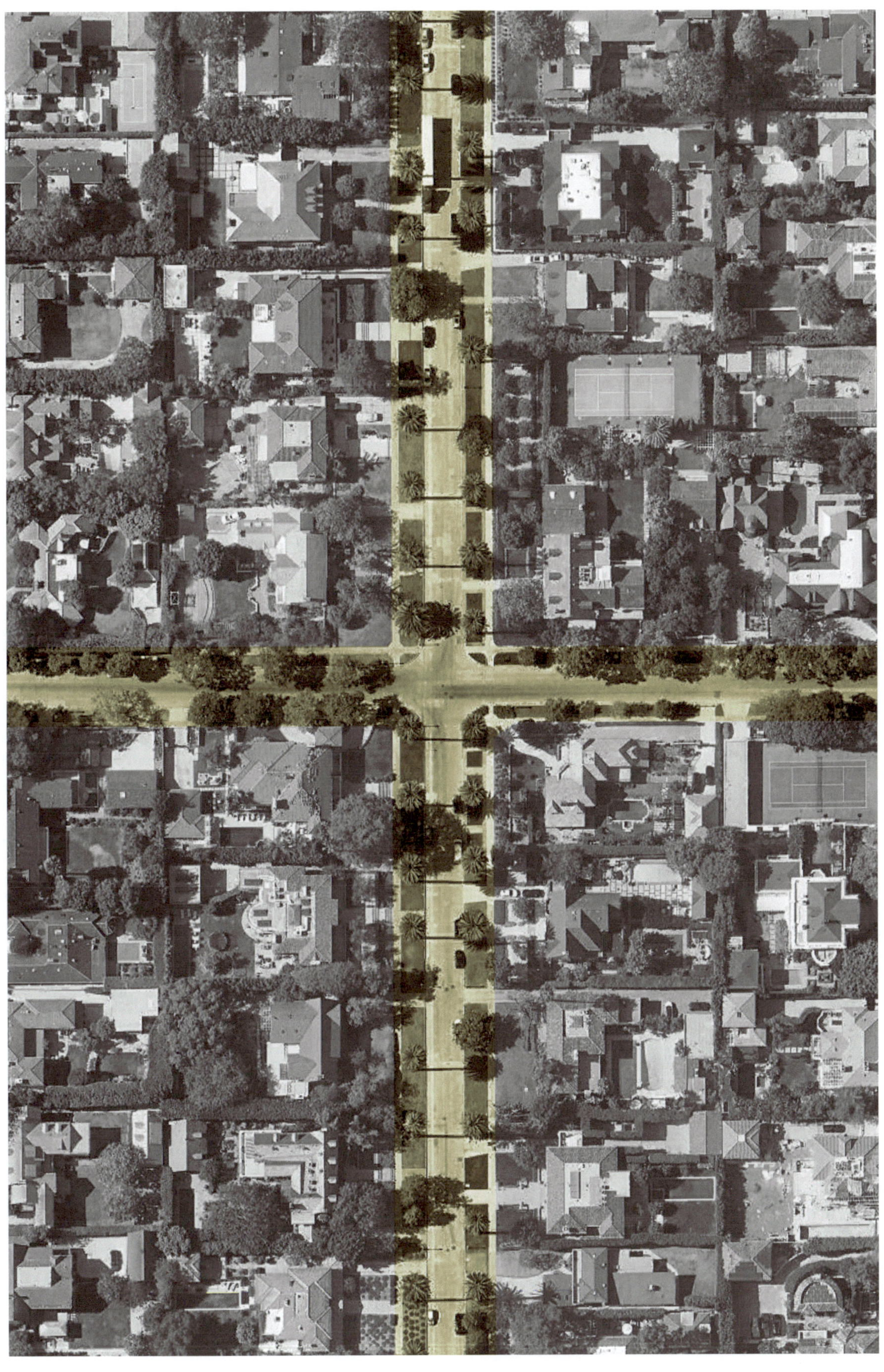

Street Perspective

Street Perspective

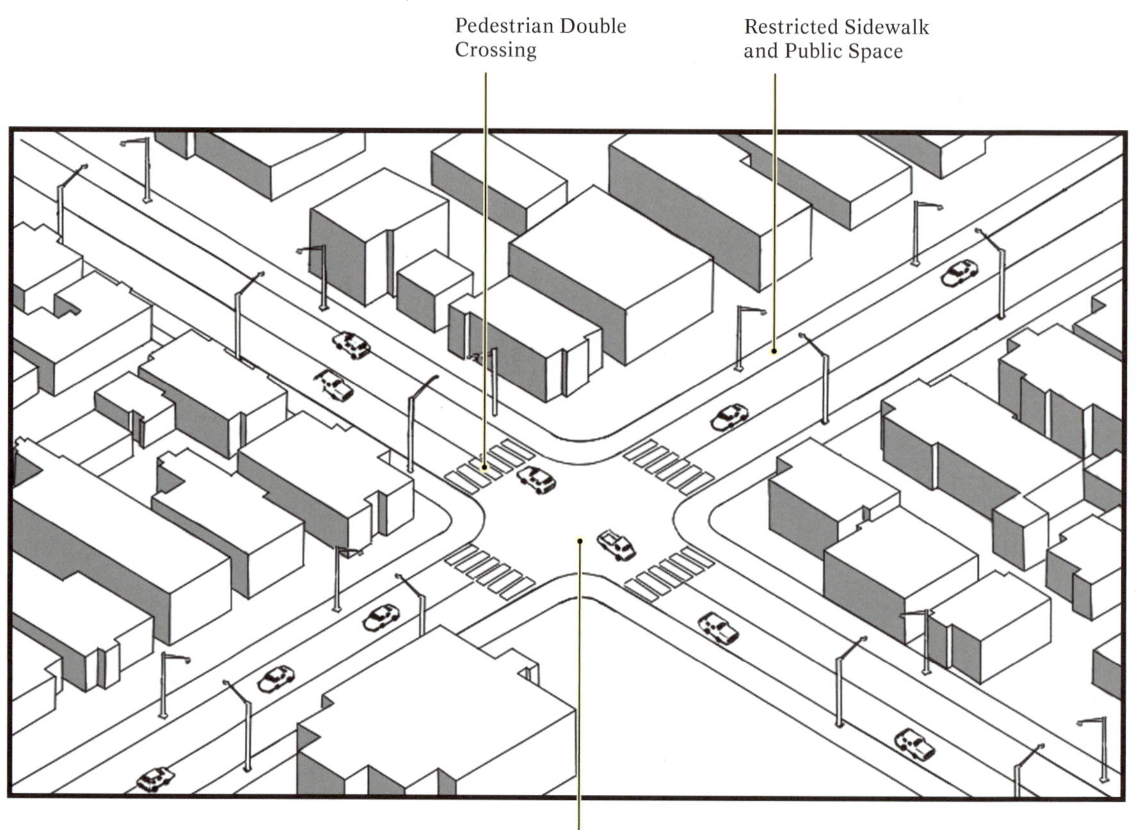
Pedestrian Double
Crossing
Restricted Sidewalk
and Public Space
Two-Way Through
Street and Intersection

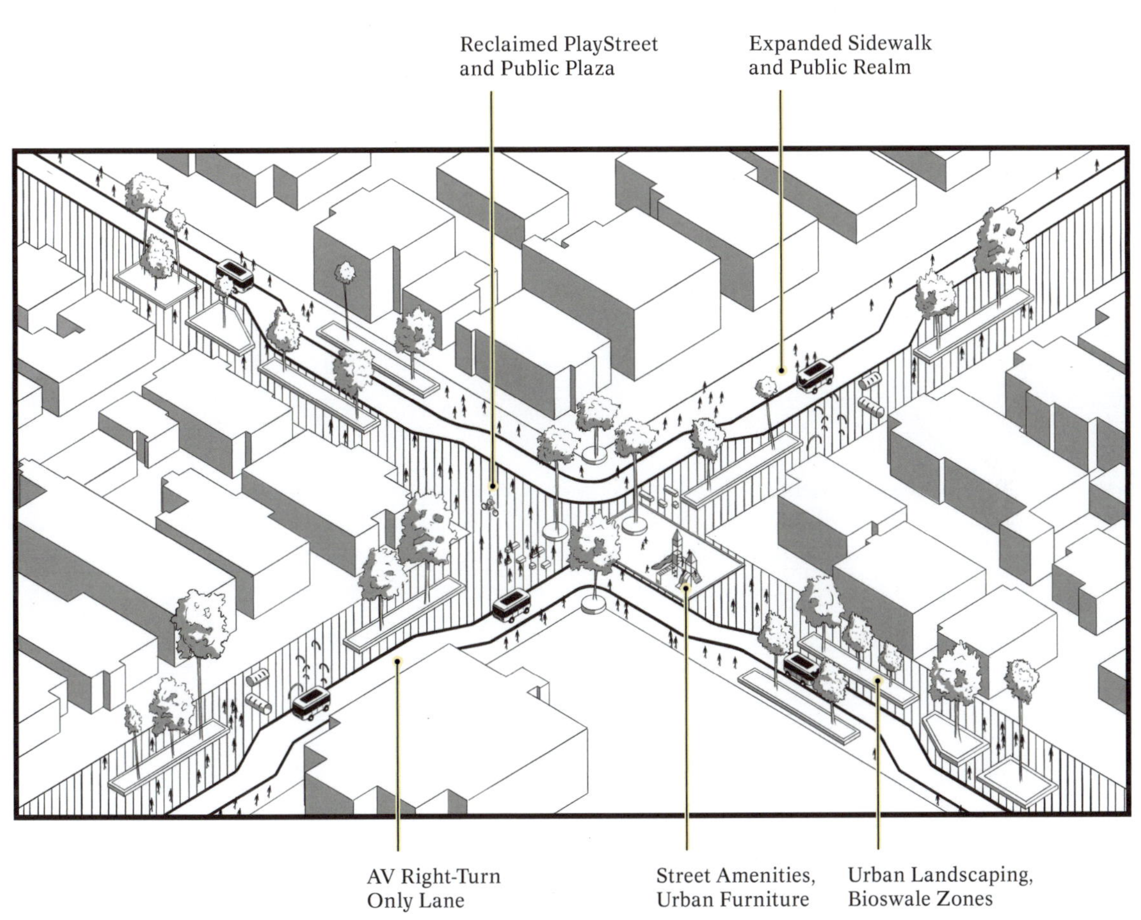
Reclaimed PlayStreet
and Public Plaza
Expanded Sidewalk
and Public Realm
AV Right-Turn
Only Lane
Street Amenities,
Urban Furniture
Urban Landscaping,
Bioswale Zones

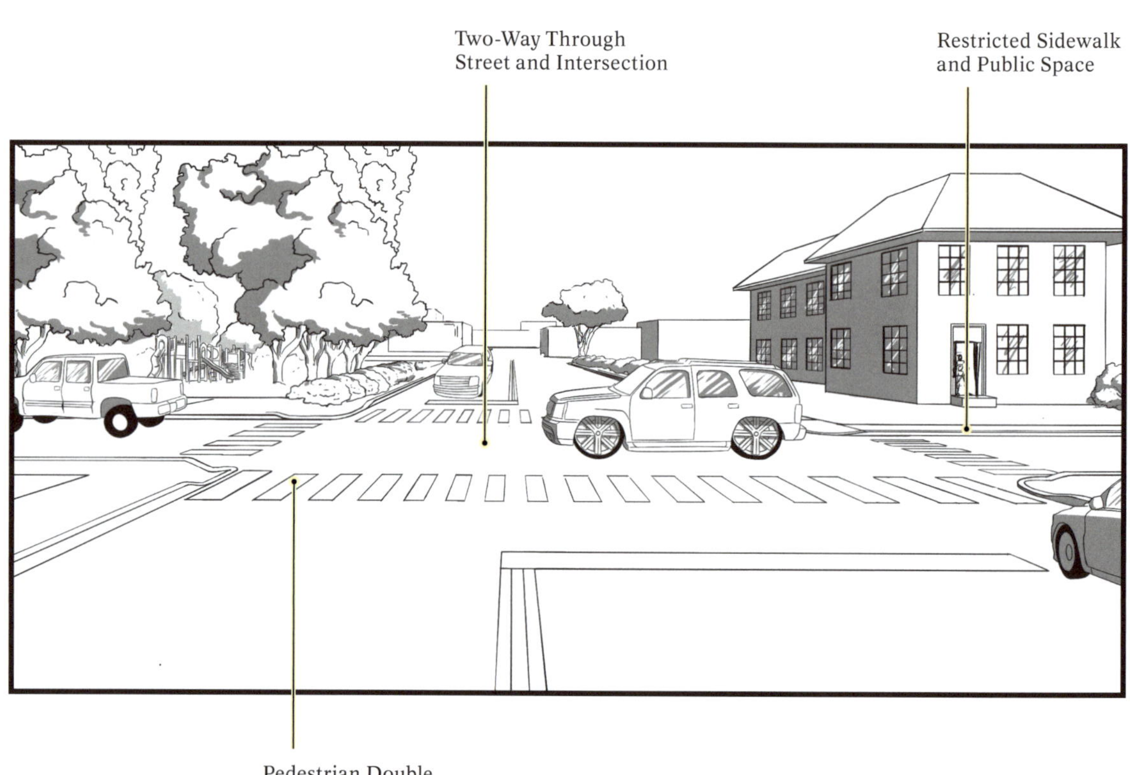
Two-Way Through
Street and Intersection
Restricted Sidewalk
and Public Space
Pedestrian Double
Crossing

Reclaimed PlayStreet and Public Plaza

Expanded Sidewalk and Public Realm

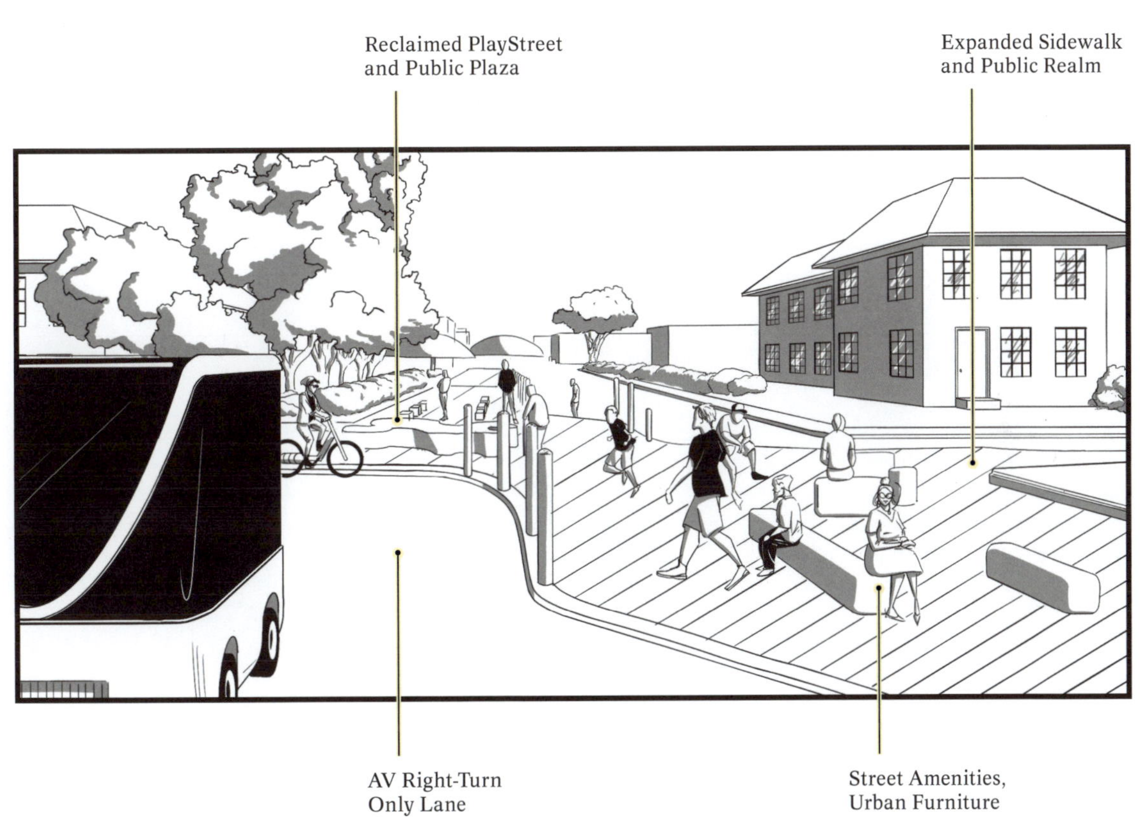

AV Right-Turn Only Lane

Street Amenities, Urban Furniture

9 M: Block-Scale Land Use and Density Transformations

Our city's urban block configurations have evolved in close calibration with the mobility systems that structure them. Defined as the space within the street pattern that is subdivided into land lots for the construction of buildings, blocks depend on the mobility access provided to them by the transportation network of the city. Besides the approximate one-fifth of urban land area dedicated to our streets and highways, the remaining majority is constituted by city blocks of a wide range of scales, dimensions, configurations, and types.

Pre-automobile, the evolution and maturation of city blocks has historically resulted in urban form that is developed to the edge of the block's lot limits. This results in a continuous street wall in which building facades form an unbroken edge that activates pedestrian sidewalks with diverse frontages and uses. While vibrant street walls and diverse urban blocks can still be found all over the world, the arrival of the automobile has impacted them in distinct ways. Required automobile parking minimums and the increased demand for parking by building inhabitants has resulted in driveways and surface lots that interrupt the perimeter of urban blocks in order to grant vehicle access to their interiors or edges. When proliferated to an extreme, in cities like Los Angeles for example, this results in porous blocks that dedicate a large percentage of their land to parking and car access. Oftentimes, these blocks no longer retain a recognizable perimeter of urban form that reinforce the edge of the street. Block back-alleys have also been altered in use by the automobile. Alleyways that, in their historical evolution, were used as secondary pedestrian accessways or functioned as a place of commons, have shifted in use in the modern block. Instead, they now service automobile access to back-loaded building garages, facilitate trash and waste removal by garbage collector vehicles, or serve as commercial loading/unloading areas for businesses.

Planned gridirons of the modern city have also been calibrated to the automobile in recent history. Certain city block dimensions cater to car efficiencies by introducing longer block lengths, for example. This is designed so that drivers don't have to stop at street intersections too frequently as they travel to the closest arterial road. At their most extreme forms, these elongated blocks and the accumulation of several blocks form the basis of the modern superblock. As has been previously discussed, the superblock concept of grouping multiple blocks originated in part to facilitate more efficient arterial trafficways for higher-speed automobiles at the periphery of a multi-block grid. This allowed the internal street divisions of superblocks to be formatted for automobile parking (as is often the case in North American public housing superblocks), lower-traffic roads, or pedestrian pathways.

The automobile has also had an effect on the block at a range of densities. Low-density block developments like the suburban single-family zoned block, for example, are subdivided and formatted around the individual front-loaded parking garage and driveway. Middle to higher density block developments, with higher density parking requirements, are often built with multi-tier parking podiums that separate urban form and land uses from the ground. These configurations result from the fact that it is typically more expensive to build parking underground than at-grade, in addition to the spatial difficulties involved with moving vehicles vertically with ramps or elevators. In all cases, the cumulative result are blocks that begin to prioritize their access and format their spatial configurations around the automobile at the ground plane, at the expense of pedestrians and other uses that are better suited to activating the street, sidewalk, and public realm.

The history of the urban block has surely evolved not just from the influence of our mobility systems but also from a variety of complex urban planning, zoning, and land use decisions. These are unique to the cultural, political, and economic contexts of their respective cities. However, it is clear that the automobile has had a distinct impact on their growth and evolution in the contemporary city. Automating those vehicles presents future opportunities to transform, evolve, and reconfigure these block arrangements to reclaim urban space. What alternative spatial futures could emerge when these blocks are allowed to densify and diversify their usages away from the spatial demands of the automobile? How can land use and zoning policy changes support a shift towards mixed-used, street-activating urban fabric at the ground plane? How can prototypical block types varying in scale and density take advantage of new spatial opportunities catalyzed by driverless technologies? How can the accumulation of blocks as an aggregate inform wider development patterns and frameworks for the future structure of cities? Let's examine these opportunities in illustrated spatial form next.

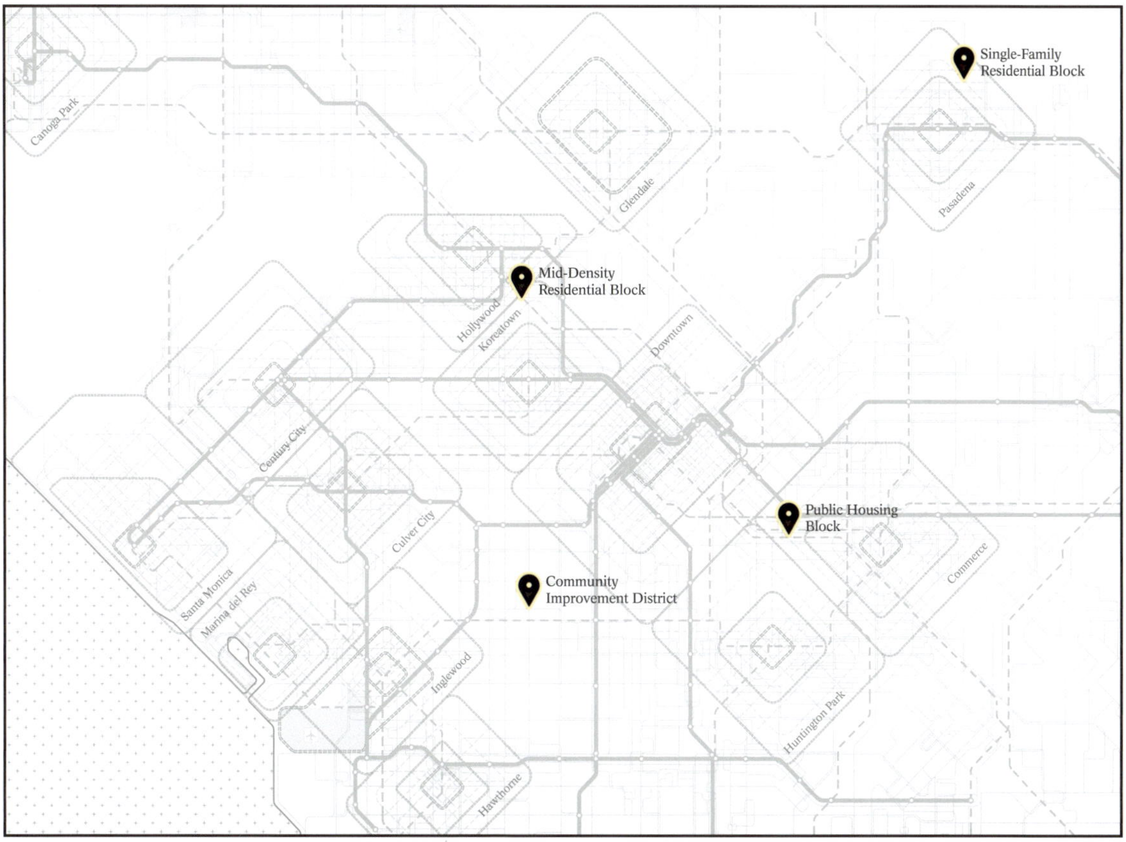

Single-Family Residential Block

Mid-Density Residential Block

Public Housing Block

Community Improvement District

Canoga Park

Glendale

Pasadena

Hollywood

Koreatown

Downtown

Century City

Culver City

Santa Monica

Marina del Rey

Inglewood

Hawthorne

Huntington Park

Commerce

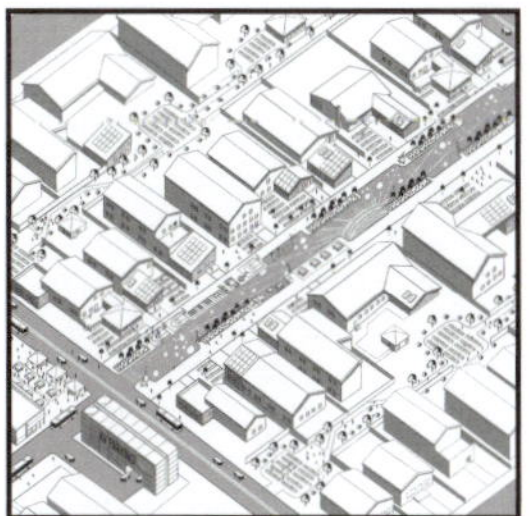

Single-Family Residential Block
Pasadena

Mid-Density Residential Block
Westlake

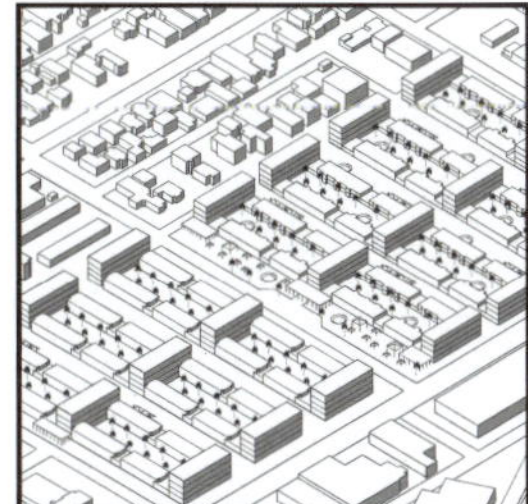

Public Housing Block
Estrada Courts, Boyle Heights

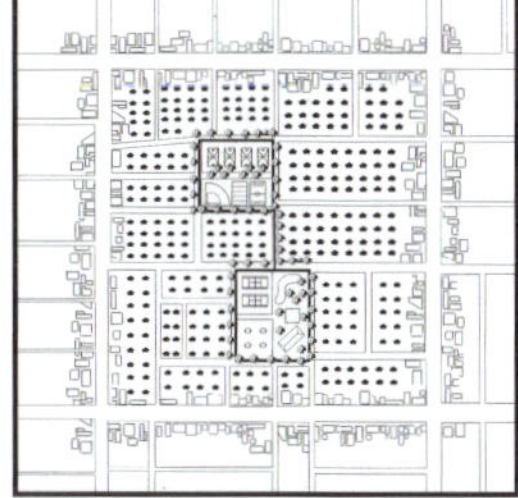

Community Improvement District
South Central

Single-Family Residential Block
Pasadena

The individually-owned automobile has exerted a strong influence on the typology of the single-family home and the suburban block that is constituted by it. Automobiles required to reach these suburban developments became incorporated into the design and footprint of individual homes, continually expanding in size over time. Today, this influence is most prominently manifested in a multi-car, private garage built into a residence that is serviced by a front- or back-facing driveway. Fully asphalted driveways at the width of these multi-car garages are often used to park additional vehicles outside. Not only do private garages and driveways consume a large amount of lot area, they also front the street's public right-of-way. Aggregated together in low-density zoned residential blocks, this results in a block configuration in which plastic garage doors, driveways, and curb cuts shape the experience and character of walking or driving by. The residential street itself remains typically underutilized, existing to facilitate individual access to each home by car and to park additional vehicles along curb edges. Los Angeles' typical block also features bisecting back-alleys that are used for vehicular access to garages at the property's rear, or for waste removal by garbage trucks.

Density and land use zoning changes in the future evolution of these blocks could unlock their latent potential. Fueled by a mobility paradigm shift, driveways and individual parking garages could become obsolete. They could be converted into expansions for existing homes or a new accessory dwelling unit (ADU)—which is a smaller, independent residential unit located on the same lot as a single-family home.[1] Garages could also be adaptively converted for small businesses or live-work units that combine residential and non-residential uses together. Driveways previously holding multiple cars and underutilized front- or side-yards could be activated by this increased density and new mix of uses. Any remaining individual parking needs could be consolidated at a block-end AV-parking structure. This would allow the underutilized residential street to be reclaimed as a PlayStreet for outdoor furniture, playgrounds, parks, and other recreational uses. Block alleyways could also become more pedestrian-friendly and communal, with collectively managed community gardens. Ride-hailed AVs and transit mini-shuttles could drop homeowners off at a block-end loading mall, which would also provide services like dry-cleaning or daycares before the short walk home. The PlayStreet would become a designated slow-zone to allow low-speed AVs to pick-up or drop-off passengers with para-transit needs at consolidated curbside zones serving several homes at once. While these changes would certainly require a cultural paradigm shift from NIMBY homeowners willing to relinquish individual automobile access and adopt land use and zoning changes, it is one that the city is already beginning to witness.[2] Thanks to changes in state and local zoning legislation, the number of ADUs in Los Angeles has exploded, more than doubling since 2017 and accounting (in 2022) for nearly a quarter of all new housing permits in the city.

1 Accessory dwelling units (ADUs) can be converted portions of existing homes (like garages), attached expansions, or detached additions on the same lot. ADUs have become increasingly popular in recent years through legislative changes in state housing policy because they have the potential to densify existing low-density residential fabric without displacing homeowners, while also having the potential to increase housing affordability for both homeowners and new ADU tenants.

2 NIMBY is an acronym for "not-in-my-backyard," a colloquialism that signifies a homeowner's opposition to the development or introduction of something new that might affect their existing quality of life or property value.

Block Elevation

Block Perspective

Single-Family Residential Block
Pasadena

2024

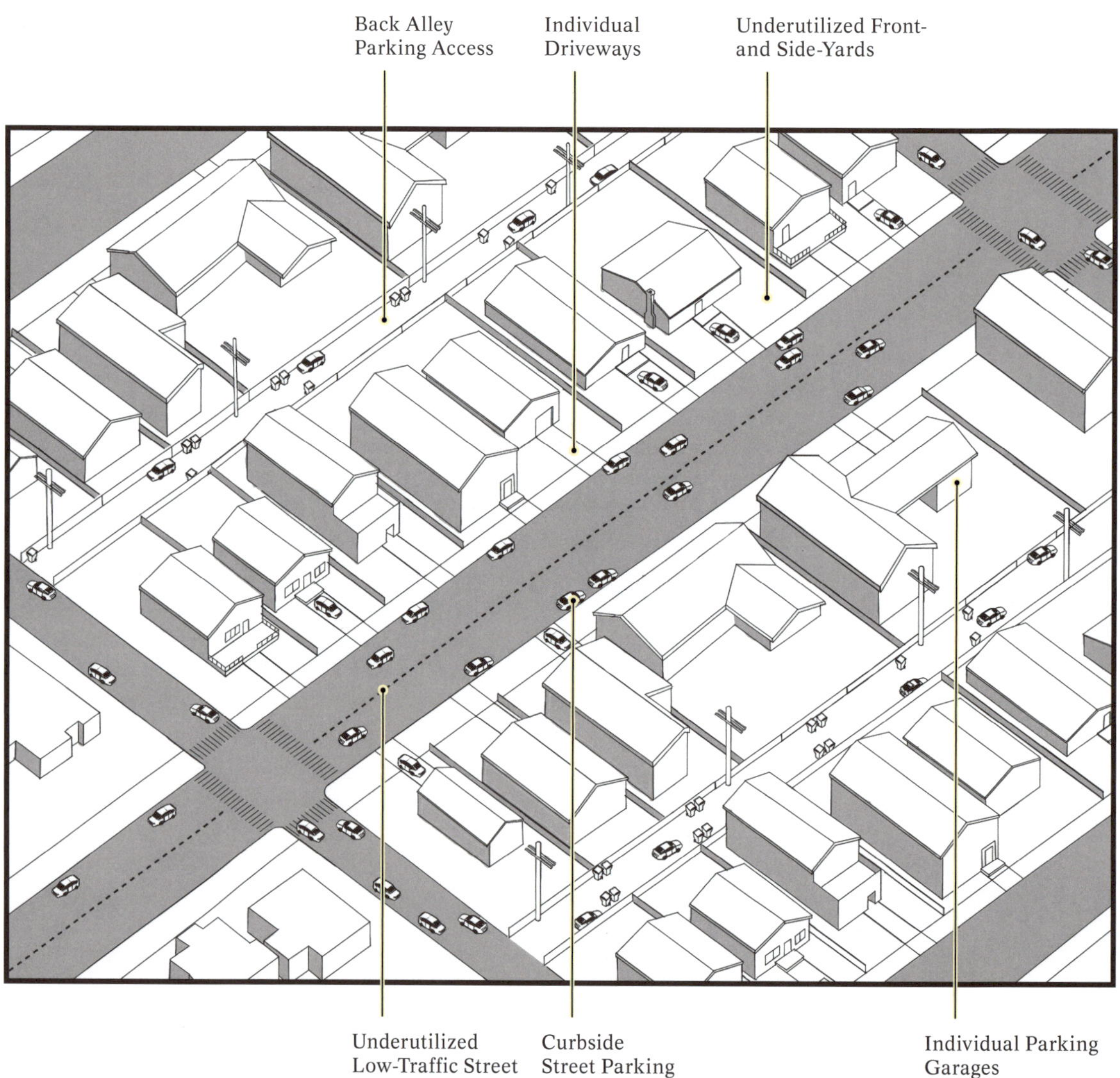

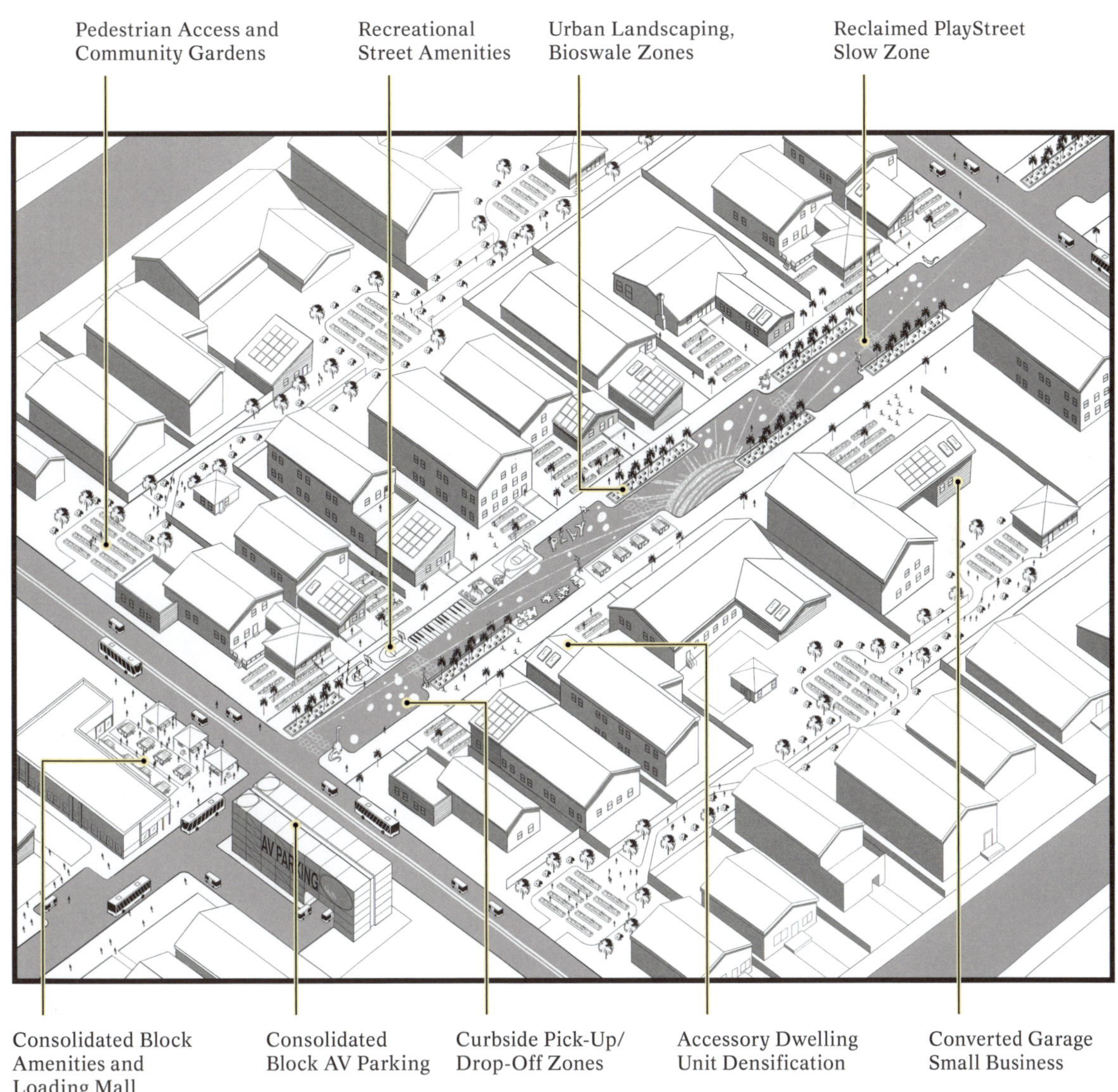
Pedestrian Access and Community Gardens
Recreational Street Amenities
Urban Landscaping, Bioswale Zones
Reclaimed PlayStreet Slow Zone
PLAY
AV PARKING
Consolidated Block Amenities and Loading Mall
Consolidated Block AV Parking
Curbside Pick-Up/ Drop-Off Zones
Accessory Dwelling Unit Densification
Converted Garage Small Business

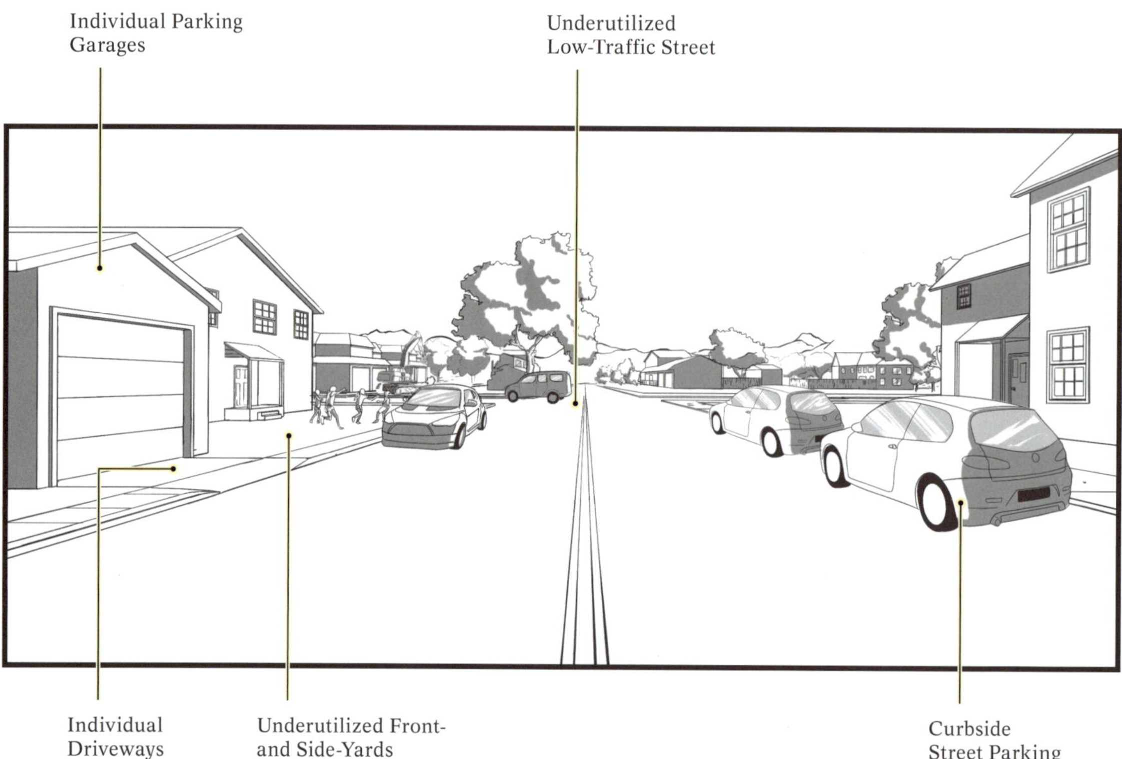
Individual Parking Garages
Underutilized Low-Traffic Street
Individual Driveways
Underutilized Front- and Side-Yards
Curbside Street Parking

2057

Converted Garage Small Business
Expanded Sidewalk and Public Realm
Urban Landscaping, Bioswale Zones
Accessory Dwelling Unit Densification
Reclaimed PlayStreet Slow Zone
Street Amenities, Urban Furniture

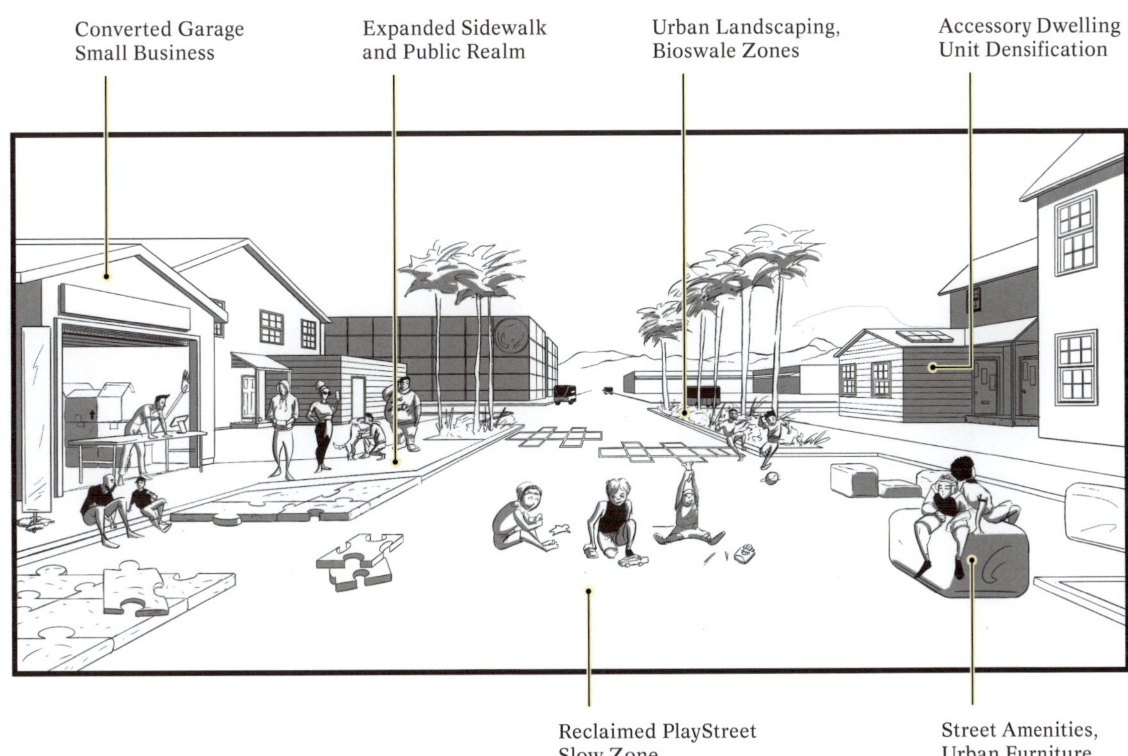

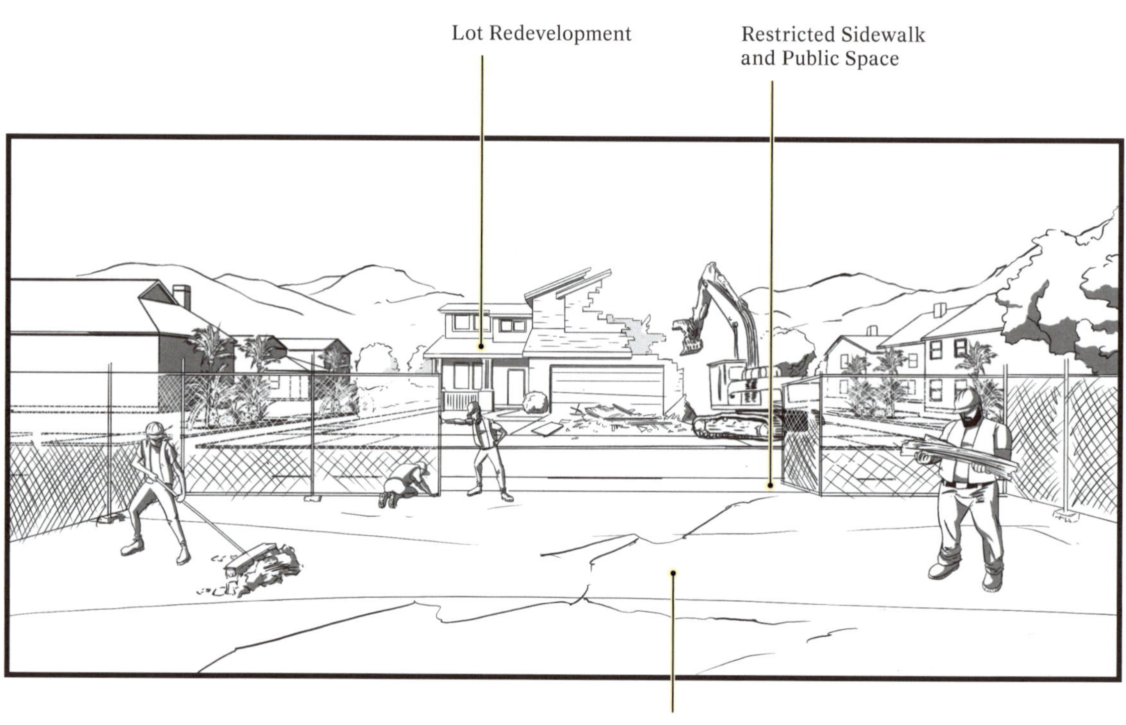
Lot Redevelopment
Restricted Sidewalk and Public Space
Vacant Lot

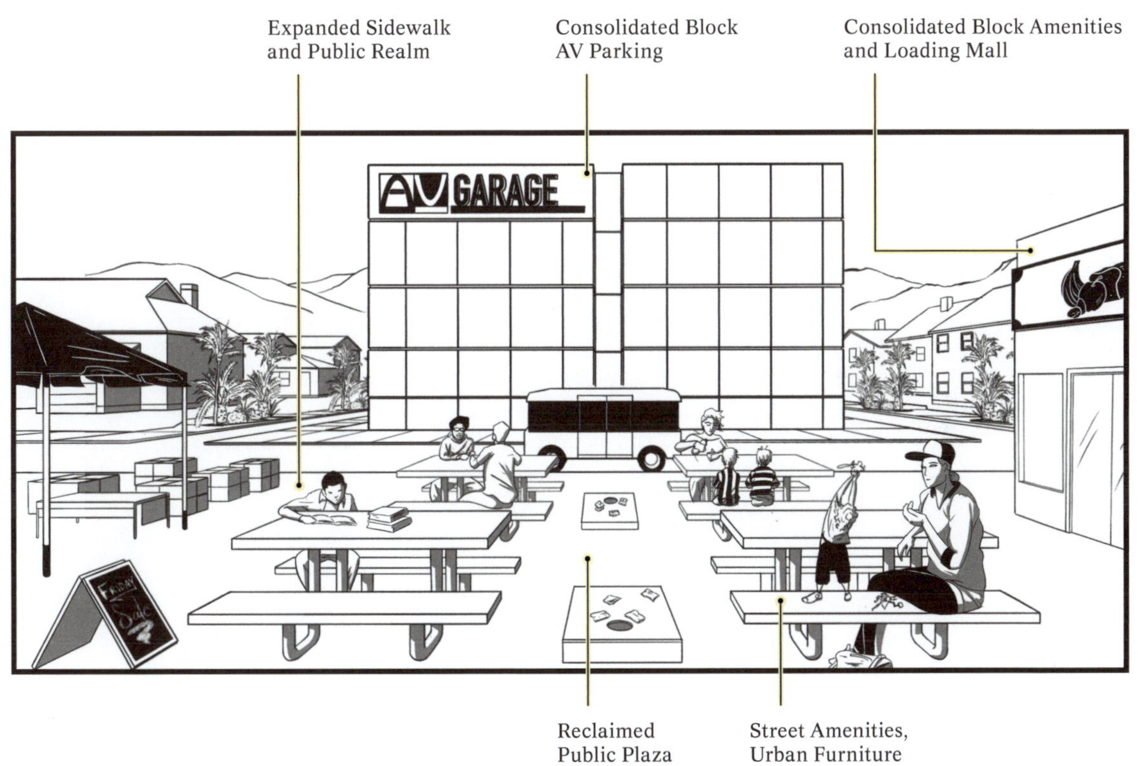
Expanded Sidewalk and Public Realm
Consolidated Block AV Parking
Consolidated Block Amenities and Loading Mall
AV GARAGE
Reclaimed Public Plaza
Street Amenities, Urban Furniture

Mid-Density Residential Block
Westlake

A common misperception of Los Angeles is that it is primarily constituted of low-density suburban blocks comprised solely of single-family homes. In truth, Los Angeles' population density metrics are on par with other major cities in the continental U.S. but this density is just expressed in different urban forms and block typologies. In the city, a variety of multi-family residential complexes ranging from three to four stories constitute a large percentage of its urban fabric. As highlighted on the right, many of these blocks are made up of large-footprint, mid-rise apartment complexes with double-loaded corridors that are organized around small interior courtyards or light wells. Due to the number of apartments that they house, these complexes are often built with a first-story residential parking garage which sits either at-grade or sunken half a level below-grade. These garage podiums are accessed by multi-entrance driveways leading down to gated parking entrances at the base of these buildings. Overfill parking can be found at surface lots located at the ends of blocks as well as through abundant street parking fronting the wide avenues that structure these mid-rises. A drive or walk down their length is void of any recreational or social spaces for communal activities to occur on sidewalks or street fronts, which are instead interrupted by wide driveways, curb cuts, and gated garage podium entrances.

The generous dimension of these mid-density residential blocks and the avenues that structure them offer great potential to transform as AVs and MaaS release demand for parking. Block-end parking could be redeveloped with new hybrid housing typologies that consolidate any reduced parking demand into AV-garages. New block amenities and commercial uses could be integrated at the block's end with an expanded bus station in conjunction with a pedestrian loading mall. This could free up lower-traffic residential streets to be pedestrianized and reclaimed for recreational and landscape uses. Instead of parking at the base of one's apartment complex, one could park or arrive at the end of a block just a short walk's distance away. This could allow the ground-floor parking podiums of mid-rise apartments to take advantage of new land use changes to become leasable commercial spaces or residential amenities. Altogether, these cumulative changes could set the stage for a more communal, recreational, and multi-use block to emerge as the foundation of the city's urban fabric.

Block Elevation

Block Perspective

Mid-Density Residential Block Westlake

2024

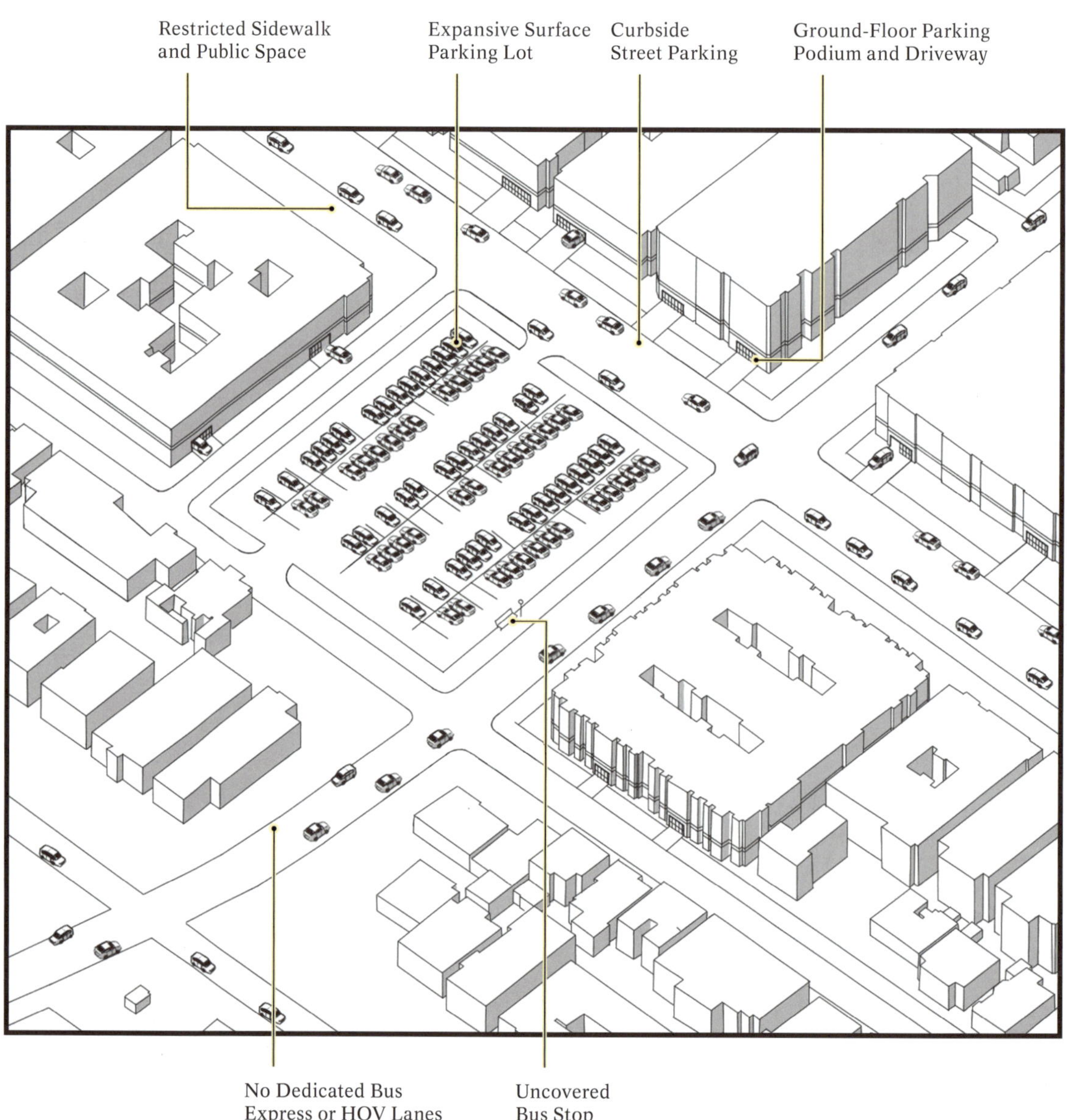

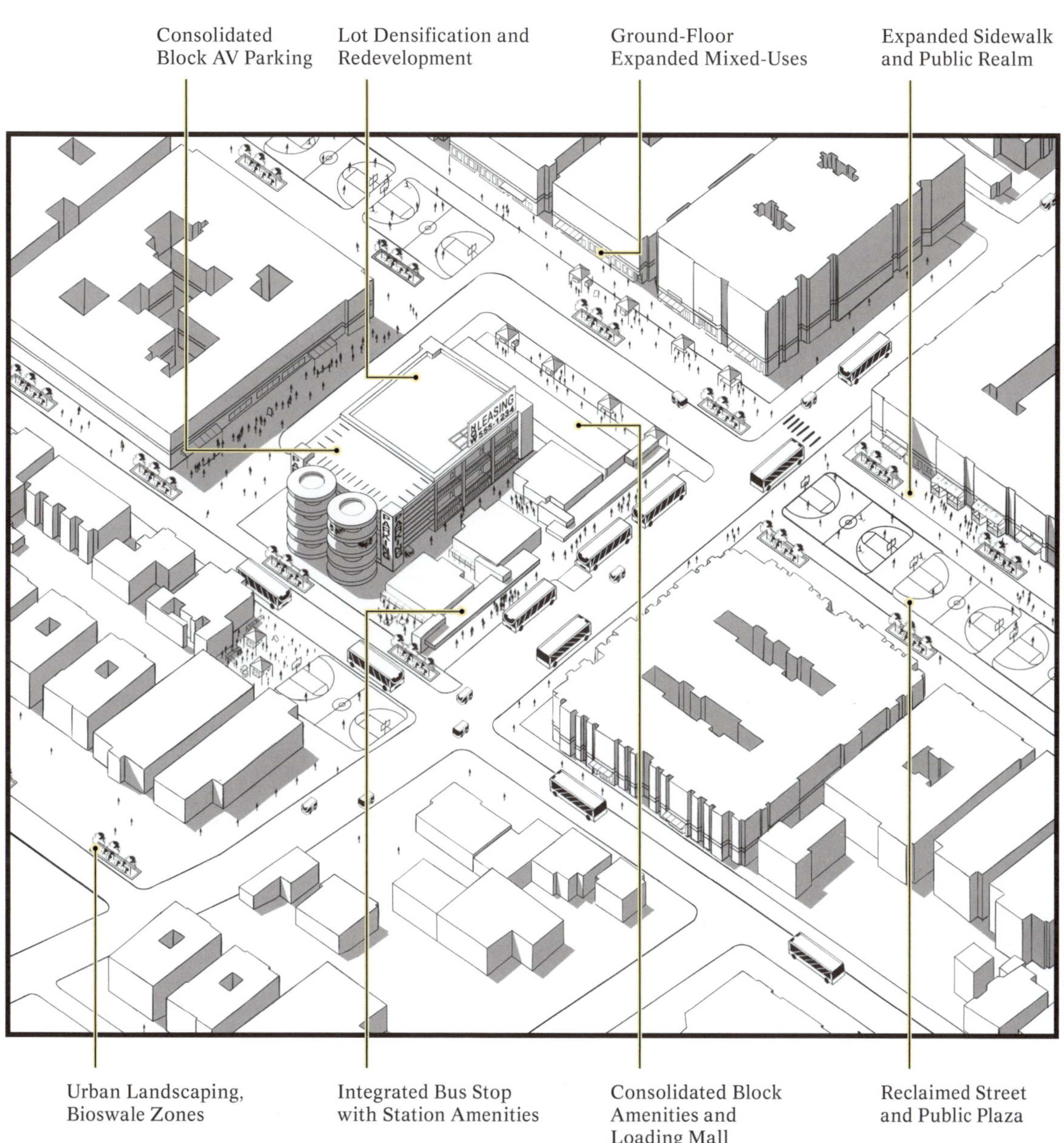
Consolidated Block AV Parking
Lot Densification and Redevelopment
Ground-Floor Expanded Mixed-Uses
Expanded Sidewalk and Public Realm
LEASING
555-1234
PARKING
Urban Landscaping, Bioswale Zones
Integrated Bus Stop with Station Amenities
Consolidated Block Amenities and Loading Mall
Reclaimed Street and Public Plaza

Public Housing Block
Estrada Courts, Boyle Heights

Public housing witnessed a boom in construction in the U.S. following the passing of the Housing Act of 1937. Often taking the form of the superblock, these developments consisted of multiple city blocks with high-density towers situated in an open landscape, or in lower-density cases a series of linear rowhouses. The construction of these large-scale developments declined in the latter half of the 20th century as the public housing model in the U.S. shifted towards policies that encouraged the provision of affordable units through the private market. However, many of these prior-era, large-scale developments still remain today, housing thousands of low-income residents on publicly-owned land administered and managed by various city's respective housing authorities.

Public housing has a long and contentious history, but let's focus on the spatial impacts that these developments have had in American cities today. Criticism of these developments stem from the lack of housing amenities they provide to residents who suffer from under-investment and under-maintenance. Tower or rowhouse typologies are typically set back from the edge of the superblock lots, rather than reinforcing or activating the street edge. This configuration—buildings sitting in a porous open landscape—means that the open spaces between building types are ill-defined, neither acting as private, sheltered courtyards nor as public-facing courts. Instead, these undefined spaces either serve as semi-private backyards, semi-public entrances, or more often than not have been claimed by surface parking lots. The large-scale superblocks that public housing developments often sit on are also often severed from the larger gridiron of the city street network, with multiple blocks circumscribed by arterial traffic corridors. Clustering of blocks means that interior streets often become cul-de-sacs or surface parking, further disconnecting residents from the continuity of the urban grid and the useful transit mobility access that that provides.

In Los Angeles, a variety of public housing superblocks in the form of rowhouses exhibit these very spatial characteristics. In particular, Estrada Courts of the Boyle Heights neighborhood is one of the city's oldest public housing projects, which today still houses over 400 units in a collection of garden-style apartment rowhouses. Managed by the Housing Authority of the City of Los Angeles (HACLA), the development is well known for its murals which are painted by local Chicano artists celebrating the working-class barrio culture. In an automated future, where the internal parking laneways that bisect these rowhouse are reclaimed, mid-rise affordable housing bars could be introduced in their place. Ground-level amenities, with uses like grocery stores, daycares, and laundromats, could be constructed in-between to transform existing rowhouse typologies into perimeter blocks. This would convert these undefined open spaces between buildings into sheltered private courtyards on the block's interior. Meanwhile, block exteriors could reintroduce through-street access with street frontages that integrate bus stops and curbside zones with street-fronting land uses. A superblock development previously interrupting the city's urban fabric could be reabsorbed in transit-oriented, pedestrian-activated ways.

Block Elevation

Block Perspective

Public Housing Block
Estrada Courts, Boyle Heights

2024

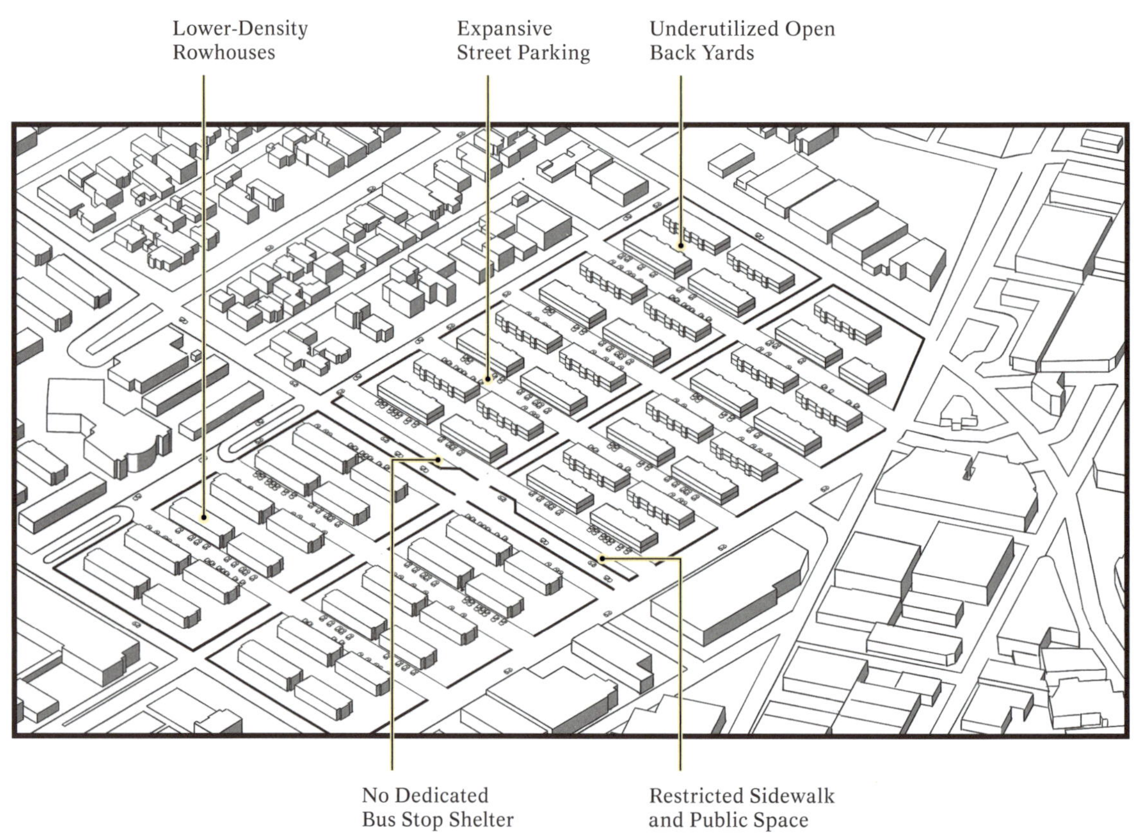

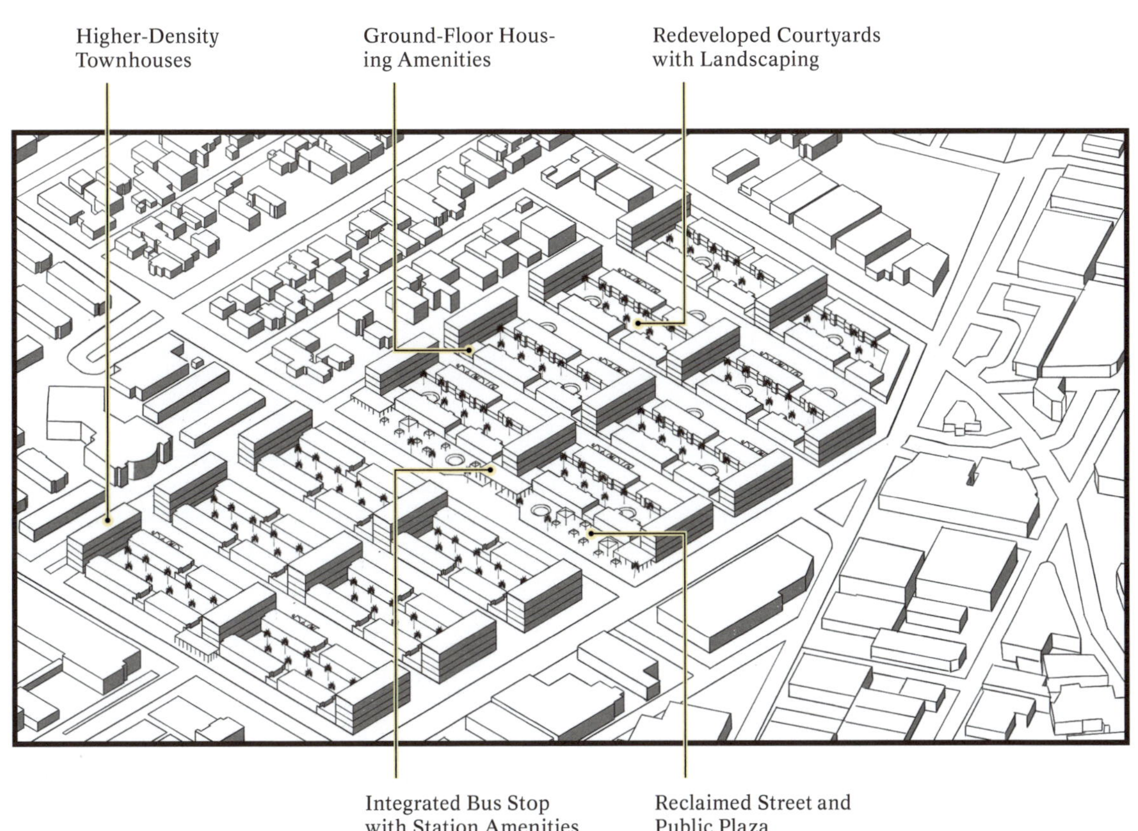
Higher-Density Townhouses
Ground-Floor Housing Amenities
Redeveloped Courtyards with Landscaping
Integrated Bus Stop with Station Amenities
Reclaimed Street and Public Plaza

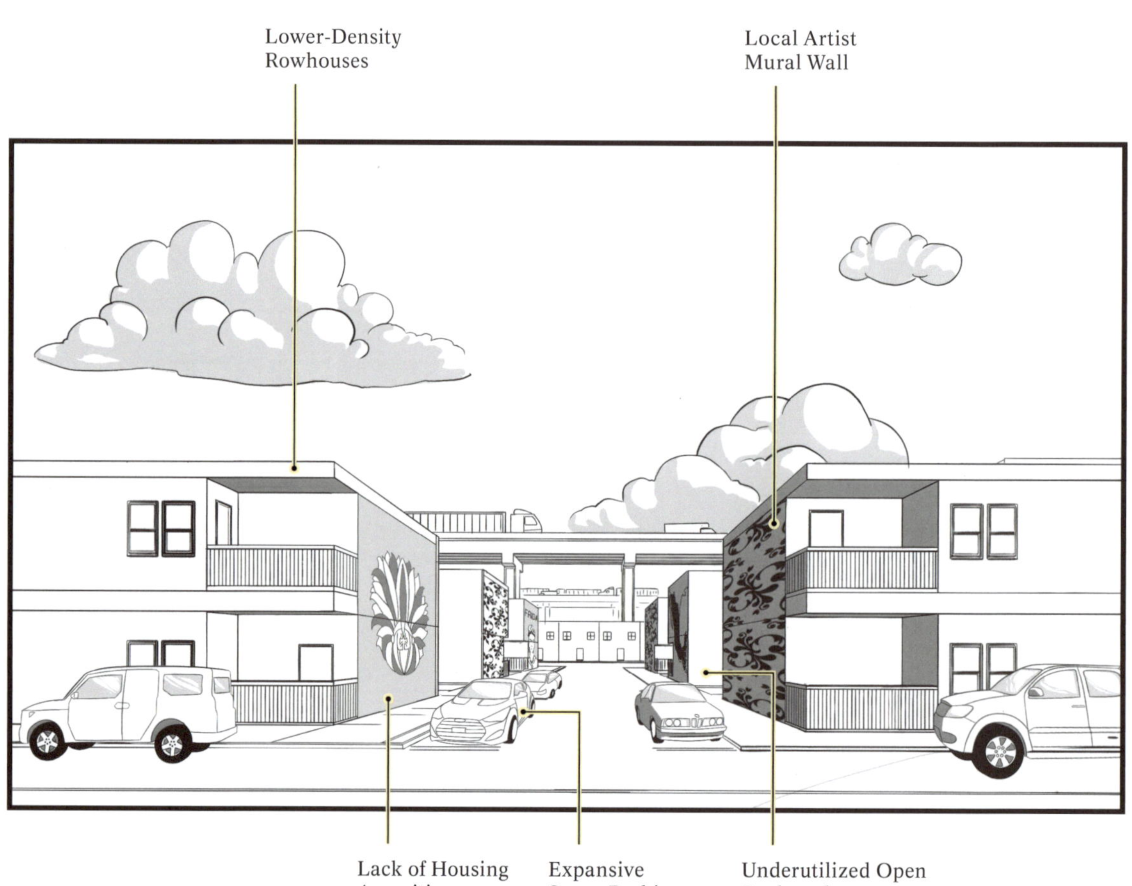
Lower-Density Rowhouses
Local Artist Mural Wall
Lack of Housing Amenities
Expansive Street Parking
Underutilized Open Backyards

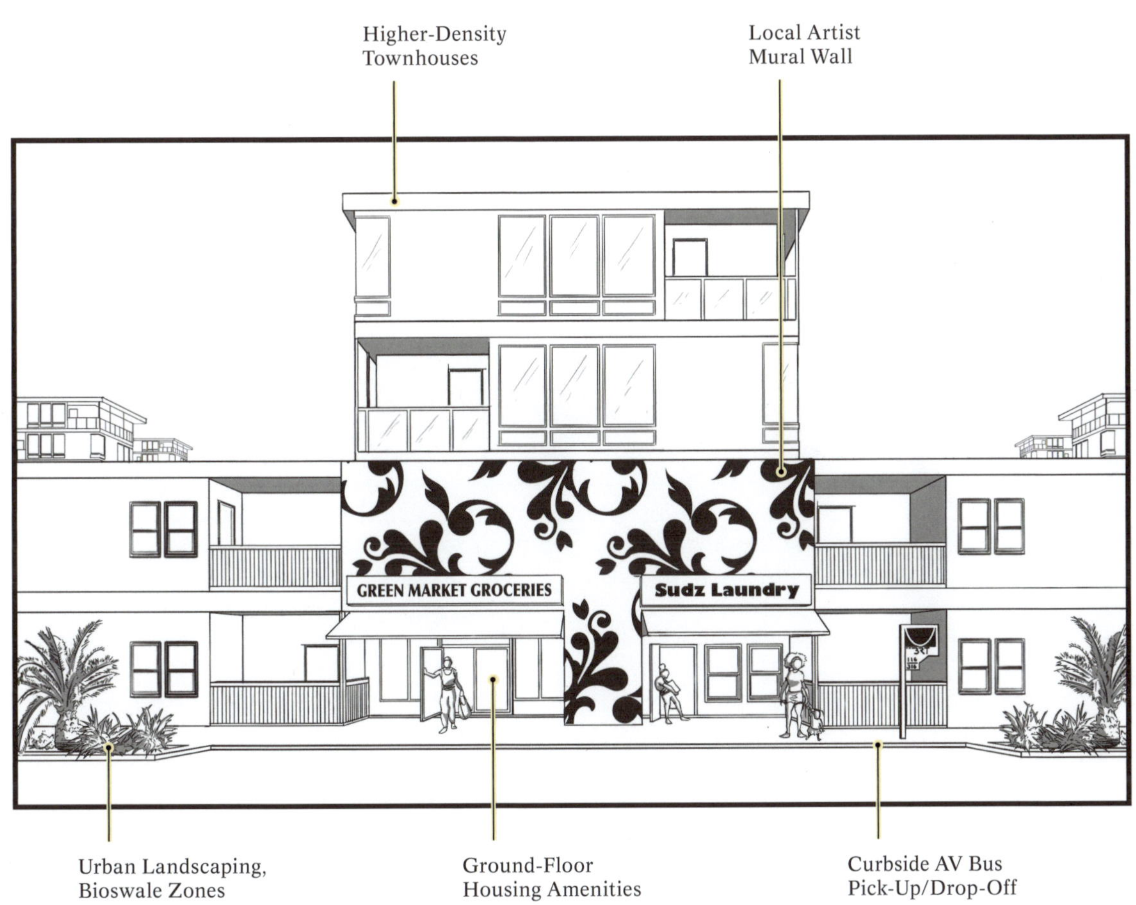
Higher-Density
Townhouses
Local Artist
Mural Wall
GREEN MARKET GROCERIES
Sudz Laundry
Urban Landscaping,
Bioswale Zones
Ground-Floor
Housing Amenities
Curbside AV Bus
Pick-Up/Drop-Off

Community Improvement District
South Central

Apart from the superblocks of its 20th century public housing developments, Los Angeles' prototypical urban fabric doesn't deploy the traditional superblock in its pattern and street network. However, if one examines the overarching structure of its fabric closely, one can see a larger hierarchy in its grid emerge. Structured upon the unrelenting Jeffersonian grid that subdivides the city, wider boulevards and avenues run at intervals of every four or five blocks wide. Organized both north-south and east-west, these corridors cluster the typical residential fabric of the city into multi-blocks that are serviced by lower-traffic local residential through streets. Multi-block avenues sometimes take on commercial or retail land uses and provide a faster thoroughfare for automobiles traveling farther distances. Meanwhile the local street grid crisscrosses the fabric in-between, stitching the residential interior of these multi-blocks together through low-traffic streets.

The implementation and expansion of automated transit sparks an opportunity to rethink the block network's hierarchical relationship to its land uses. The concept of the community improvement district (CID) is overlaid on this structure, which describes a model for self-taxing a district that collects revenue on property owners within defined geographic boundaries. These funds are then reinvested to pay for district upgrades, infrastructure investments, public space programs, and management services for that specific area. Diagrammed in his section, CIDs could be overlaid on these multi-block clusters as a blueprint for reclaiming internal residential streets for alternative uses, while allowing the border of these multi-blocks to densify. Up-zoning and land use changes could establish commercial corridors of higher density development on the border, while protecting the existing interior lower-density residential blocks from displacement. Internal community loops can be implemented in the center of these multi-blocks, in which a perimeter of low-traffic streets are reclaimed as shared PlayStreets for public community use. Adjacent lots to the community loop could expand PlayStreets into parks, playgrounds, and recreational uses. These loops would be free of any vehicular traffic, deploying the right-turn only street intersection typology illustrated in the previous chapter.

Funds from taxing new developments on the multi-block exterior corridors would go directly to implementing these spatial and infrastructural changes on the interior. Based on the business improvement district (BID) model that is used in many cities today, CIDs could have elected boards—made up of owners and renters of the district's properties—in order to ensure that new investments are in the interest of the existing residents of the area. Together, this model could establish a template for rethinking the structural relationship of the city's multi-block Jeffersonian grid to its street, land use, and density hierarchies.

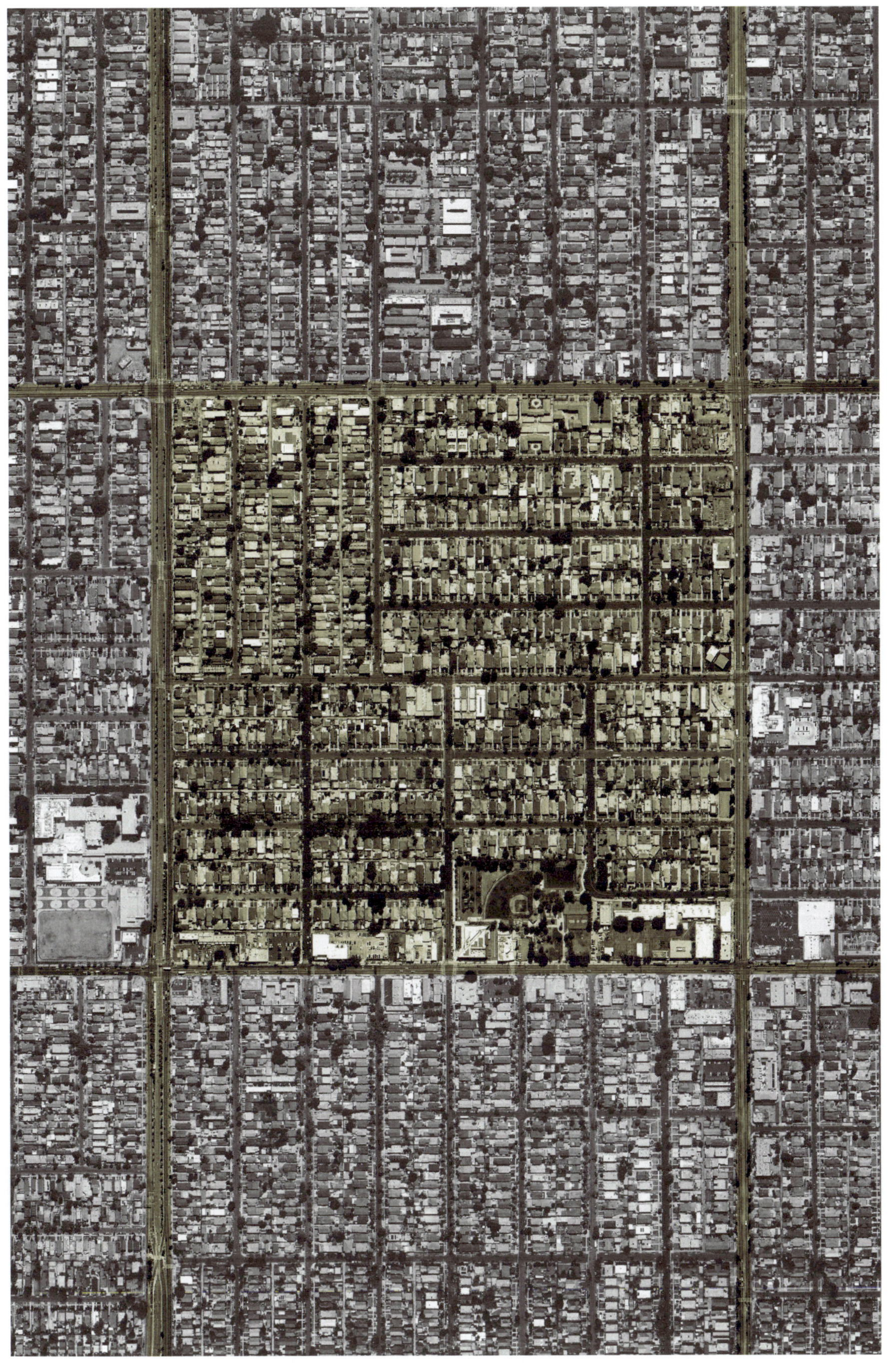

iPHONE XR SCREEN REPLACEMENT
$75
iPhone 11
Lcd Repair
$85
iPhone 11 Pro Max
Lcd Repair
$95
boostmobile
iPhone XS MAX
Lcd Repair
$80
iPHONE X SCREEN REPLACEMENT
$75
boostmobile

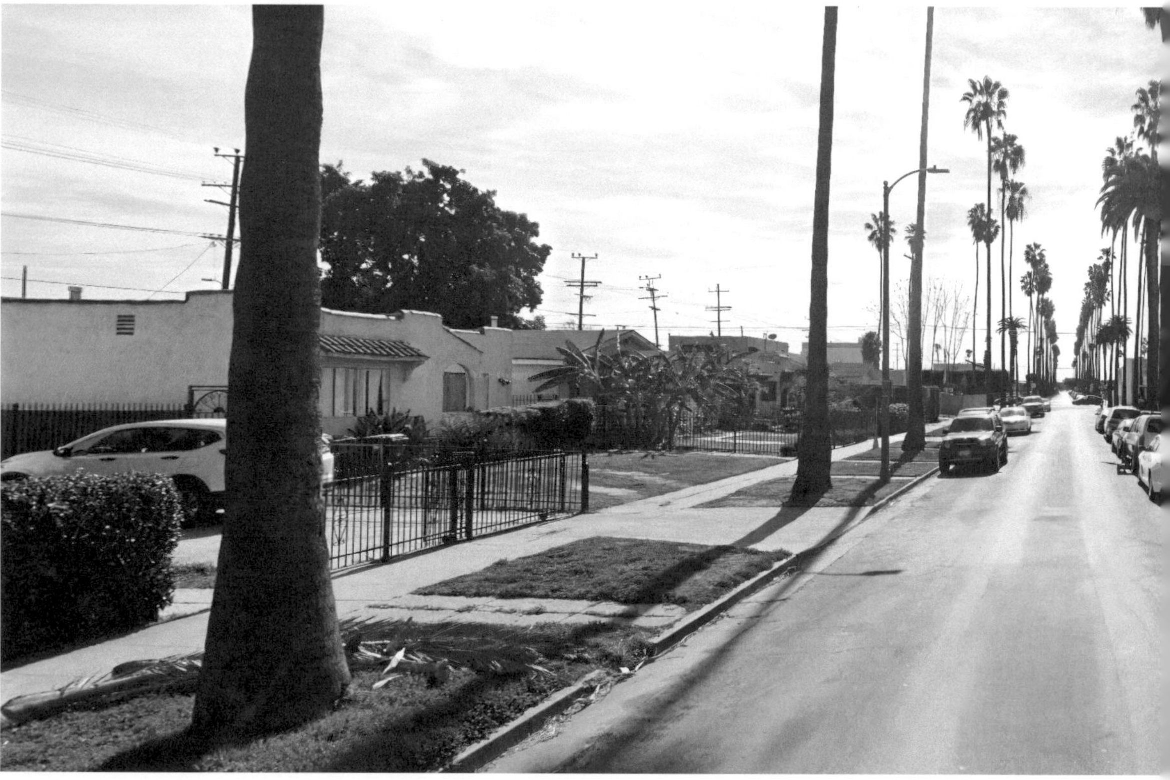

Block Commercial Exterior

Block Residential Interior

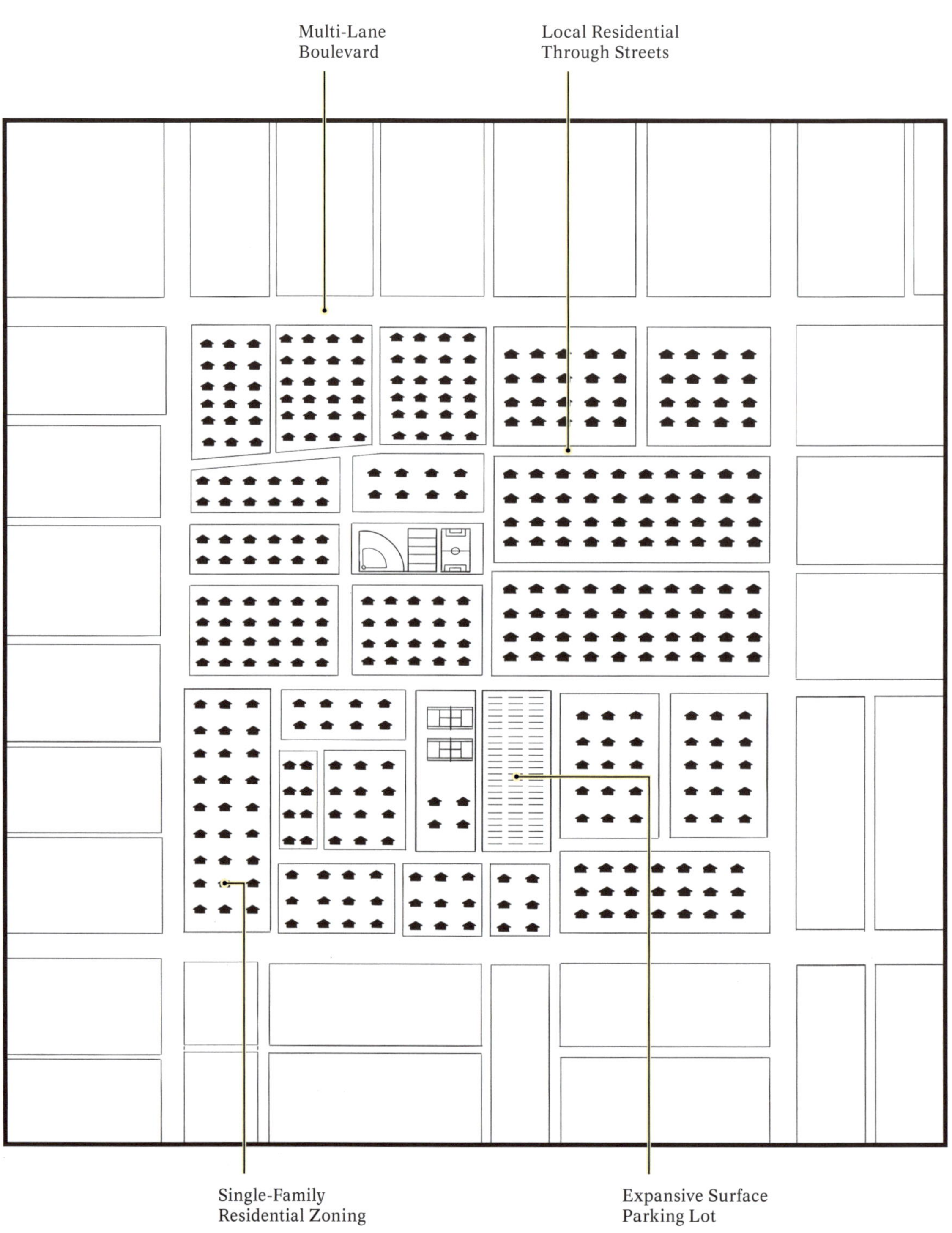
Multi-Lane
Boulevard
Local Residential
Through Streets
Single-Family
Residential Zoning
Expansive Surface
Parking Lot

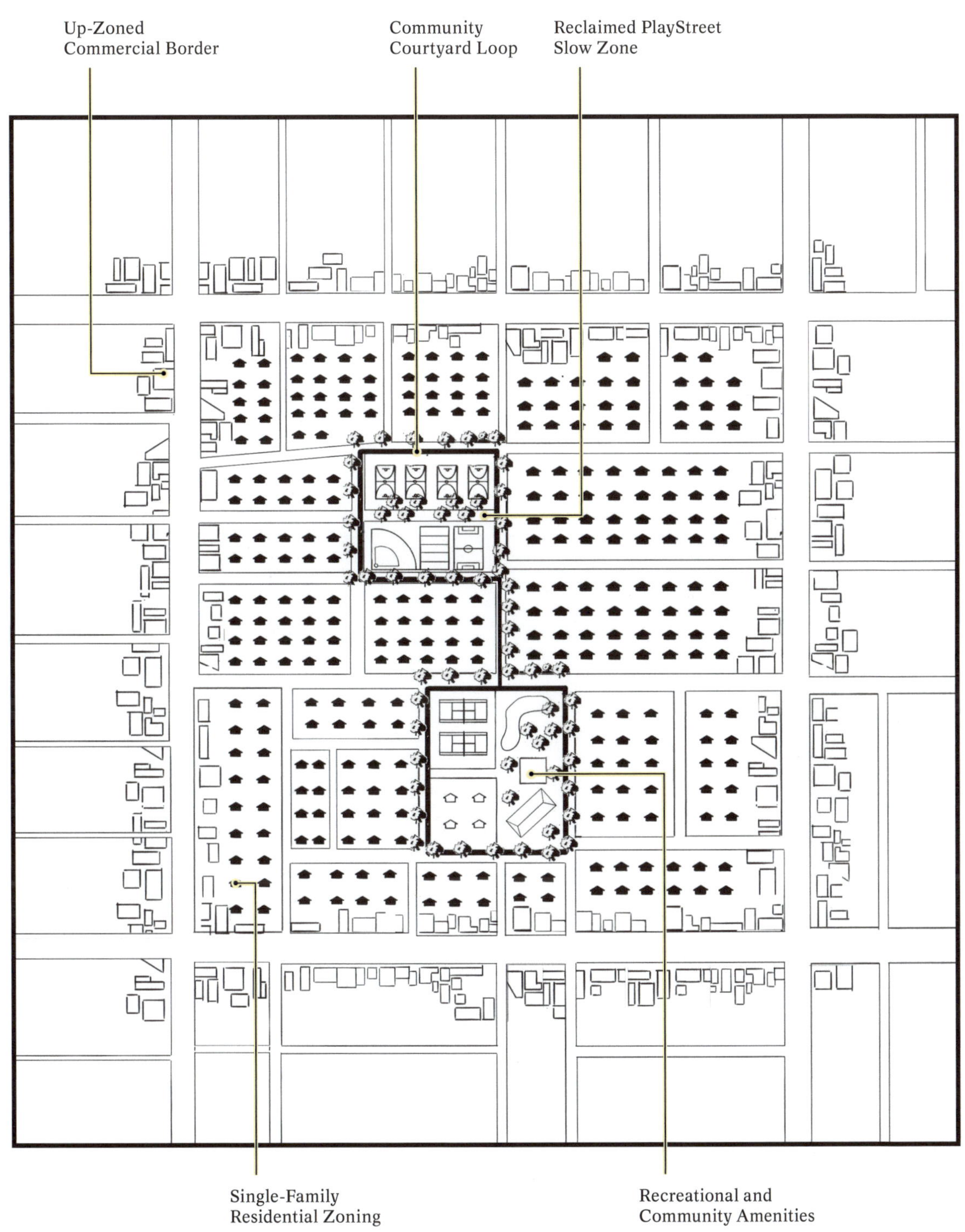
2057
Up-Zoned
Commercial Border
Community
Courtyard Loop
Reclaimed PlayStreet
Slow Zone
Single-Family
Residential Zoning
Recreational and
Community Amenities

10 S: Adaptive Conversion of Car-Centric Building Types

It is estimated that on average, as much as 50% of modern America's urban land area is dedicated to storing, facilitating the movement, and orienting around automobile use.[1] If on average, our streets and roads take up 18% of the total land area of cities, this means that almost one-third (32%) of our urban land is dedicated to parking automobiles, auto-oriented land uses, and car-centric building forms. This percentage includes not only the vast amount of surface parking dedicated to storing cars—estimated on average to take about 22% of urban land area—but also a plethora of architectural building that contribute to the built environment of our cities.

1 Melosi, "The Automobile Shapes the City," Automobile in American Life and Society, 2005.

These car-centric building types have evolved in response to the auto-dependency traits that many of our cities exhibit today, producing land use patterns and urban form that favor automobile access and parking over alternative uses of valuable urban space. In many cases, this involves the aforementioned incorporation of expansive parking space in front of buildings that cater to patrons arriving by car. Ranging in scale from retail strip malls and motels, all the way to huge entertainment venues like sports stadiums, these typologies are spatially calibrated to the number of patrons they can serve and the number of cars that their patrons arrive in. This has resulted in spatial building configurations that either retreat from fronting the street, or introduce vehicular access that interrupts a continuous street wall. Both spatial moves are in service of dedicating significant area of their private lots in the front, back, or side to vehicular parking. They also introduce large curb-cuts that segment pedestrian sidewalks with parking lot access. In some cases, like gas stations, these curb-cuts run the full length of the building lot so that vehicles can enter and exit along any point of the property. Auto-oriented buildings also reconfigure in other ways, including introducing over-scaled signage advertising to the speed of passing cars. The result of all these design decisions are public realms that are less vibrant, activated, and pedestrian-oriented due to a decrease in sidewalk-activating building fronts.

Besides the impact that lot parking has had on the design of buildings, some building types are literally designed around the spatial dimensions and movement radii of the individual automobile. Most noticeably, is the proliferation of parking garage structures and buildings whose sole function is to move and store cars in dense vertical arrangements. Because of the wide turning radii of automobiles, parking garages have emerged around a variety of vertical ramping systems, wide footprints, and parking configurations (single-row single-loaded, double-row double-loaded, etc.) prioritizing mono-functional efficiency above all else. Built with stacked concrete slabs, these buildings have been generally stripped of any aesthetic contributions to the public realm in their historical evolution. Other building typologies, like drive-throughs or drive-ins, distinctly format their architectures around the turning radii and queuing constraints of the automobile. In the case of drive-through restaurants, the majority of pedestrian access to the building is severed by an automobile queuing lane circling the building on multiple sides. Even the internal layout and organization of these building types are configured to service patrons from their cars through drive-through windows.

With the arrival of driverless vehicles, all of these spatial calibrations should be questioned again. When these buildings have been configured to the spatial operations and dimensions of the car, what happens when that vehicle becomes automated? Furthermore, the predicted decrease in demand for parking space will open up transformational opportunities for reclaiming, redesigning, and densifying these auto-oriented typologies as they evolve. These building types have proliferated all over cities, holding vast amounts of embodied carbon in their construction and use. Rather than completely demolishing and building anew with the potential to incur far greater carbon costs, many of them could instead be adaptively converted to incorporate AVs. In particular, how can typologies that have catered to the car be redesigned to contribute to the public realm and respond to transit and pedestrians? What alternative land uses could exist for parking structures when future demand for their current mono-use decreases? What hybrid building typologies could emerge that incorporate the benefits of AVs while minimizing its externalities? These building types are assets, holding the potential for new inventive urban futures that contribute to a paradigm shift for redesigning our cities.

This book takes the position that adaptive conversion strategies must be prioritized with any driverless technological changes. It is generally wary of new typological inventions that are solely predicated on the arrival of new technology. This is because, as has been roundly discussed, our urban histories have borne witness to urban infrastructures, fabric, and form that were predicated on the singular efficiencies of the technologies of their time, resulting in a plethora of aforementioned negative externalities. Might we be repeating the same mistake of the past, in attempting to design new building types solely around the constraints and requirements of autonomous vehicles? Instead, any emergent building typologies must establish principles prioritizing human-centric urban form that actively engages with the street and the city's public realm. Only then, can new technologies be used as a tool to advance towards these goals. Some of these new tools include AV's potential ability to move forward and backward regardless of direction, since their sensors operate multidirectionally rather than being constrained by forward-facing drivers. Might this be utilized to produce fleet-parking in vertical orientations that are far less footprint consumptive, with layouts that don't require wide turning radii to account for human error or ramps to accommodate human driving? How can AVs work with other technological innovations like dynamic lane management? How can these tools be deployed to minimize the footprint of vehicles on urban space, reclaiming lot space for expanding commercial, recreational, and environmental uses?

Any future transformations of privately-owned buildings and lots certainly require the right economic incentives. Developers and property owners must find it more profitable to adaptively convert car-centric buildings for expanded business opportunities than to completely demolish and rebuild. These design and development decisions must be incentivized by the right regulatory policies or abatements, including the removal of parking minimum requirements. If done so in conjunction though, the very tone and texture of our cities could transform in architecturally exciting ways.

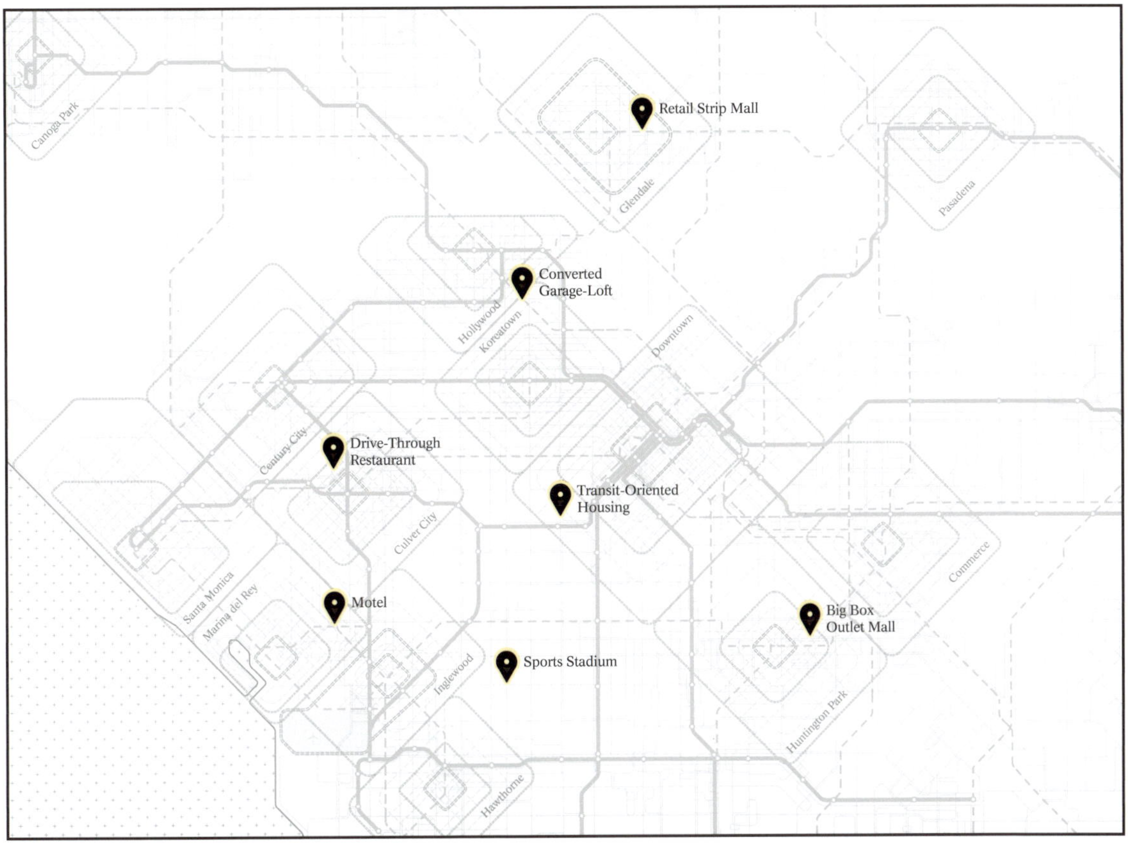

Retail Strip Mall
Converted Garage-Loft
Drive-Through Restaurant
Transit-Oriented Housing
Motel
Sports Stadium
Big Box Outlet Mall
Canoga Park
Glendale
Pasadena
Hollywood
Koreatown
Downtown
Century City
Culver City
Santa Monica
Marina del Rey
Inglewood
Commerce
Huntington Park
Hawthorne

Converted Garage-Loft
Koreatown

Drive-Through Restaurant
In-N-Out

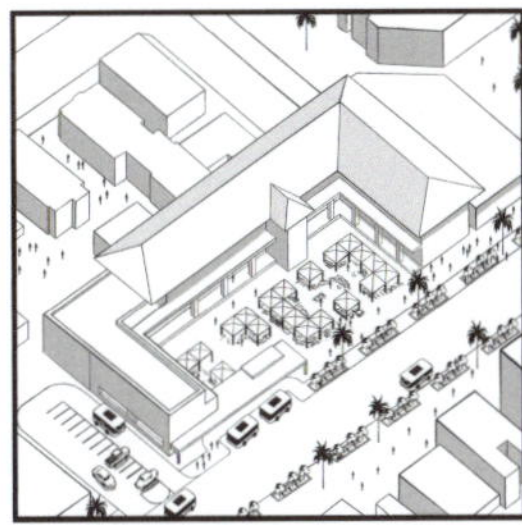

Retail Strip Mall
Glendale

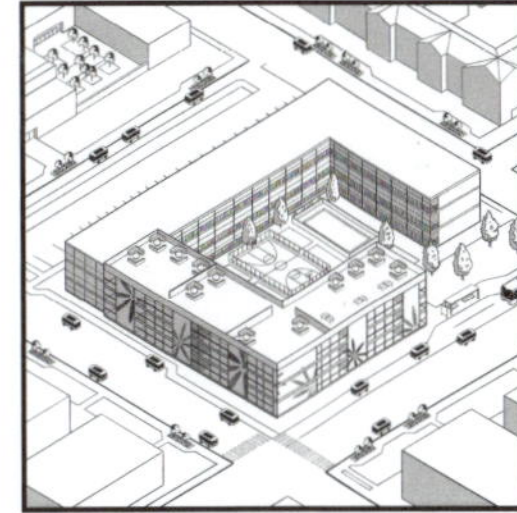

Motel
Playa Vista

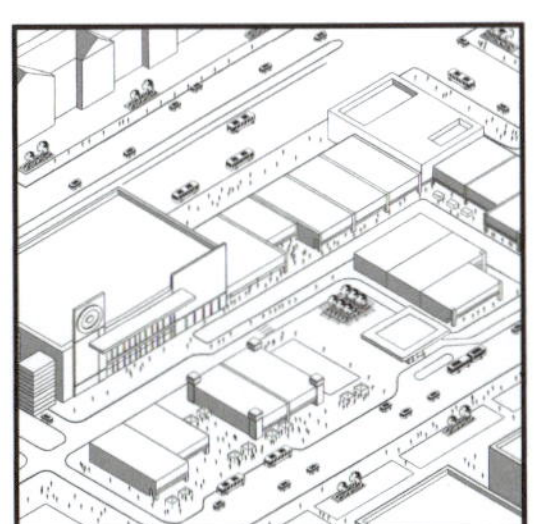

Big Box Outlet Mall
Huntington Park

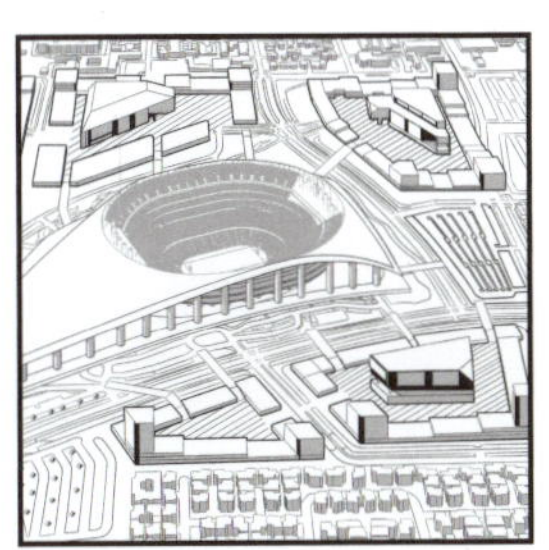

Sports Stadium
Sofi Stadium, Inglewood

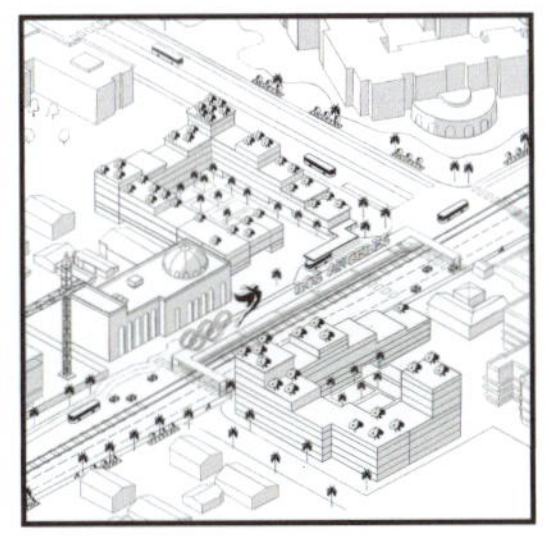

Transit-Oriented Housing
Exposition Boulevard

Converted Garage-Loft
Koreatown

The most prevalent building type literally designed for the singular use and spatial dimensions of housing automobiles is the parking garage. The burgeoning rise of automobile ownership in the 20th century came with a steep demand for housing those cars, with streets and lots unable to keep up with this demand. In their early histories, garages were designed with aesthetic ornamentation intended to blend them with their context and contribute to the public realm of the city. Over time, however, the typology of the vertical parking garage has been stripped down to its bare minimum for cost-efficiencies sake. Formed by thinly stacked concrete slabs that are open to the elements, these buildings are designed around the minimum floor-to-ceiling height building code requirements required by cars, which is typically seven feet for standard parking spaces or around eight feet for van-accessible spaces. Modern parking garages are either of the mechanized variety (employing car lifts serviced by valets) or are ramped, with the latter being built ubiquitously since the 1950s to allow for self-parking and all-day access. With ramped garages, a variety of layouts and configurations have emerged, ranging from the double helix spiral ramps to interior inclined parking ramps. The footprint of garages are fairly substantial due to the dimensions and radii of the ramps needed to move cars vertically. Parking garages in the urban cores of cities can be highly profitable ventures, requiring low labor and maintenance costs to manage. They are also often built in tandem with other high-density land uses like residential and office towers, separating these urban forms from the ground through parking podiums that are less expensive to build above-ground than sinking the garage underground.

While adaptive conversion of this building type could take many forms, it is imagined here in this exploration to incorporate residential uses, a persistent need facing cities as they densify. However, the particular challenge that adaptive reuse of parking garage structures present is their low floor-to-floor heights that were not built to service human-oriented uses and spatial comforts. Along with structural retrofitting and inclined ramp surfaces, low ceilings make introducing the ceiling-mounted mechanical ducting required for human occupation an interesting design challenge. With the right economic incentives, however, removing portions of every other floor can allow for residential units like loft-style apartments to be introduced, which could comfortably fit in the non-standard floor heights of these structures. Hybridizing these buildings with apartments, while still maintaining a smaller portion of their floor area for parking demands, could also allow a resident to park directly in front of their unit which is a convenience and feature not typically found in higher-density living. Given the vast quantities of parking garages found all over our urban landscape, converting even some of them to residential uses could put a significant dent in the housing crisis facing cities today.

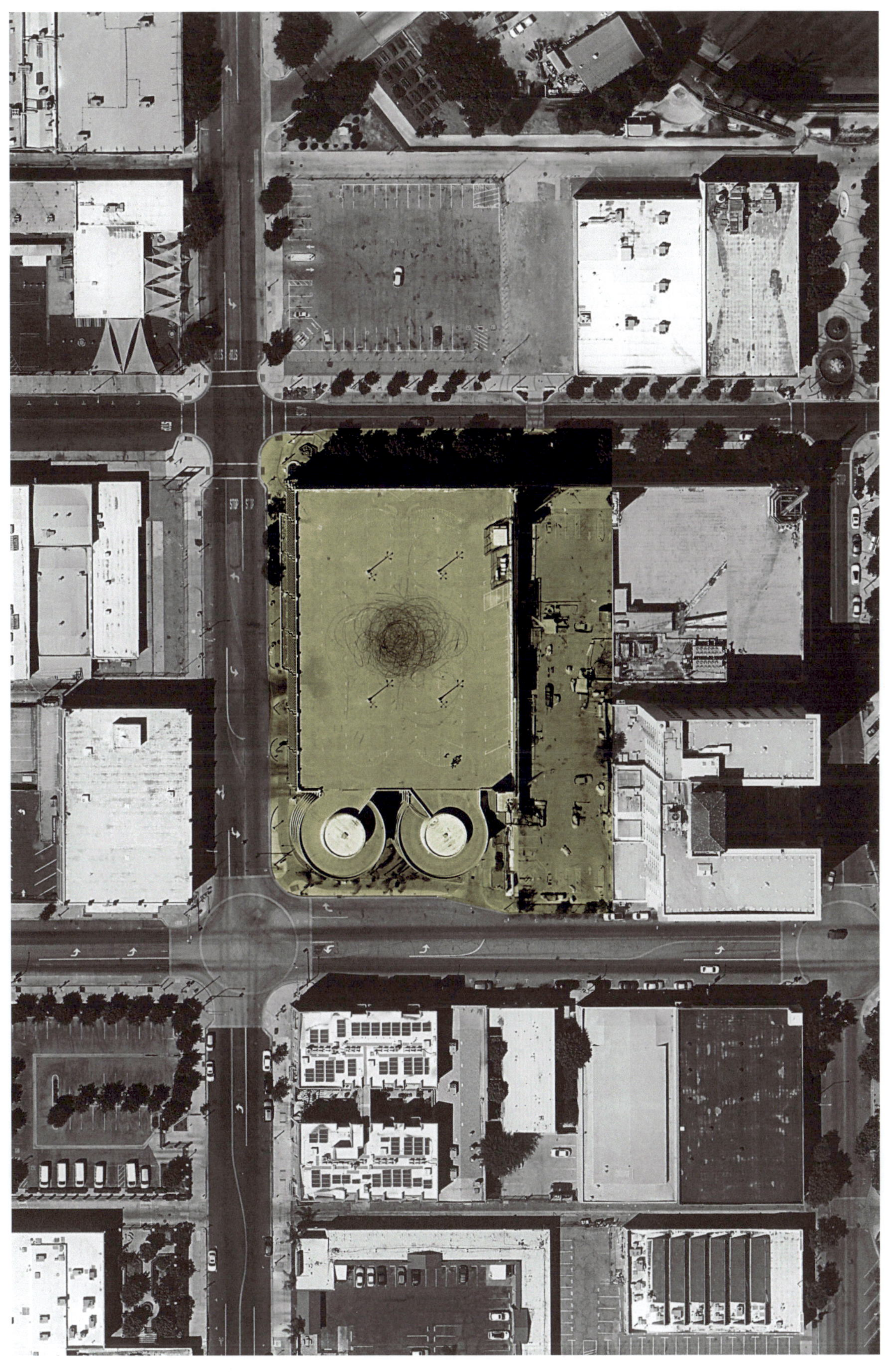

Building Elevation

Building Perspective

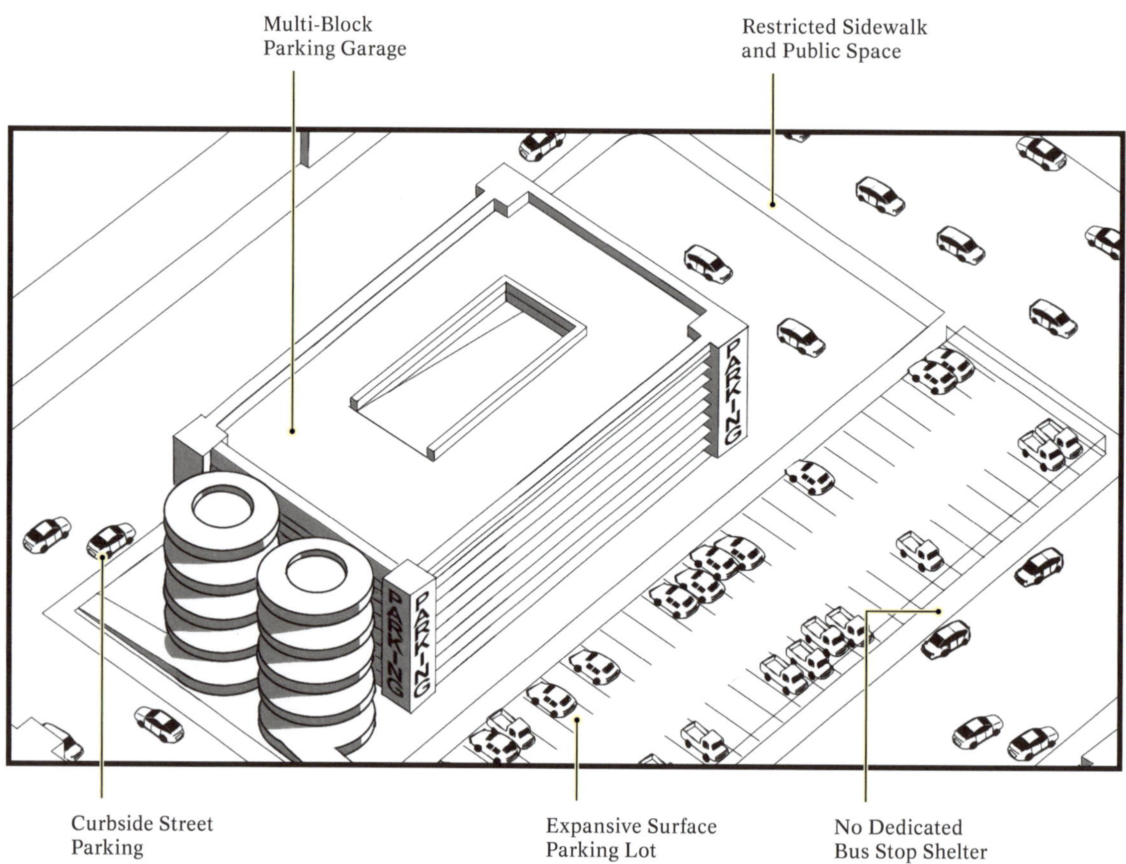
Multi-Block
Parking Garage
Restricted Sidewalk
and Public Space
PARKING
PARKING
Curbside Street
Parking
Expansive Surface
Parking Lot
No Dedicated
Bus Stop Shelter

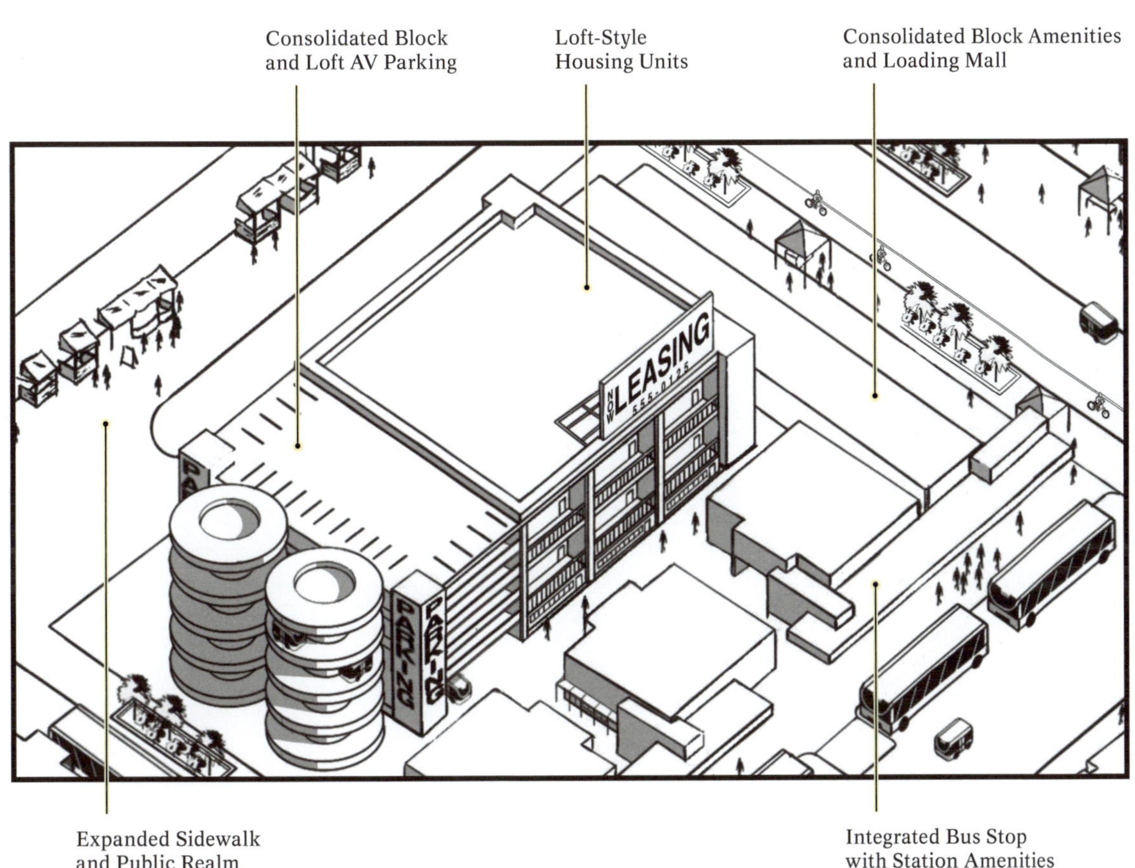
Consolidated Block
and Loft AV Parking
Loft-Style
Housing Units
Consolidated Block Amenities
and Loading Mall
NOW
LEASING
555-0125
PARKING
Expanded Sidewalk
and Public Realm
Integrated Bus Stop
with Station Amenities

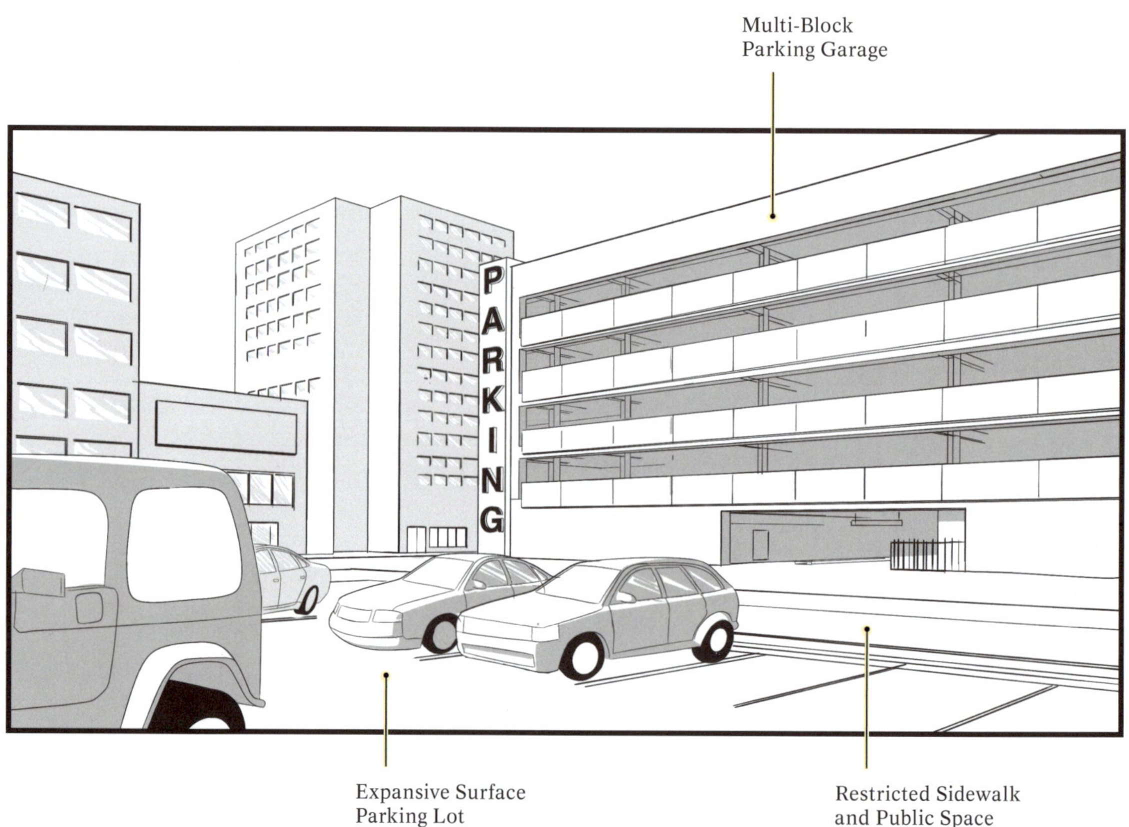
Multi-Block
Parking Garage
PARKING
Expansive Surface
Parking Lot
Restricted Sidewalk
and Public Space

2057

Consolidated Block and Loft AV Parking
Loft-Style Housing Units
LOFT HOUSING
NOW LEASING
PARKING
LOFT HOUSING
NOW LEASING
AV Curbside Package Delivery
Expanded Sidewalk and Public Realm

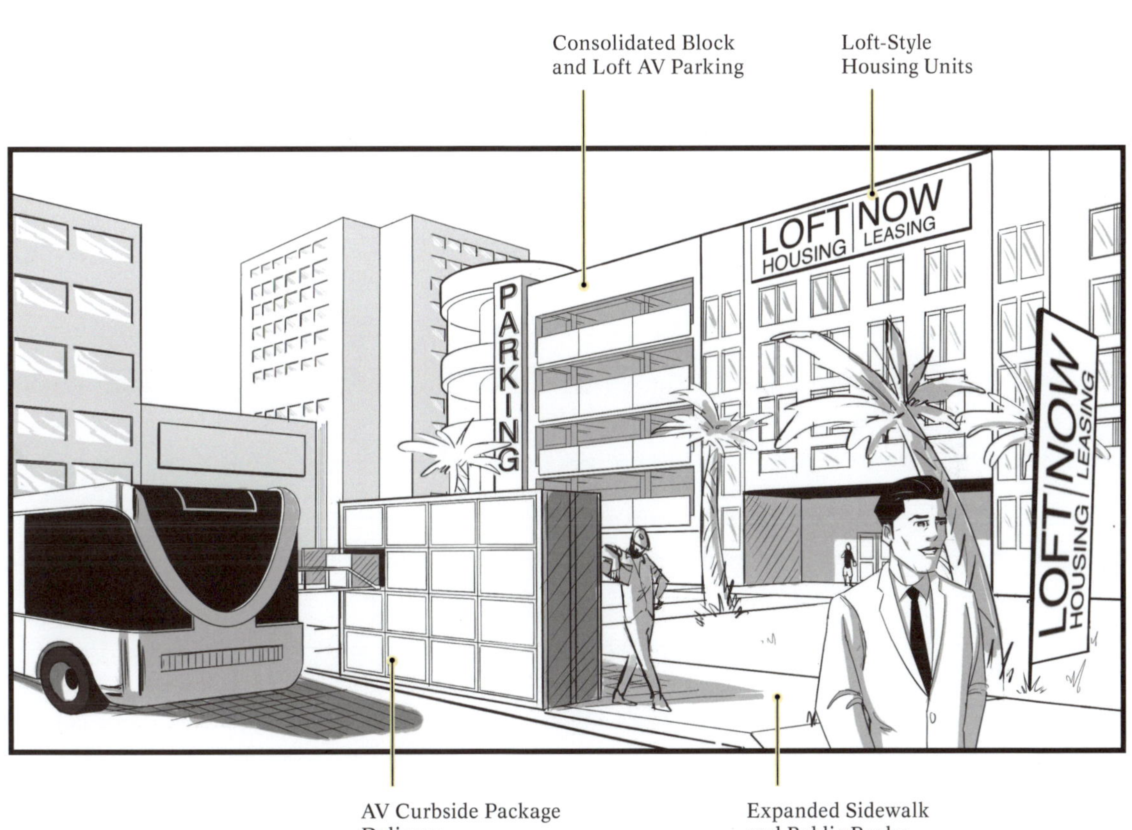

Drive-Through Restaurant
In-N-Out

The drive-through restaurant is a quintessential example of a building that has been designed around the spatial dimensions and queuing constraints of an automobile. Its key design and operational feature is one, or sometimes even two, retail counters that face the building's exterior, oriented towards the position and height of a driver's window. Its site plan is configured to allow a line of vehicles to queue around the building, with driving patrons placing an order at the beginning of the queue while their order is assembled by the time they reach the drive-through window. These queuing lanes snake through and bifurcate a parking lot that services patrons who want to dine-in, with both the queuing lane and parking lot encircling the restaurant in asphalt. Due to the amount of space needed to park and queue cars, there is limited space leftover on the site for any other public uses like outdoor dining and seating. Although prevalent in many other car-dependent cities, this building type is especially ubiquitous in Los Angeles and has become synonymously identified with the fast-food industry including businesses like In-N-Out and McDonald's.

With a decreased demand for parking, reclaimed parking lots could be transformed for uses that better service the patron or the business owner. Portions of that space could be freed for restaurant expansions, additions, or new business opportunities and leasable areas for the lot owner. In parallel, outdoor dining or other publicly-accessible spaces could be expanded or introduced to service more patrons. These changes could activate adjacent sidewalks of the street and public realm, taking advantage of Los Angeles' temperate climate. With the majority of patrons arriving through ride-hailed AVs and automated shuttles, dedicated and covered curbside drop-off or pick-up zones could be designed into the form of the building's front. Additionally, autonomous, on-demand, food-delivery vehicles and services could be integrated into the building type in more efficient ways than the current configuration of queued, human-driven vehicles that circle the lot. This could have design implications on a flexible zone that allows exchanges between driverless food delivery vehicles and the building to occur.

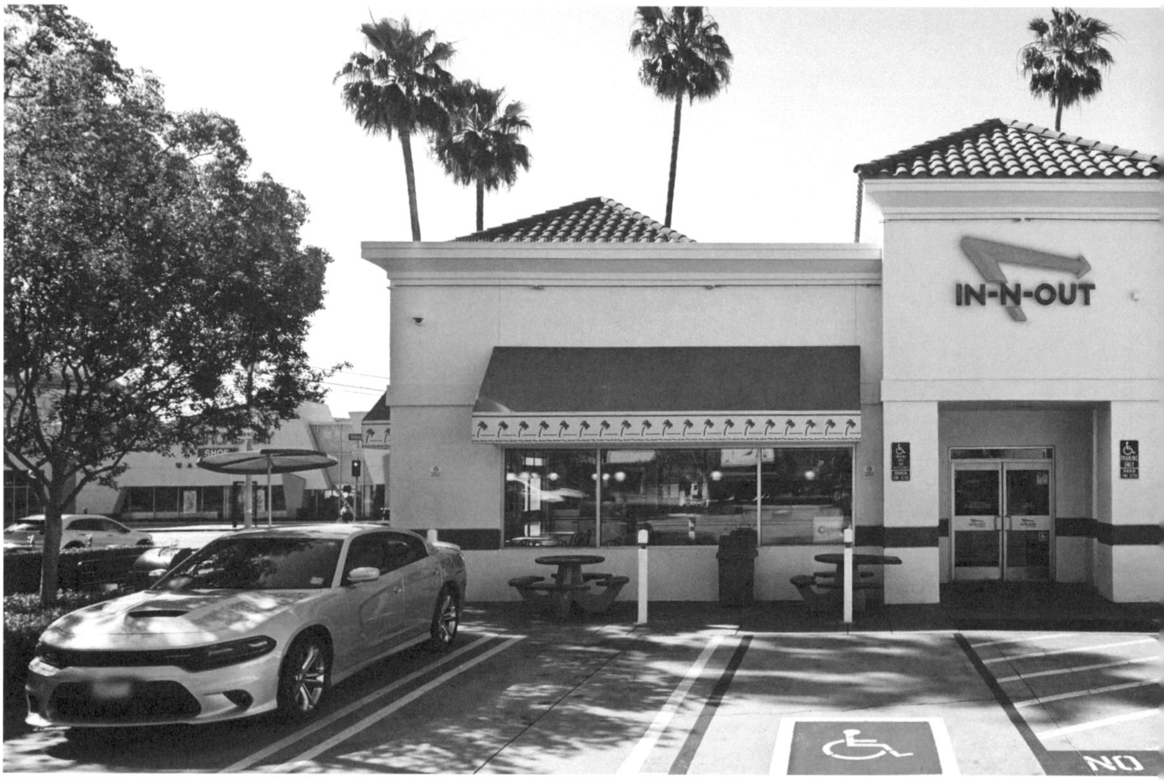
IN-N-OUT
NO

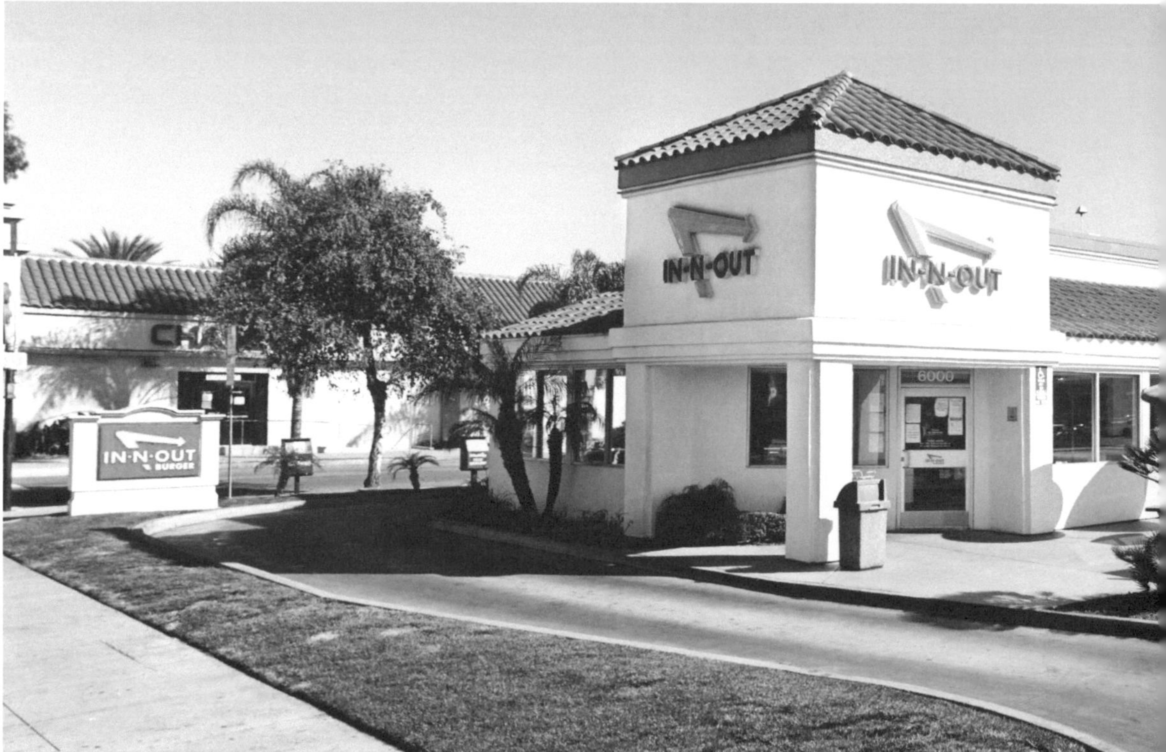
IN-N-OUT
IN-N-OUT
IN-N-OUT
BURGER
6000

Building Elevation

Building Perspective

Drive-Through Restaurant In-N-Out

2024

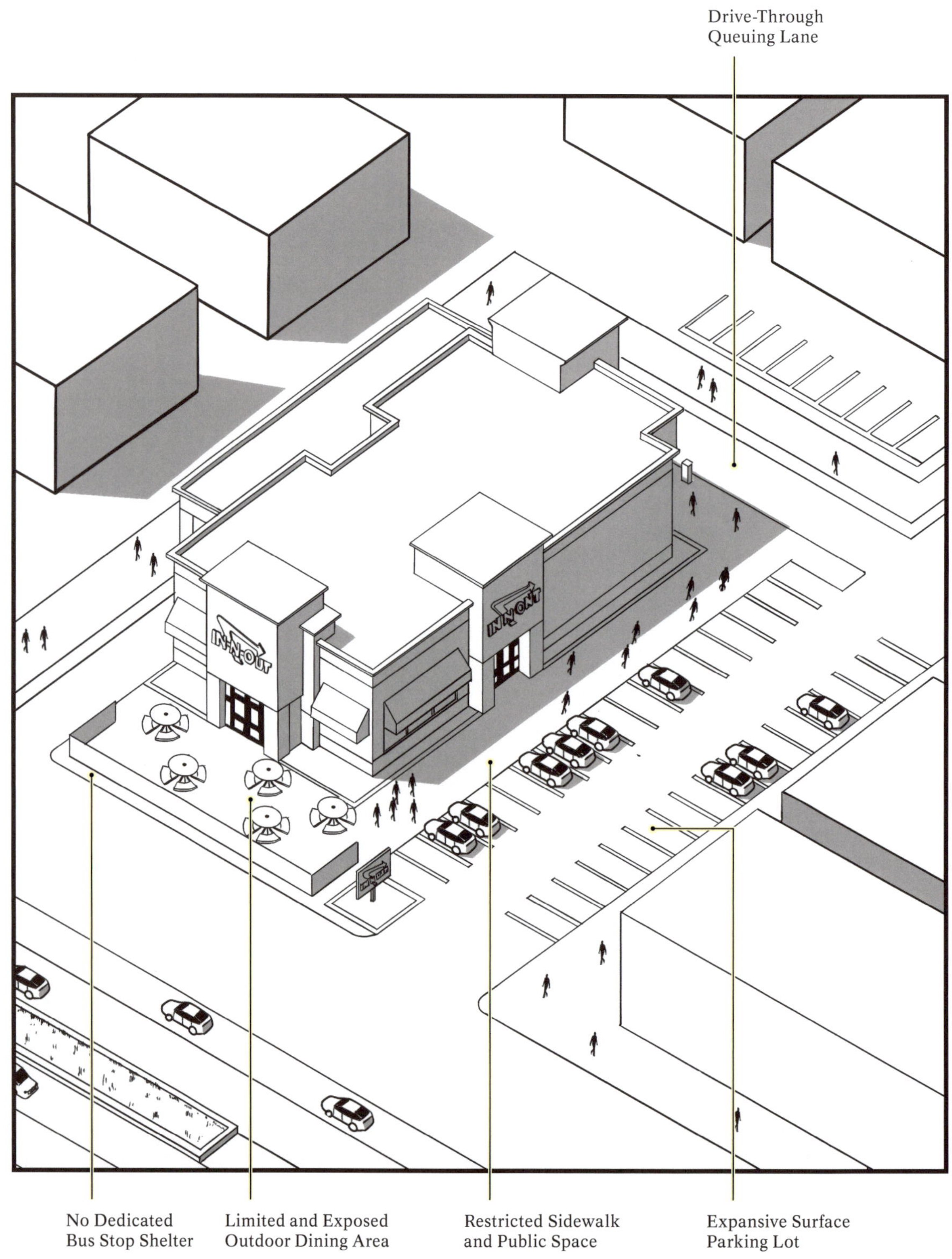

2057

Restaurant Expansion
or Addition
Integrated AV
Food Delivery
IN-N-OUT
IN-N-OUT
Integrated Curbside
Pick-Up/Drop-Off Zones
Expanded Outdoor
Dining Area
Expanded Sidewalk
and Public Realm

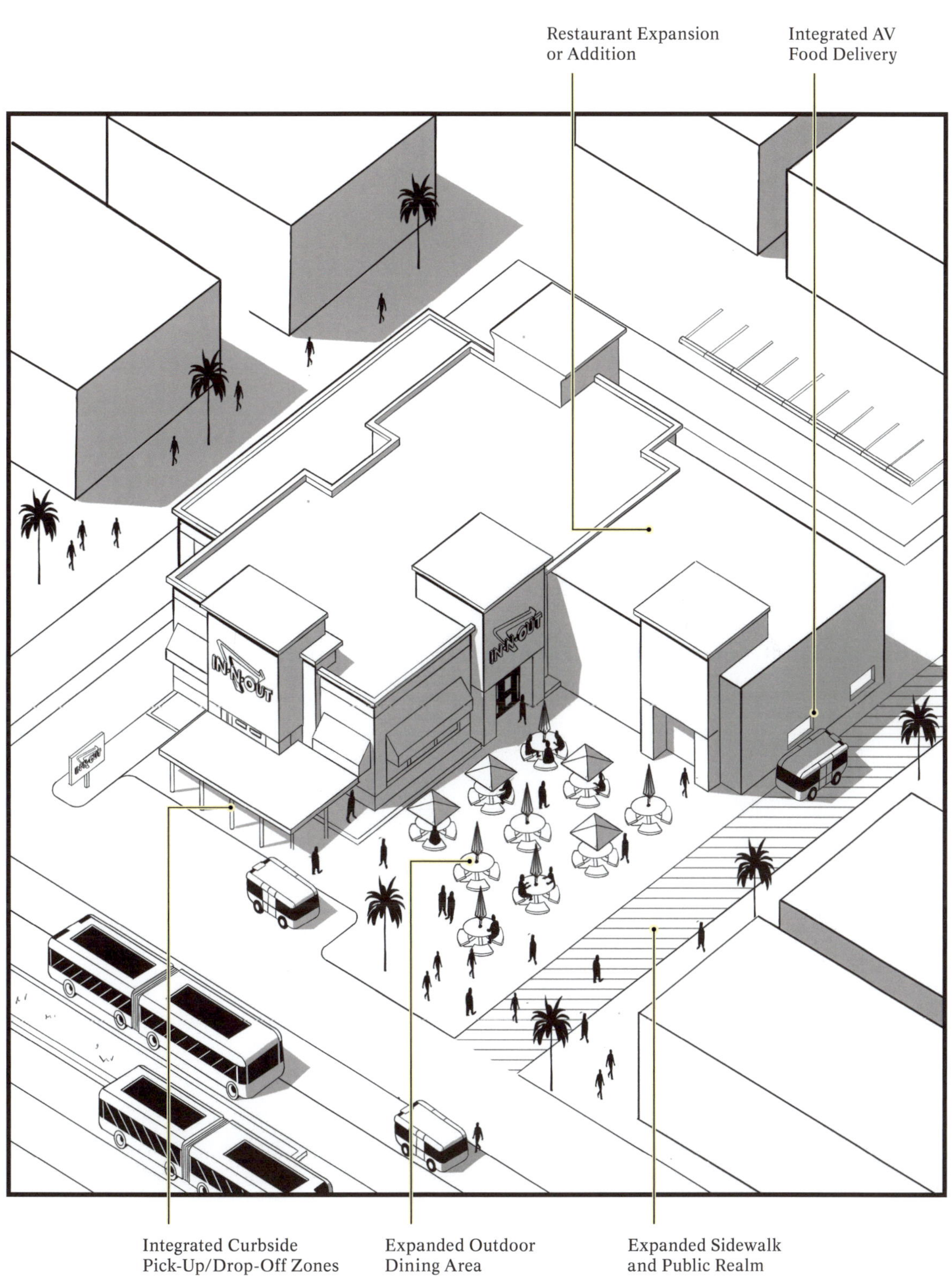

Retail Strip Mall
Glendale

A building type that is common to Los Angeles is the L-shaped strip mall which has evolved and proliferated throughout many neighborhoods of the city. Typically located along major commercial or arterial roads, it is a two-story retail structure that is shaped in the footprint of an "L" around a parking lot. This particular footprint evolved from the calibration between the amount of commercial retail area the building can hold, to the amount of parking needed by patrons to access those stores. Typically, this is a ratio of 4 to 5 parking spaces per 1,000 square feet of retail space. The typological form of this building—set back from the sidewalk with an arm that is perpendicular to oncoming vehicular traffic—features shopfronts and oversized signage that advertise towards the speed of passing vehicles. Despite the significant area of the site dedicated to parking, during peak hours the lot is still often overcapacity, requiring businesses to enlist valet services to double or triple park the surface lot and its driving lanes to the brim. With little to no space leftover for patrons besides a small outdoor circulation corridor to access stores, one can often find restaurant patrons milling about in the parking lot as they wait to be seated. In denser parts of the city, informal street vendors often set up stalls on the street-facing sidewalks of these lots to sell their goods to waiting patrons.

With driverless vehicles and a decreased demand for parking them, these spatial calibrations will be disrupted. As the building's footprint will no longer be tied to a minimum required parking ratio, the strip mall could be expanded in length to accommodate new retail businesses and stores. This expansion could take the form of an open courtyard, as illustrated in this exploration, which could shelter and activate a central plaza on the lot reconfigured for more pedestrian-oriented uses. Street vendors that previously crowded sidewalks could have a dedicated space for selling their goods, or a formal farmers market could function in the space. Nearby bus stops could be integrated into the design of the building expansion, with covered shelters and curbside loading zones for patrons. Patrons previously waiting on crowded building edges and asphalt could have public seating and space dedicated to them. Alternatively, this reclaimed open courtyard space could also be converted into a pocket park with urban landscaping or other potential uses.

DENTISTA
VERMONT DENTAL
NAILS
ENVIOS DE DINERO

VERMONT DENTAL
LON
UNISEX
5GB
CENTRO DE PAGOS
HOLLY NAILS
Cell: 714-717-1899
ENVIOS DE DINERO
cricket
GUNACO
DO NOT ENTER

Building Elevation

Building Perspective

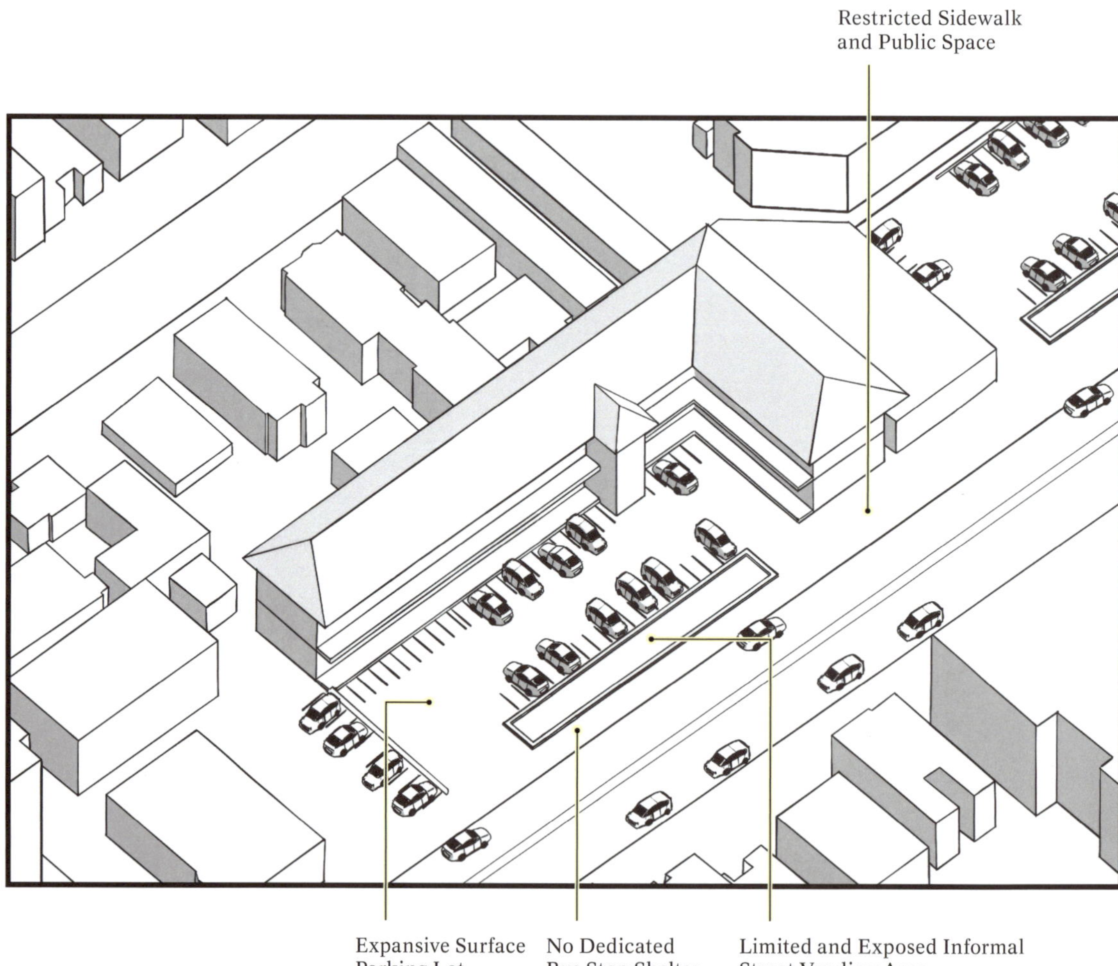
Restricted Sidewalk
and Public Space
Expansive Surface
Parking Lot
No Dedicated
Bus Stop Shelter
Limited and Exposed Informal
Street Vending Area

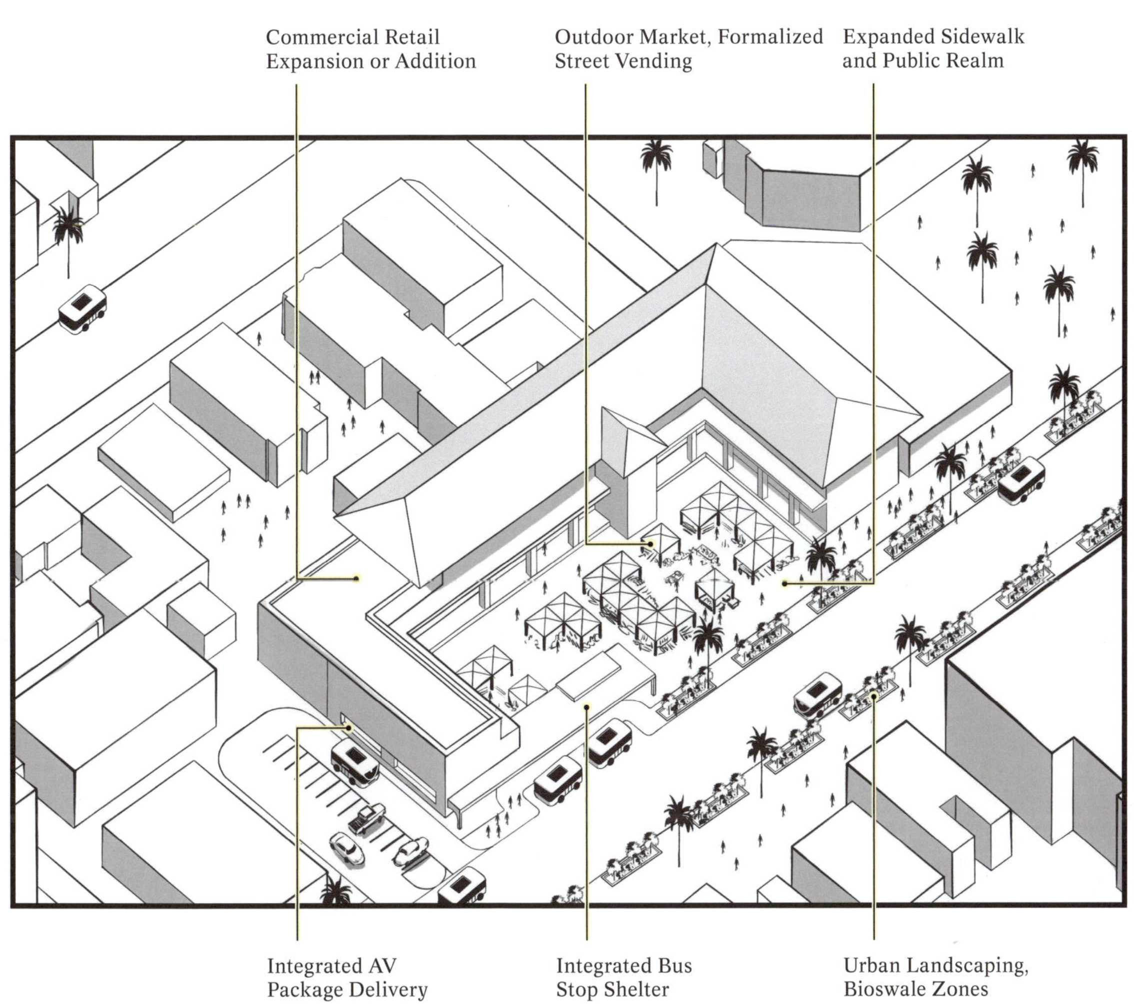
Commercial Retail
Expansion or Addition
Outdoor Market, Formalized
Street Vending
Expanded Sidewalk
and Public Realm
Integrated AV
Package Delivery
Integrated Bus
Stop Shelter
Urban Landscaping,
Bioswale Zones

Motel
Playa Vista

The motel, also known as a motor hotel, motor inn, or motor lodge, is a typology of hotel that is designed to typically house motorists. Usually these building types are low-rise (one to three stories), constructed in an “L”, “U”, or “I” shaped layout of single-loaded rooms that are directly accessed from an exterior circulation corridor facing a parking lot. Motel room doors typically open directly onto an adjacent parking space, making it easy to unload suitcases from the trunk of a vehicle. This typology of building proliferated widely in the 1960s as highway systems and major car corridors spread over territories. The need for inexpensive, easily-accessible, road-side, overnight accommodations close to these routes led to the growth of the motel concept which pepper the urban fabric of car-centric cities, often in close proximity to major transportation hubs or intersections. While sometimes designed with accommodation amenities, over time many of these amenities, which include pools, outdoors spaces, or other uses to service the hotel patron, are no longer provided in favor of space for more cars or more rooms. Motels also feature visually distinctive neon-lit signage or other oversized icons to advertise vacancies to the fast-traveling passing vehicle.

“L”, “U”, or “I” shaped motels could expand their footprints to reclaim surface parking for the construction of new rooms and other hotel amenities. These additions or expansions could take the typological form of a courtyard, which reinforce the street edge while sheltering the interior open space of the motel. Instead of having rooms open onto an unsightly surface lot, rooms could face a space dedicated to hotel recreational amenities like a pool, ball court, or landscaping. Motel pools or outdoor amenities that were previously exposed to adjacent moving and parking vehicles could now become more patron-friendly spaces protected by the new building addition. In other footprint configurations, the building could expand from exterior single-loaded room arrangements to double-loaded interior circulation corridors through the construction of a second row of rooms. Finally, integrated AV bus stops or curbside zones could facilitate a patron’s arrival by automated transit vehicles.

Building Elevation

Building Perspective

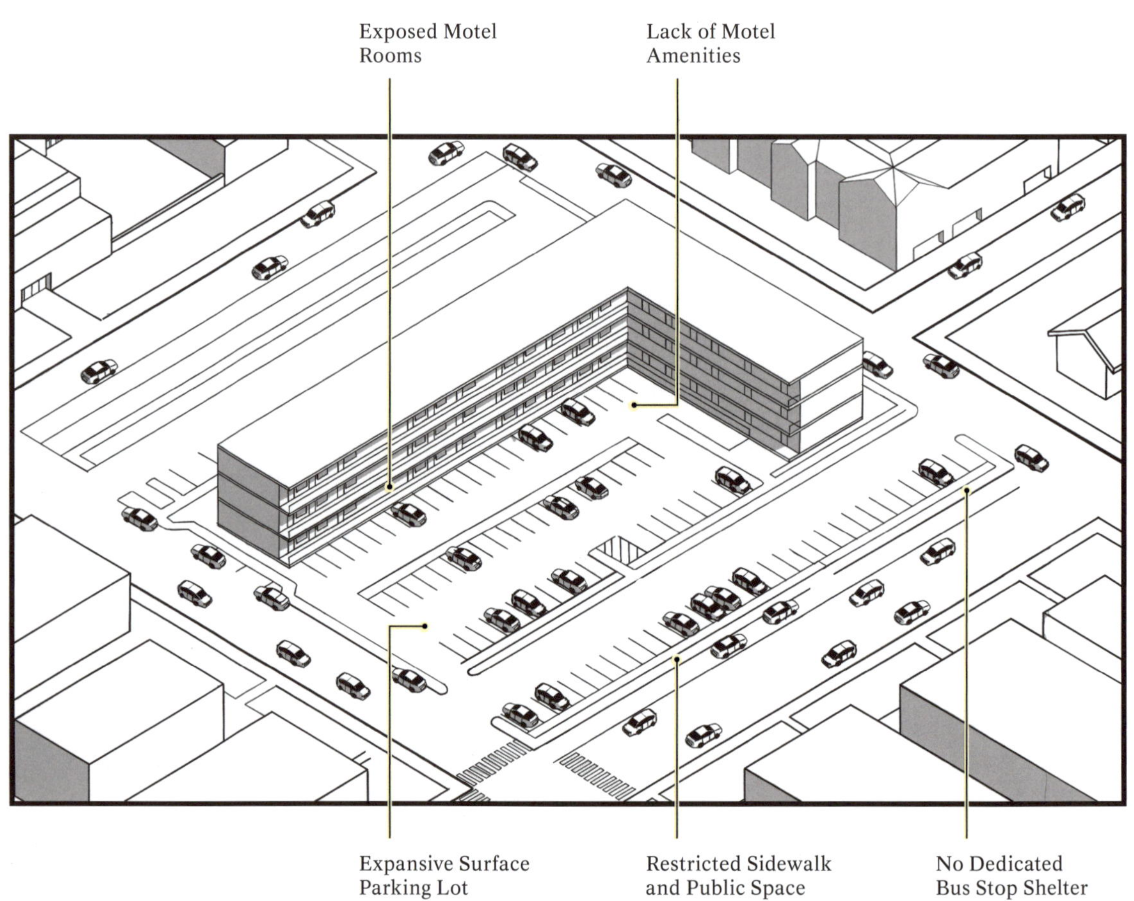
Exposed Motel
Rooms
Lack of Motel
Amenities
Expansive Surface
Parking Lot
Restricted Sidewalk
and Public Space
No Dedicated
Bus Stop Shelter

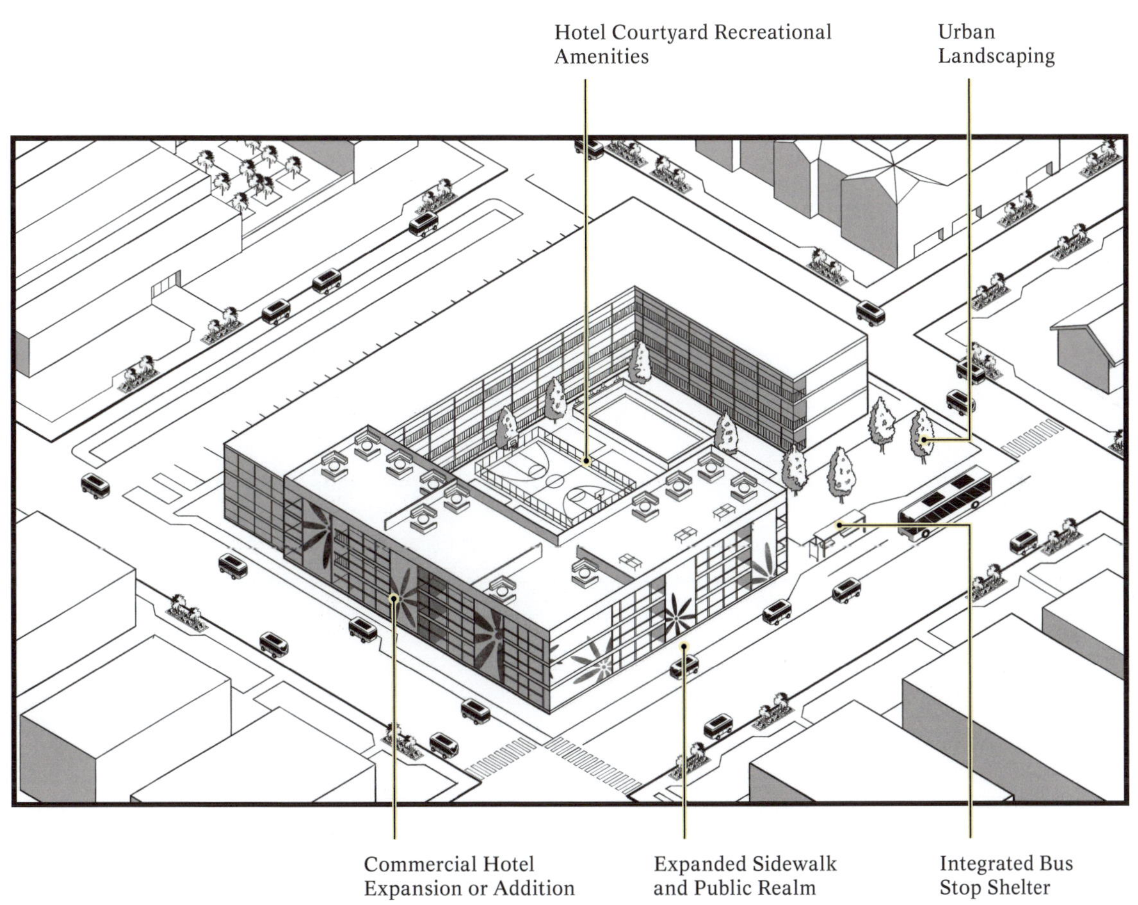
Hotel Courtyard Recreational Amenities
Urban Landscaping
Commercial Hotel Expansion or Addition
Expanded Sidewalk and Public Realm
Integrated Bus Stop Shelter

Big Box Outlet Mall
Huntington Park

The big box store, also known as the superstore, megastore, or supercenter, is a large footprint retail establishment that, along with other large footprint stores, form a retail mall or outlet. The term "big box" refers to the general appearance of these buildings that are large, freestanding, single-story warehouses. Constructed of four load-bearing walls roofed by exposed trusses, this form minimizes the number of interior columns and partition walls to allow for ultimate merchandise display flexibility and storage within. Typically, big box stores sit in the middle of an expansive paved parking lot able to hold hundreds of cars for patrons to access the tens of thousands of square feet of retail space. Retailer consumers are typically expected to arrive by automobile, rather than through walking or public transit, and so the building's exterior signage, scale, and form are designed for efficient access by the motor vehicle with no accommodations for the pedestrian. Big box stores, and their accumulation into an outdoor retail mall, are typically located outside of urban centers where land is less expensive. They take up huge swaths of a block, of which much area is dedicated to paved parking in the front and loading/unloading zones for trucks in the back.

In a driverless future, the expansive surface lot required to service arriving consumers by individual vehicles will see reduced demand and could therefore be consolidated into a minimal footprint, vertically-serviced AV-parking structure. The majority of remaining patrons arriving by mass transit or ride-hailed AVs could be dropped off at a central public plaza on the lot. This reclaimed public space could orient the expansion and construction of new smaller-scale retail additions and commercial subdivisions. A previously single-loaded circulation sidewalk that divided parking from retail could be transformed into a double-loaded pedestrian street mall that is activated on both of its edges. The big box outlet mall could transform from a single-stop use into a block that becomes an active destination outside of traditional shopping hours, with uses that support new pedestrian-oriented public spaces and plazas.

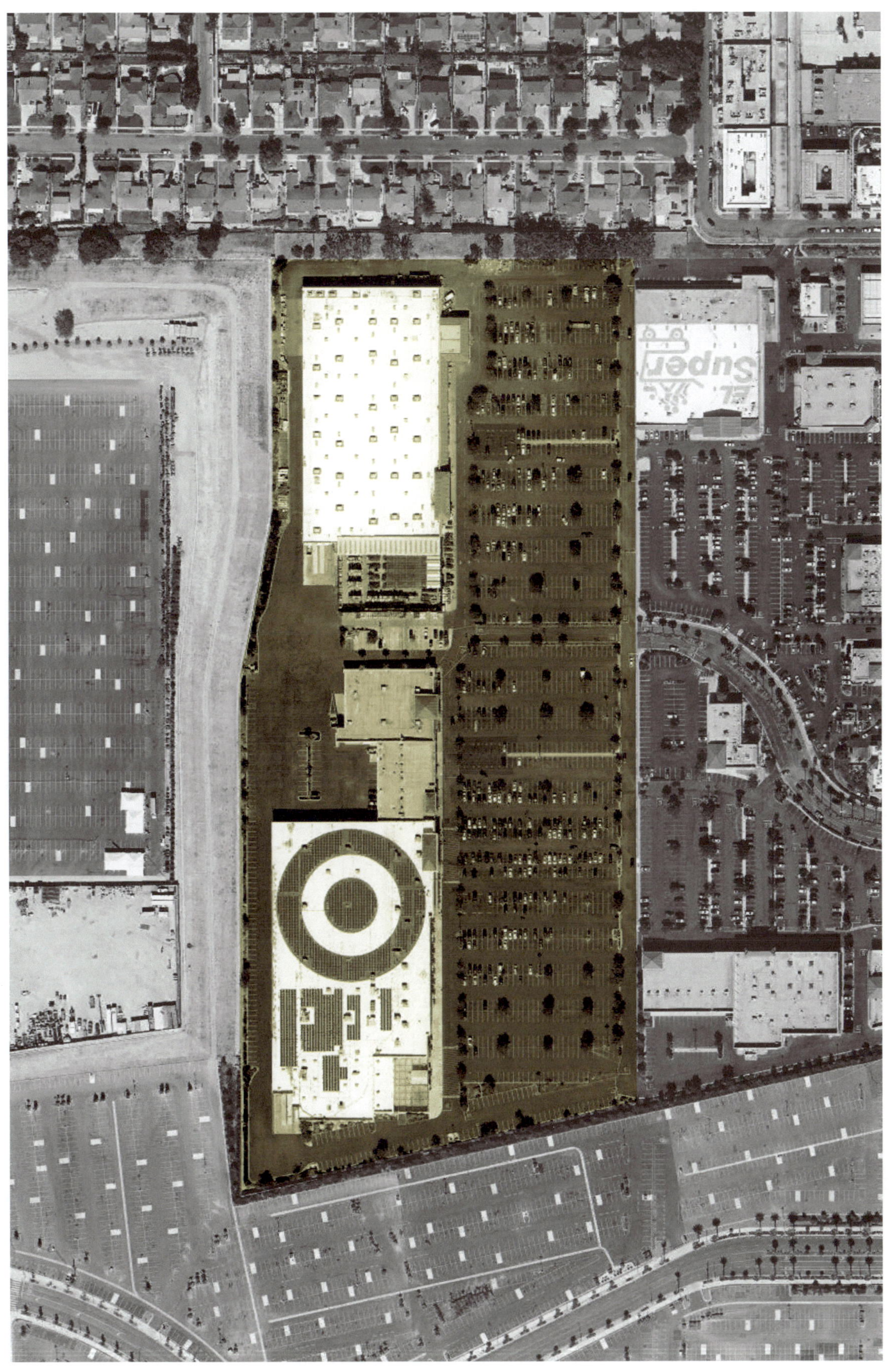
El Super

Building Elevation

Building Perspective

Big Box Outlet Mall
Huntington Park

2024

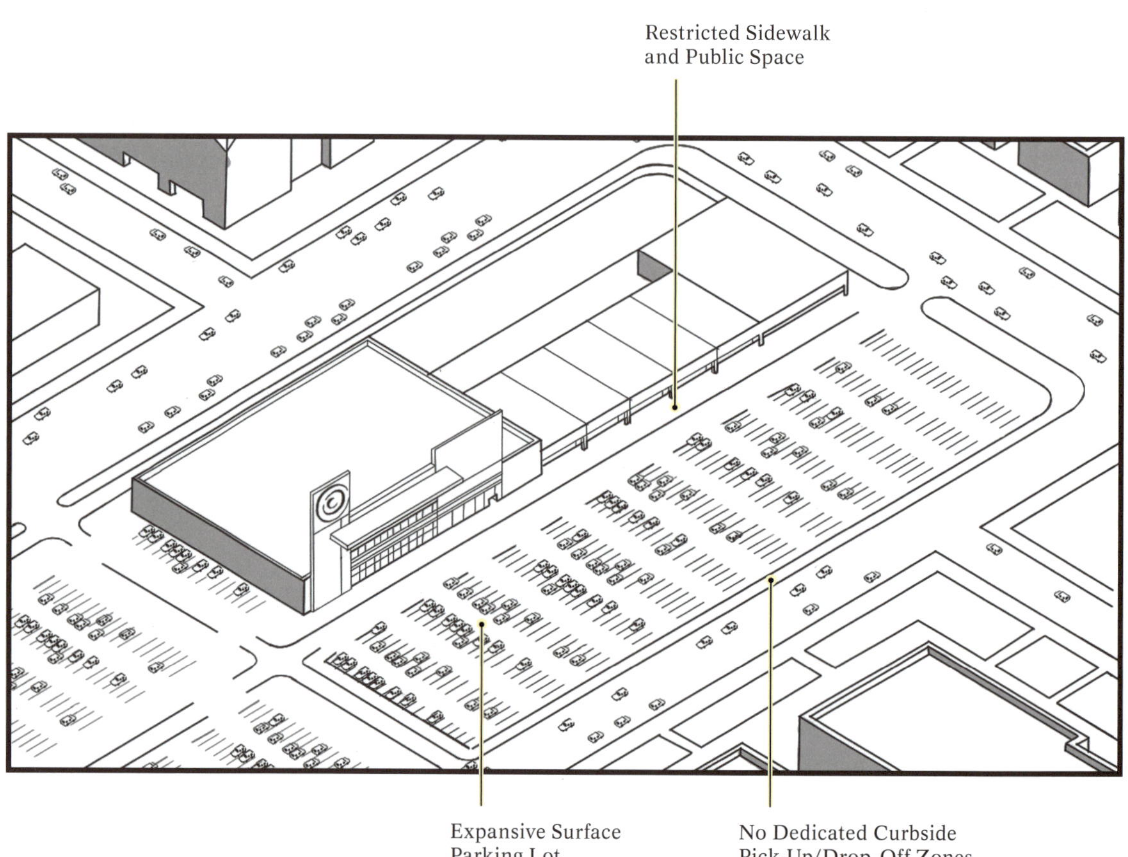

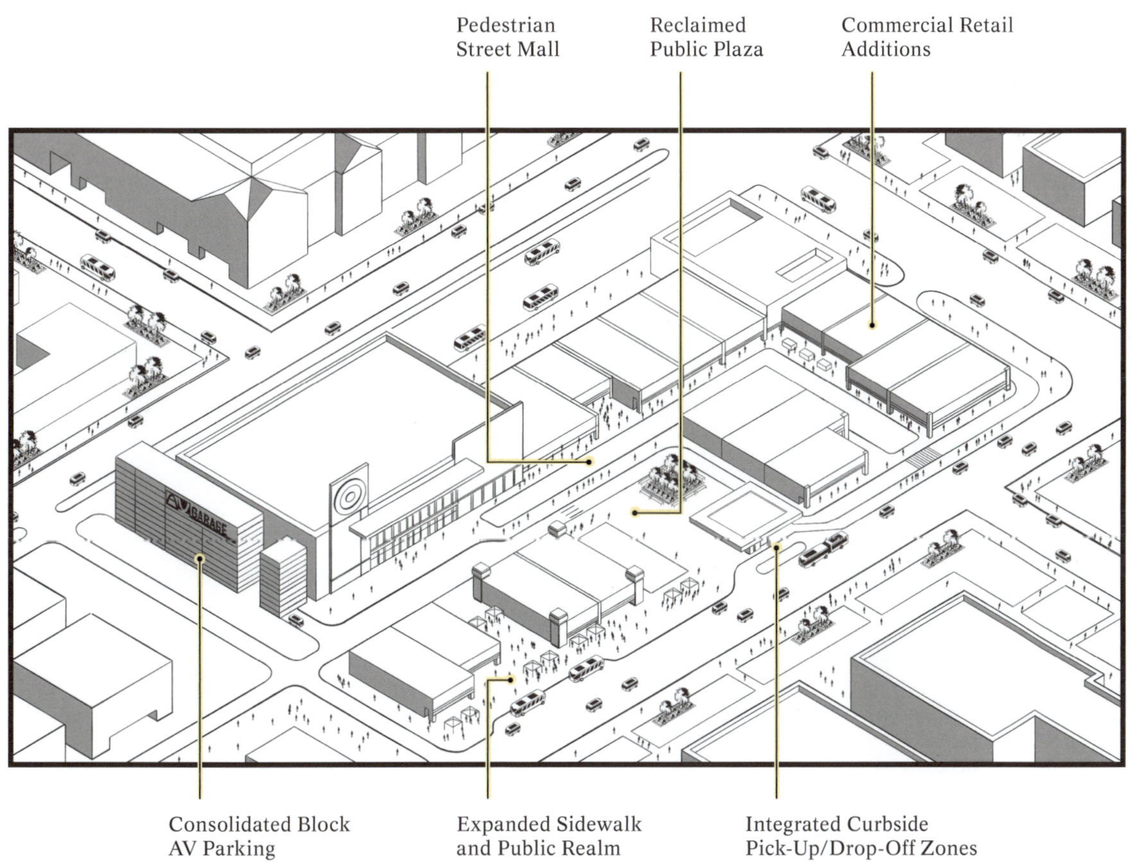
Pedestrian
Street Mall
Reclaimed
Public Plaza
Commercial Retail
Additions
AV GARAGE
Consolidated Block
AV Parking
Expanded Sidewalk
and Public Realm
Integrated Curbside
Pick-Up/Drop-Off Zones

Expansive Surface
Parking Lot
Restricted Sidewalk
and Public Space

2057

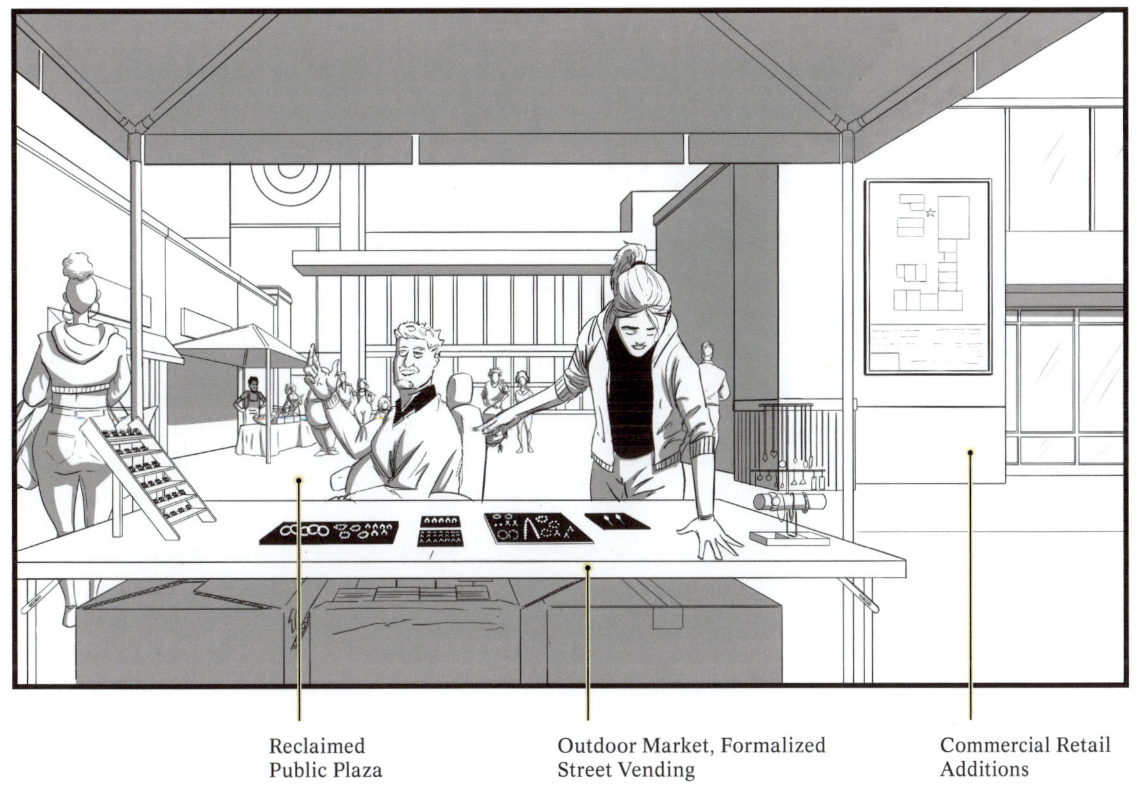

Reclaimed Public Plaza

Outdoor Market, Formalized Street Vending

Commercial Retail Additions

Sports Stadium
Sofi Stadium, Inglewood

Large-scale stadiums that house sports, concerts, or other entertainment events can be found in major metropolitan centers, often designed as architectural destinations and aesthetic icons for their respective cities. However, many of these large footprint structures are also designed to sit in a vast expanse of asphalted surface lots that house parking for tens of thousands of patrons arriving by automobile. The challenge with these typologies is that they must facilitate a large influx of people with cars arriving and leaving during peak demand, resulting in major traffic gridlock in the neighborhoods surrounding these venues. However, during the majority of the day when the stadium is inactive, these asphalted surfaces remain mostly empty, devoid of any use. Designed as monoliths, these typologies are often at a scale that is vastly incongruous with the lower-density neighborhoods around it. In recent years, the ubiquitous use of ride-hailing services also further complicates and contributes to these traffic jams, not to mention the near impossibility for a patron to locate a ride-hailed vehicle in a sea of cars. This results in a variety of detrimental effects on the urban environment, which not only include traffic congestion but also scale disjunctions, environmental noise, and exhaust pollution effects on adjacent neighborhoods.

In Los Angeles, the variety of sports team stadiums and entertainment venues in the city face no exception to these challenges. Illustrated in this section is the recently constructed Sofi Stadium, home to the Los Angeles Rams football team. With a shift towards automated mass transit use, vast areas surrounding stadiums could be reclaimed for new development ranging from commercial retail blocks, entertainment plazas, public spaces, or parks. Higher-density housing and a range of mixed uses could activate the neighborhood at more times of the week, while pedestrian connections could stitch these block developments back to the stadium. These public and economic uses could better support the stadium itself, extending the daily and weekly lifecycle of the building and better integrating it into the scale of the urban fabric around it. Additionally, both automated mass transit and ride-hailed AVs could have a zone dedicated to organized queuing during the stadium's peak demand. Organized pick-up/drop-off lanes within designated boundaries or zones could be supported by digital queue management systems. These types of hard and soft infrastructure coordination have already been previously implemented in some airport ride-hailing zones. For example, Uber's "Airport Queue" system matches exiting passengers en masse to arriving ride-hailed vehicles, designating specific curbside zones by number or lane where passengers meet their vehicles to reduce congestion and confusion.

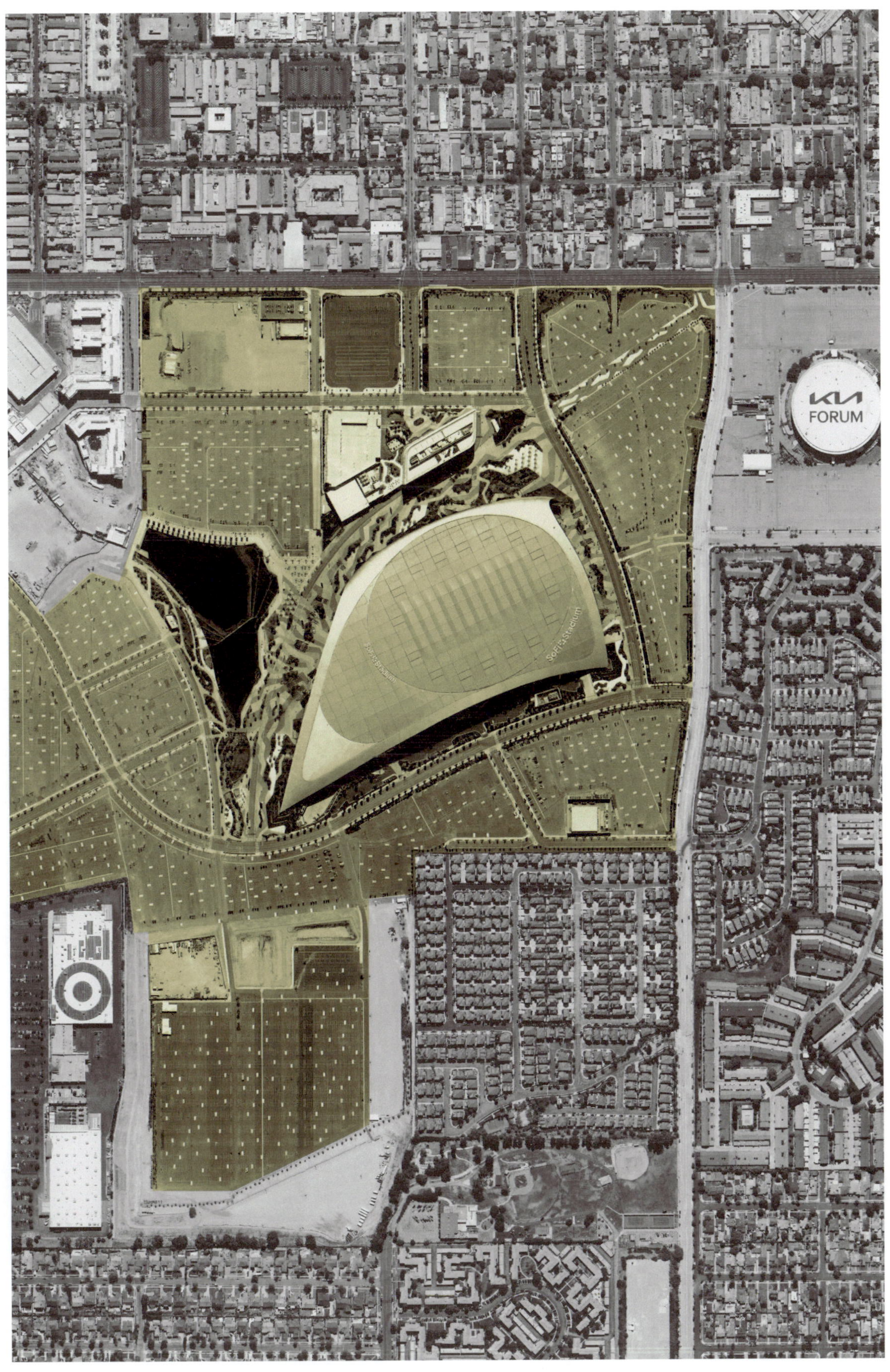
FORUM
SoFi Stadium

Building Aerial View

Building Perspective

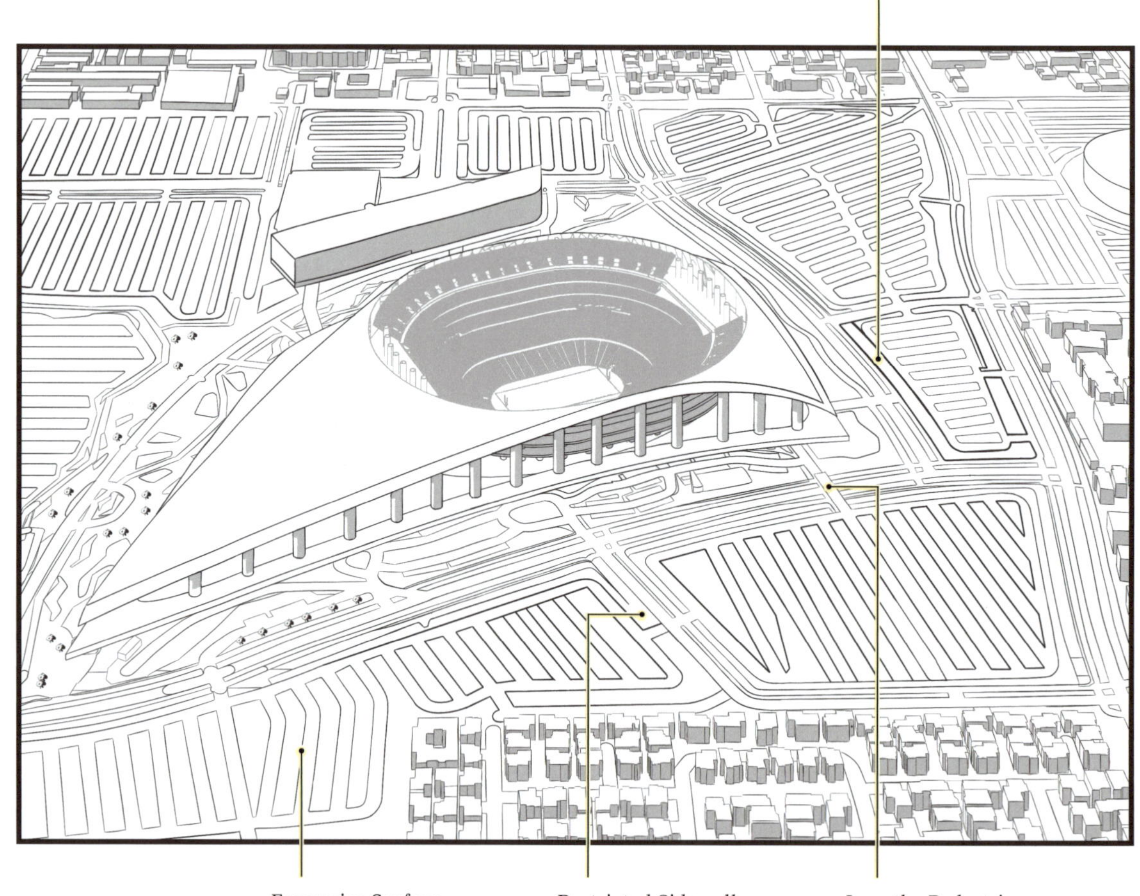
No Dedicated Curbside
Pick-Up/Drop-Off Zones
Expansive Surface
Parking Lot
Restricted Sidewalk
and Public Space
Lengthy Pedestrian
Crossing

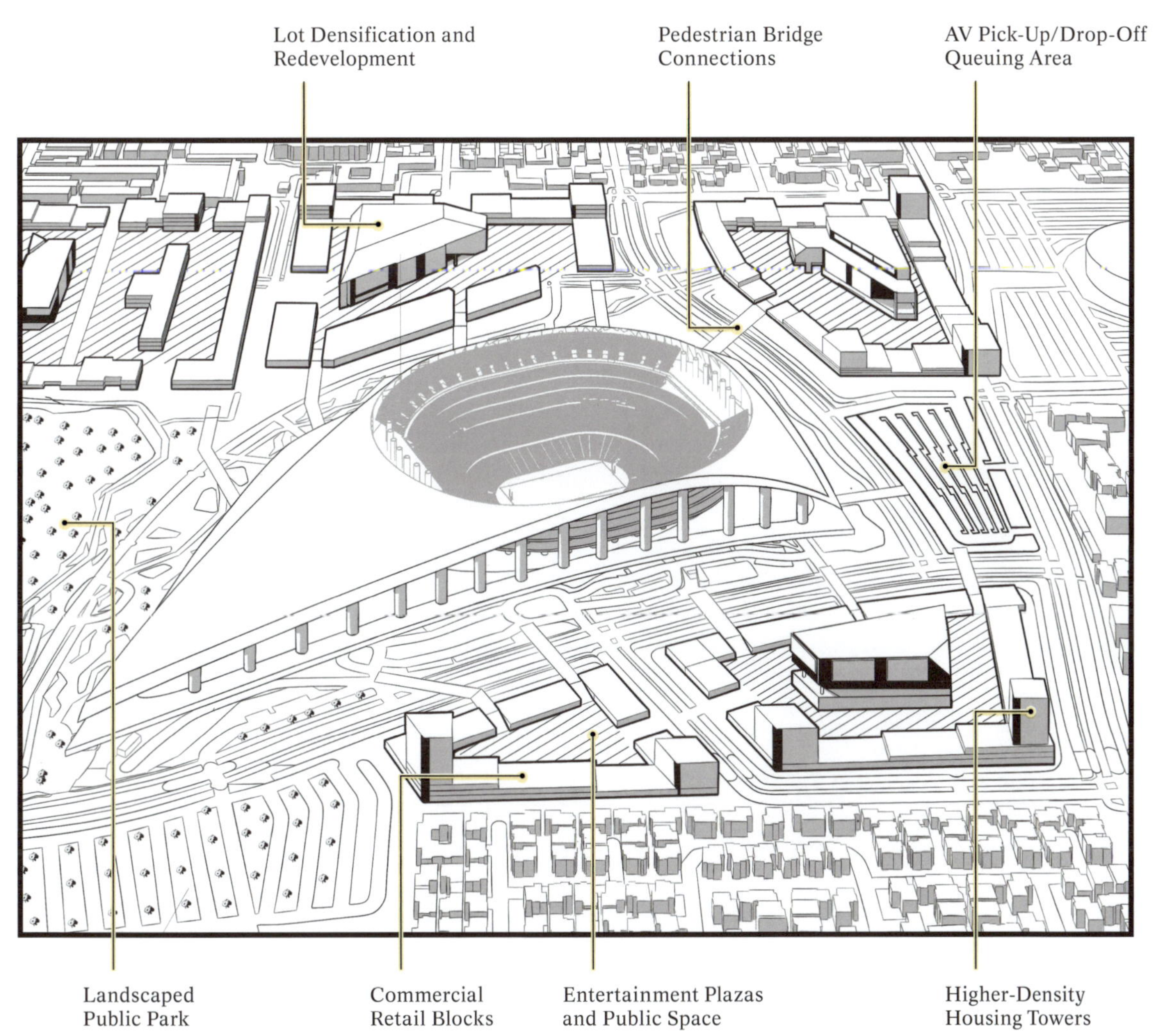
Lot Densification and
Redevelopment
Pedestrian Bridge
Connections
AV Pick-Up/Drop-Off
Queuing Area
Landscaped
Public Park
Commercial
Retail Blocks
Entertainment Plazas
and Public Space
Higher-Density
Housing Towers

Transit-Oriented Housing
Exposition Boulevard

As mentioned in this chapter's introduction, these typological explorations focus primarily on the adaptive conversion of existing car-oriented buildings. In the case of new construction, the opportunity presents itself to explore new typologies catalyzed by driverless technologies. Examples of this include the AV-only garages and dynamic lane-managed streets that have been previously illustrated in prior typologies. Any new building types formatted around the AV, however, must prioritize human-centric uses that contribute to the urban environment. In particular, opportunities must be taken to densify around new or expanded transit hubs and stations in Los Angeles, many of which currently exist as park-and-ride typologies surrounded by surface parking or empty lots. Many of the city's existing light rail stations are built at-grade due to the increased cost of tunneling underground. This means that subway and rail vehicles must share their right-of-way with the street section, with many of their station platforms located in the center of boulevards on isolated islands bordered by multi-lane vehicles. Pedestrians arriving or exiting the platform must contend with major crosswalks in order to reach sidewalks, while the station platform itself is often only lightly or not-at-all covered, exposing waiting passengers to the elements.

Illustrated here, new transit-oriented housing typologies can address these issues by integrating with expanded transit rail station through covered walkways, interior waiting areas, station amenities, and new bus stop canopies. The building typology itself, built on prior park-and-ride surface lots, can be designed around an open central courtyard that creates a public space for transit users waiting for a transfer. The station—and the new, denser, multi-family housing adjacent to it—is activated by mixed-use ground-floor retail and other amenities. A rare but good example of this transit-oriented development in Los Angeles is the Wilshire-Vermont transit village project located at the corner of Wilshire Boulevard and Vermont Avenue. Designed by the firm Arquitectonica, this multi-family apartment complex and its central courtyard is designed to be the arrival point of the Wilshire-Vermont stop of the city's Metro B Line station that lies directly underneath the property. The building facilitates a pedestrian connection to a high-capacity bus stop on the Wilshire corridor, while ground-floor retail, amenities, and educational spaces activate the exterior street wall and interior public courtyard of the building. 20% of its units are designed for low- and moderate-income residents, further supporting the city's transit ridership demographic. Altogether, this case study and the illustrated exploration in this section provide a typological blueprint for mixed-use, transit-oriented housing in a city that must densify around its transit stations and corridors.

Station Elevation

Station Perspective

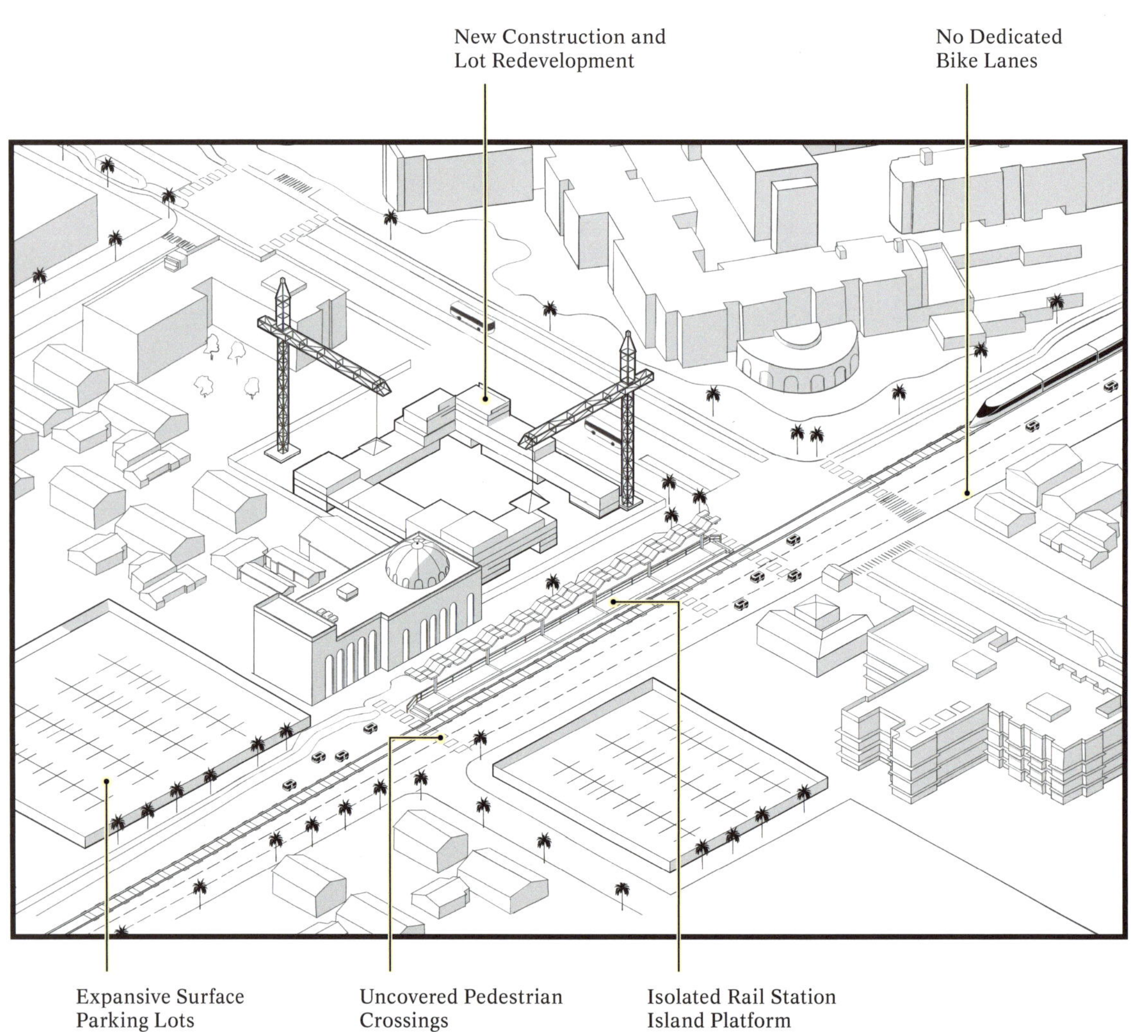
New Construction and
Lot Redevelopment
No Dedicated
Bike Lanes
Expansive Surface
Parking Lots
Uncovered Pedestrian
Crossings
Isolated Rail Station
Island Platform

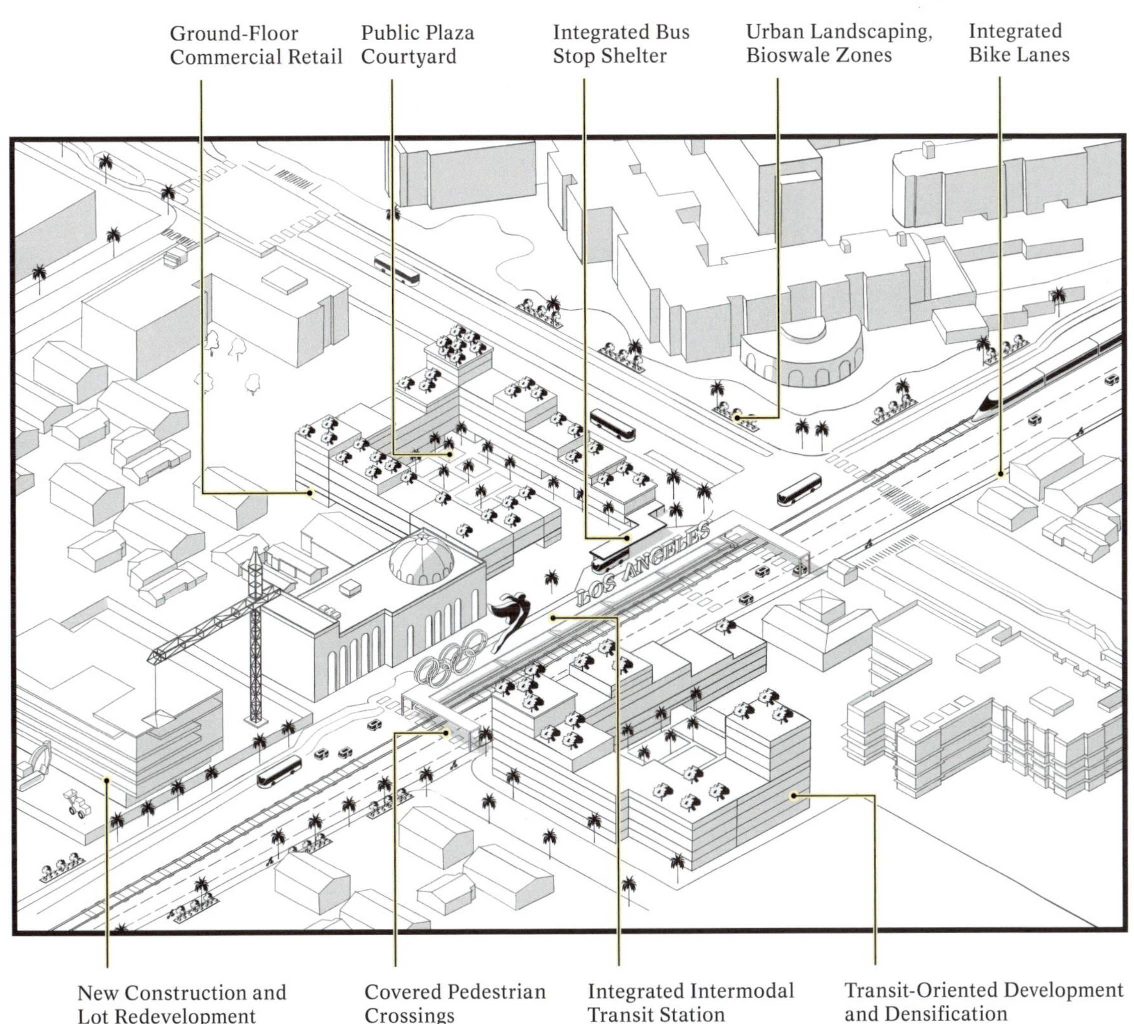
Ground-Floor Commercial Retail
Public Plaza Courtyard
Integrated Bus Stop Shelter
Urban Landscaping, Bioswale Zones
Integrated Bike Lanes
LOS ANGELES
New Construction and Lot Redevelopment
Covered Pedestrian Crossings
Integrated Intermodal Transit Station
Transit-Oriented Development and Densification

11 Paradigm Shifts

Paradigm Shifts and Societal Ambitions for the Future City

The spatial and infrastructural transformations imagined in this book require several paradigmatic shifts to occur that will inevitably disrupt the status quo operations of many contemporary cities. As first formalized by philosopher and scientist Thomas Kuhn, a paradigm shift refers to a major change in the worldview, concepts, and practices of how something works or is accomplished.[1] It implies a new way of thinking and acting that upends and replaces a prior established way of doing things. Often, these shifts are catalyzed by a revolutionary innovation or with a technological disruptor, which can also encounter resistance from incumbent operational modes and those unwilling to adapt. Within the context of urban planning and design, a paradigm shift entails a focal change in how, to what ends, and for whom our cities are designed.

1 Kuhn, *The Structure of Scientific Revolutions*, 1962.

The case for a mobility paradigm shift has been made distinctly clear, having been discussed at length as the focus of this book. Our mobility systems and transportation networks must be designed to move people around, rather than cars. Rather than measuring our current systems by vehicle throughput and speed, we need to understand them as multipurpose systems for the movement of people through space. This would present a far more complete picture of the relationship between urban space and urban mobility. Transit and other shared modes are far more efficient, sustainable, and pluralistic ways of utilizing the contested and valuable public space of cities. Multi-modal, diverse options for accommodating a range of trip types, city residents, and mobility needs must be offered. Cities must free themselves from car dependency. AVs must be deployed to assist and catalyze such transportation shifts.

Additionally, any paradigm shifts to the mobility systems of our cities necessarily implies a paradigm shift in the other status quos that are implicated. A transportation paradigm shift cannot be instigated without an accompanying cultural paradigm shift. Changing the way one moves around a city means changing the transportation behaviors, motivations, lifestyles, and experiences that contribute to those mobility decisions. This also implies a cultural change in the way we collectively view and promote certain mobility choices over others through media and messaging. No matter what transportation modes and driverless technologies are provided, if society and individuals do not adopt them, then these larger shifts will not happen. If deployed correctly, driverless vehicles can act as a useful disruptor to challenge prior mobility habits that have been taken for granted or choices that have been presupposed.

A mobility paradigm shift also entails a change in the way we design around, develop, and plan cities. This development paradigm shift involves expanding how we think about land use, zoning, and parking in coordination with transportation choices. Increasing development density around major transit corridors and hubs, introducing more mixes of land use types, and decreasing development area dedicated to parking are important decisions that both support and are supported by the flows of people and goods to these destinations. Moreover, the design of our mobility infrastructures, block structures, and architecture also shape and are shaped by these principles. Using AVs as an emerging tool, these multi-scalar design and development decisions form the blueprint of a paradigm shift for the urban structure and future evolution of cities.

This design and development shift also implies a critical policy and planning paradigm shift to accompany it. Automation must be incorporated into urban transportation planning models and networks to take advantage of their useful opportunities. Our transportation policies must support and enable designing for new paradigms of moving around. Cities must remove regulatory parking minimums, implement policies that induce multi-modality and transit, and redistribute the legal use-rights to public space and infrastructure for more pluralistic uses. Policy tools are just as critical in this shift as physical and spatial design adaptations. In fact, most of the design and development transformations imagined within this book can only be catalyzed by these policy and planning shifts. The arrival of AVs represent an important moment to implement these regulatory paradigms that might not have been politically or culturally tenable without the impetus of a new technology to upend the status quo.

In the end, none of these paradigm shifts can take place if the economic and fiscal logics driving them don't make sense. That is just the reality of the current neoliberal socioeconomic systems that define our public-private divisions. While not the focus of this book, it is a reality that must be acknowledged, and one that driverless vehicles could provide advantageous opportunities given the potential cost savings associated with them. They must be leveraged to conduce new modes of operating through public-private partnerships that align economic and social outcomes together. This requires visionaries and disciplines to work towards a common goal to align priorities. This leads to a final point, which is to highlight the complicity of these paradigm shifts to produce each other. These overlapping paradigms are co-dependent, and they must enable each other in a cyclical feedback loop. Changing the way one moves around a city, changes the city that emerges culturally, developmentally, economically, and more. This book is hopeful that this emergence can be instigated by rethinking our transportation choices, including the driverless disruptors landing in cities today.

To encapsulate this discussion on paradigm shifts, let's resurface a pivotal precedent previously discussed—the Cheonggyecheon Stream Restoration Project, which saw the dismantling and removal of an elevated freeway and the restoration of a historic stream in Seoul's urban heart. What began as a local project catalyzed by a seemingly simple objective—to remove the Cheonggye highway—came to symbolize a major paradigm shift in Korean urbanization in the early turn of the 21st century. In *A City and Its Stream*, academic and historian Peter Rowe describes how Cheonggyecheon "signaled the finality of a turning point in discourse about the city—official and otherwise—from an ambivalence about public projects to an optimistic forward-looking attitude, from a focus on relatively functional matters to broader issues about lifestyle and livability, and from simply managing urban affairs to actively promoting public good."[2]

2 Rowe, *A City and Its Stream*, 2011.

In addition to becoming a vehicle for environmental, social, and economic revitalizations, the project also indicated a shift in Korean planning priorities both top-down and bottom-up—with city agencies, officials, and residents now placing emphasis on health, sustainability, and social responsibility. Cheonggyecheon became a paradigmatic template for planning intention and action, as well as governance and political will-power. Initiated by Seoul Mayor Lee Myung-bak, the quick and ambitious execution of his promised platform enabled him to leverage its success on a national level, vaulting him into the country's presidency just a few years later in 2008. As a paradigmatic shifting project, the city also went on to demolish nearly 15 other expressways in the time since its documented and overwhelming success. In a similar vein to Seoul's own paradigm shifts, this book calls on us a society to reexamine the interdependent relationship between our mobility and our infrastructure, our mobility and urbanization, and most importantly, our mobility and our *societal ambitions*.

Large-scale infrastructural transformations as turning points in urbanization paradigms can also be found in Los Angeles' own history, which serves as an important historical precedent to frame the new infrastructural paradigms visioned in this book. From 1936 to 1959, Los Angeles County channelized 52 miles of the Los Angeles River after a series of devastating floods had occurred leading up to that point. The river is a major hydrological thoroughfare that runs north-south from the Santa Susana Mountains, through east Downtown, all the way down to Long Beach where it discharges into the San Pedro Bay and the Pacific Ocean. Led by the U.S. Army Corps of Engineers and backed by federal funding, this immense large-scale infrastructural act was conceived of solely as a flood-control measure, despite the other diverse uses and ecologies that the alluvial river held. For over a thousand years, the Los Angeles River and its rich plant and animal habitats provided a livelihood for indigenous Native American settlements and then European colonizers in the region. Throughout its history, the river was the primary water source for the Los Angeles Basin and, up until the opening of the Los Angeles Aqueduct in 1913, an important hydrological source for agriculture and economy. Through the 19th and up until the early 20th centuries, the river powered the city's industry and also served as an important transportation corridor, creating economic value and growth. When the Army Corps of Engineers channelized the river, completely encasing the river's bed and banks in concrete with only a trickle of water usually flowing down the middle, the river ceased to provide many of its prior important functions. It was turned into an engineered mono-functional infrastructure that did one thing well, controlling floods, at the expense of all else. It became representative of the city's sprawling hyper-urbanization, entombing its landscape in steel and cement into an infrastructure that the city has turned its back—so much so that most residents today are even unaware of its very existence and the indispensable role it used to play. The river's channelization signified a paradigm shift in Angeleno urbanization and industrialization at that time, redefining urban priorities and societal attitudes about a city and its citizen's relationship with infrastructure and urban development. Along with the city's construction of mono-functional highways that began crisscrossing Los Angeles in that era, ideas around engineered efficiency and techno-centric urban aspirations became synonymously defined.

In recent years, Los Angeles has excitingly rediscovered its river and has begun redefining its potential future relationship to

Aerial view of Seoul's Cheonggyecheon Stream Restoration Project.

Aerial view of the fully channelized Los Angeles River.

it. Initiated in a 2016 River Ecosystem Restoration study, plans have been put forth by the city to reimagine this infrastructure. The study aims to restore and reestablish the historic riparian and freshwater marsh habitat of the river as a greenway to support increased wildlife habitats, passive recreation, and public access, while still maintaining existing levels of flood risk management. Restoration includes reintroducing ecological and fluvial processes through a more "natural" hydrologic regime to reconnect the river to its historic floodplains and tributaries. 2016 was the same year that Angelenos also voted to approve Measure M, the city's expansive investment into expanding its transit mobility. Both are not only huge infrastructural undertakings, but also indicate a shift in *societal ambitions* for the city—the idea that our urban infrastructures, mobility and ecological landscapes, and public realm must be rethought in more comprehensive, collective, and multi-performative ways.

In the footsteps of this lineage, as Los Angeles turns inward again and away from the outward expansion of its first and second eras, future paradigms in the third iteration of the city must emerge. Developmentally, denser urban clusters and mixed land uses must form around new transit hierarchies, arteries, and poly-nucleated centers. Infrastructurally, numerous mono-functional highways and wide street right-of-ways that have prioritized the passage of cars must be reclaimed. Architecturally, urban form must stop catering to parking and instead engage with and contribute to the city's public realm. Politically, our policies and economic incentives must align to spark these spatial changes in the built environment. Transportation-wise, Los Angeles must transition from a mono-centric, downtown-focused network towards a poly-centric web of driverless, multi-occupancy vehicles offering a range of mobility options. The city's mobility network could become neither a true hub-and-spoke, emblematic of its first era defined by streetcars, nor a grid paradigm, emblematic of its second era defined by automobiles, but one that takes advantages of elements of both models. If this were to occur, a new blueprint for the city would be laid down, unleashing the full potential of its poly-centric metropolitan structure. The city would transition from an individualistic car-oriented city into one that is more diverse, more dense, more walkable, and more accessible. If done in this manner, Los Angeles could reinvent its DNA yet again for decades to come.

In the exciting case of Angeleno mobility, various tools, extensive funding, media attention, political willpower, and public approval for change have already begun to take place. However, driverless vehicles are landing in a city that has yet to put plans in place to not only take advantage of these fast-landing disruptors, but to importantly ensure that they do not erase the hard-fought gains the city has set in motion. How can new mobility technologies and renewed appetites for more diverse mobility choices catalyze an Angeleno paradigm shift? How can the evolving struggles and lessons of doing so inform paradigm shifts for other car-dependent cities around the globe? And lastly, how can the infrastructural project excitingly redefine a city's urban aspirations, ecological relationships, and societal ambitions?

Metrics of Success

If the mobility paradigm shift that this book endeavors to push cities towards were to occur, how would we know that we have reached it? How should cities mark their incremental successes towards these goals? What benchmarks should be set for cities and citizens to aim towards? How should we measure the potential arrival to the Los Angeles of 2057 imagined within?

At various points in this text, statistical metrics—the area dedicated to parking in cities, vehicle miles traveled by cars, traffic congestion metrics, etc.—have been used to quantifiably capture and explain the lengths to which our cities have been shaped by automobile dependency, often at the cost of other transportation modes like transit. This next section outlines some important metrics that indicate a shift away from car dependency and towards more shared and multi-modal cities. They can be thought of as key performance indicators (KPIs) that are quantifiable, measurable gauges of progress towards an intended result. KPIs can provide a focus and analytical basis for decision making. They are not only indicators of a hopeful and eventual mobility paradigm shift, but should also be used as targets to hold cities accountable in the interim. Acting as lodestars, KPIs can guide city agencies, urban transportation planners, policy-makers, and other various stakeholders that have an influential hand in helping them come to pass.

Using again the city of Los Angeles as a starting point, the set of statistical figures illustrated diagrammatically in this section takes key data metrics of the city from recent years, and extrapolates them towards the future Measure M benchmark year of 2057—of which this book has set as a soft target. This extrapolation is purely speculative, and as such can be taken with a grain of salt, but it is contextually informed by real metrics exhibited by transit-rich cities elsewhere around the world. In particular, the mobility metrics of another American city, New York, served as reference benchmarks. Therefore, while seemingly idealistic—particularly in the context of auto-dependent Los Angeles—it is certainly not out of the realm of possibility and in fact could very well be achieved if the steps outlined in this book are heeded.

3 All figures based on Los Angeles metrics unless otherwise indicated.

4 Estimates based on latest year prior that data was available.

5 L.A. drivers had an average commute time of 33 minutes, versus 54 minutes for those who took transit.

6 This figure is 17% in New York City.

7 For the median U.S. city, this figure is 30%.

8 This figure is 1.3 billion in New York City.

Key Performance Indicators[3]	2024[4]	2057
Transit Mode Share Primary transport mode taken for work commute	7%	50%
Car Ownership Share Percentage of households that do not own a personal automobile	8%	70%
Households Served by Transit Percentage of households that are serviced by minimum adequate transit service	29%	60%
Number of Transit Trips Taken Average number of transit trips taken per year	21	90
Vehicle Miles Traveled Average weekday vehicle miles traveled by personal automobile per resident	23	15
Carbon Emissions by Mode Total global carbon emissions output (in billion tonnes) by transport mode per year	2.8	0.5
Travel Time Multiplier[5] Average time multiplier it takes to reach a destination by transit versus automobile	1.6x	0.8x
Job Access Penalty by Mode[6] Jobs accessible by transit as a percentage of jobs accessible by automobile	3%	20%
Congestion Penalty by Job Access[7] Percentage of jobs inaccessible due to automobile traffic congestion	62%	15%
Annual Passenger Boardings[8] Total number (in millions) of passenger boardings on transit modes yearly	253	950

Transit Mode Share

Only 7% of Angelenos use transit for their work commute. In a driverless future, all multi-modal options expand in use while mode dependency on personal cars decreases significantly.

9 New York City's transit modal share is 56%.

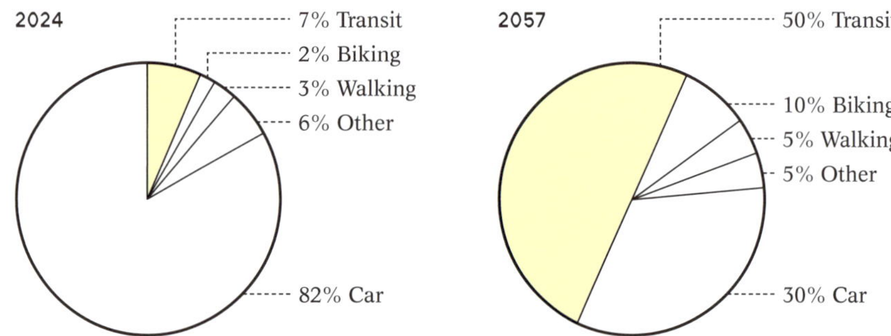

Car Ownership Share

Only 8% of Angeleno households do not own a personal automobile. In a driverless future, most households have sold their personal cars, instead consuming automated mobility-as-a-service.

10 54% of New York City households do not own a car.

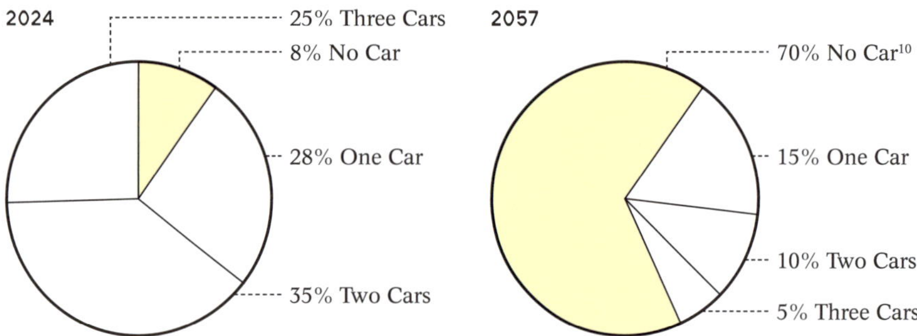

Households Served by Transit

In Los Angeles, only 29% of households are geographically met by a minimum benchmark of adequate transit service. The rest fall in transit gaps, defined as a mismatch in range between transit demand and transit service availability.

11 63% of New York City households are serviced by adequate transit service.

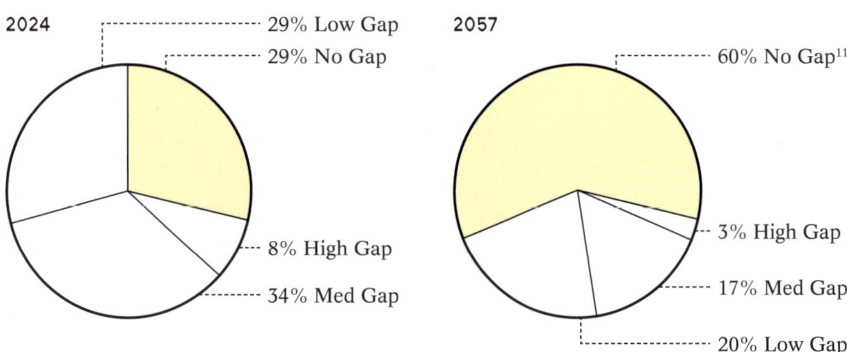

Number of Transit Trips Taken

Angelenos take only 21 transit trips on average per year, per resident, a figure that has decreased in recent years. In a driverless future, this trend reverses and transit trips taken greatly increases.

12 New Yorkers take on average, 86 transit trips per year, per resident.

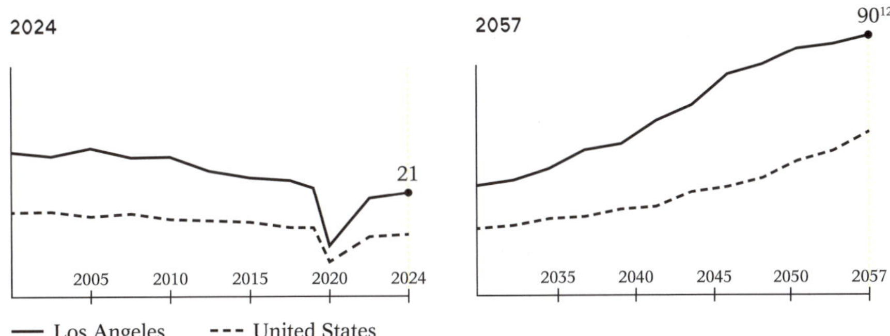

Vehicle Miles Traveled (VMT)

Angelenos travel 23 miles per resident, per weekday on average in a personal car. In a driverless future, miles traveled in a personal AV decreases with VMT shifting into transit and other modes.

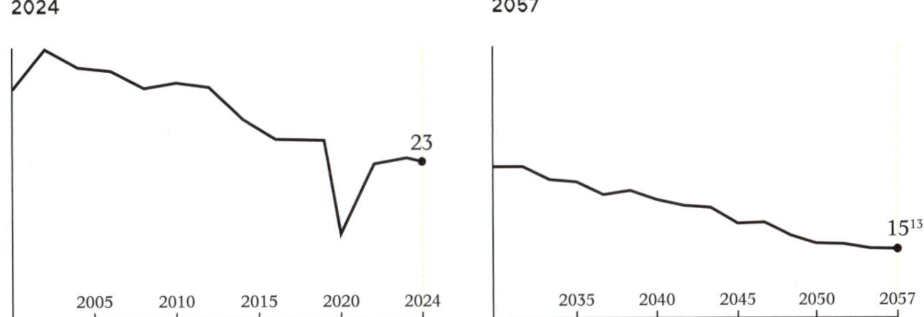

13 New Yorkers drive 17 miles per resident, per day.

Carbon Emissions by Mode

Total global carbon emissions output in most transport modes has risen drastically in recent years.[14] Due to a shift towards shared, electric mobility and transit, this trend reverses significantly.[15]

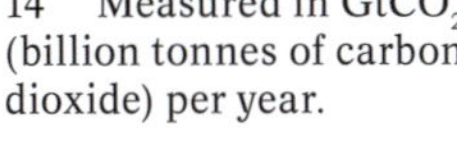
14 Measured in $GtCO_2$ (billion tonnes of carbon dioxide) per year.

15 Based on the International Energy Agency's 2070 Sustainable Development Scenario.

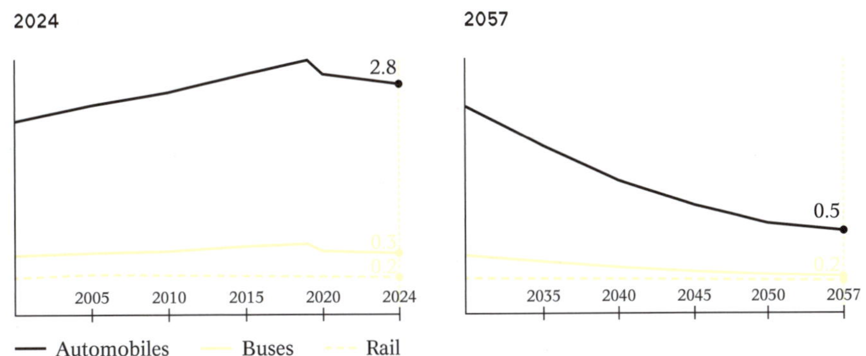

Language and Media as a Choice

Instigating these compounded paradigm shifts may seem daunting and overly idealistic to many. This is because the current paradigm of 20th century-based automobility has completely and overwhelming subsumed our assumptions and conventions today. The initial opportunities that automobiles brought led to an overreliance on those same automobiles. Automobile reliance has become automobile dependency. Automobile dependency has become automobile culture. And car culture has become a lifestyle choice, a love affair, and an obsession that has culminated in societal rites of passage ranging from the celebration of earning one's driver's license, to the purchasing of one's first car, to the customizing and showcasing of one's car as a quintessential form of identity and desire. Particularly

1951 Advertisement for the Chevrolet Skyline Deluxe automobile.

in the context of the U.S., this culture has merged seamlessly with American ideals of freedom, exploration, self-expression, and personal choice, becoming seemingly inseparable.

In *Carjacked: The Culture of the Automobile and Its Effect on Our Lives*, authors Catherine Lutz and Anne Lutz Fernandez define car culture as "a society built around private modes of transportation, but with massive public investment in the infrastructure that allows those private uses"[16] When framed this way, this definition reveals how shockingly routinized we have accepted this tradeoff—one that takes public tax funding and allocates it towards private uses. This is a result of political and economic decisions that not only encourage certain modes of operation, but also how we view and define them.

16 Lutz and Fernandez, *Carjacked*, 2010.

Changing this culture and instigating this paradigm shift begins with language and media. Since the 1930s, North American transportation policy, public agencies, and private media strove to redefine the values and responsibilities surrounding our mobility choices. In *Autonorama: The Illusory Promise of High-Tech Driving*, author Peter Norton describes how during this time, "pedestrian safety, once defined as the responsibility of drivers, was redefined to prioritize drivers. Traffic congestion, once taken to be a symptom of excessive reliance on cars to move people, was redefined as a problem of inadequate streets."[17] He goes on to note that "street railways, once justified as spatially efficient passenger vehicles, were faulted as obstructers of automobiles… [and] the delay that motorists once caused one another by choosing to drive was redefined as a public responsibility to be relieved only through new road capacity, which in turn gradually changed driving from a choice among alternatives to a practical necessity."[18]

17 Norton, *Autonorama*, 2021.

18 Ibid.

The language used in redefining automobile acceptance, use, and subsequent dependency can also be traced to the fact that by and large, the most audible historians of the automobile in American society was Motordom. Celebrating *only* America's love affair with the automobile through defining legal use-rights, car adverts, movies, and social norms at the expense of all else in our societal memory, means reinforcing and giving in to the idea that those choice were the will of the people. Selectively left out of Motordom's historical accounting is the varied resistances to automobile domination and hard-fought struggles against Motordom-defined language that this book, and others, have highlighted. This omission, as Norton notes, "enables and protects assumptions that car domination is the popular will" of society and that any alternative efforts to travel and operate differently will fail.[19]

19 Ibid.

The justification by Motordom that there are no other alternatives to automobility because society has made that decision itself is categorically false. Car dependency was not primarily the product of mass preferences but rather a result of decades of political and economic choices that proliferated into our built environment. Those decisions were made and won by Motordom in history, and consumer mass behavior dutifully fell into line. When no alternative choices to driving exist, then *choosing* to drive is not the same as a *preference* to drive. When alternatives to driving have been offered that are economically, experientially, and performatively better, then consumers and populations have taken advantage of them.

The language of Motordom has been so assimilated in society that we not only accept but propagate it when new technological disruptors like AVs arrive, using new technologies to perpetuate the status quo. And that same rhetoric is also used to caricature critics of car dependency, accusing them of wanting to deprive consumers from the freedoms afforded by car ownership and use. To be clear, this book—and the transportation models and policies it envisions—does not demand that we give up car ownership and use altogether. Cars, driverless or not, can be one of many useful mobility tools. There are trips for which a car is necessary—for the mobility challenged, or when transporting goods or items exceeds what one can carry by hand, for example. However, cars as the only available all-purpose tool, used regardless of trip suitability, especially in cities where urban space is contested and valuable, is the only modal choice available in many contemporary cities. This is what defines car culture and car *dependency*: where no other mobility choices exist, where our urban form has shaped itself completely around the car, and where society has decided to subsidize and prioritize this mode at the expense of all others.

The irony of our current transportation paradigm is that public transit is frequently condemned as a misuse of public taxes and subsidies, even though it cannot even begin to compare to the length with which private automobile use and infrastructure has been subsidized in cities. This is despite the fact that almost all alternatives to automobility are less costly—spatially, environmentally, and fiscally—when considering all of the embodied spatial demands and infrastructural affordances offered to drivers. It is only because a world built for cars has become so routinized that we barely notice it anymore. Expanding highways with more lanes is far more expensive than repainting an existing lane for bus rapid transit. According to studies, "highways do not—and except for brief periods in our nation's history—never have paid for themselves through taxes."[20] Building thousands of acres of parking in cities is far more spatially costly than using that space to house people and address the urban housing crises. The average embodied carbon emissions by transport type on a per passenger-mile basis is far more costly by cars than it is for rail, buses, or bikes—ranging from 5 to 26 times more. Norton argues that even if only 10% of the funds now going to building, maintaining, and policing roads were to shift towards investing in basic transit, cycling lanes, and pedestrian movement, that we could start to see beneficial trends within a year or two. Therefore, the potential economic and spatial gains that driverless technologies offer must be taken advantage of to invest in new paradigms that incur far less costs on cities.

20 Dutzik and Davis, *Do Roads Pay for Themselves?* Maryland PIRG Foundation, 2011.

What these various discoveries uncover is that firstly, choices still exist for us today as a society. We must insist on the provision of diverse modal choices. We must insist that alternative paradigms to the mobility, development, design, and policy status quos be instigated. We must choose to reexamine our own mobility choices, behaviors, and their complicity in perpetuating the existing state of affairs. Secondly, it is up to us as a collective to redefine a new language and value system for guiding and implementing driverless technologies. It is only by reexamining how we talk about old and new mobility technologies, questioning language-conditioned assumptions, and defining new language and visualizations for projecting alternative futures, that a paradigm shift can be instigated. There are moments in our urban histories in which opportunities to change the narrative and set our cities on alternate evolutionary paths have presented themselves. Can we make the arrival of driverless technologies one of those defining moments?

Technological Tools and a Call to Action

In 1966, the late British academic and architect Cedric Price was once quoted as stating, "technology is the answer, but what was the question?" Price asked this question as a provocation, in order to interrogate and reconsider the impact of technological progress and innovation on the design and architecture fields of that time. Today, that same question takes on the embedded significance of contemporary technologies' ability to address—and complicity in contributing to—the various urban and environmental issues facing cities.

To conclude this book, I resurface Price's question in order to contextualize a contemporary moment in which many AV proponents have arrived at the conclusion that driverless technologies are the answer to our urban mobility woes, including the negative externalities that auto-based urbanism has wrought on cities and their societies. Instead of trying to answer Price's question, this book instead calls on us as a society to ask different questions altogether, and to think and act accordingly.

The broad question that this two-volume set asks on the front cover of its slipcase is, "how will autonomous vehicles impact the future of cities?" In order to answer that initial question, other embedded questions have been surfaced throughout the course of this book. How do we envision moving around our cities, and in what modes and methods? What principles do we value in our transportation systems and the choices, behaviors, and lifestyles they engender? What design and planning outcomes for our cities and built environments emerge from those values and principles? What tools can we use to achieve such visions for our future cities?

Is driverless technology *an* answer to these questions? It could be, as this book has hopefully made a strong case for. AVs are not, however, *the only* answer. Progressive cities and leaders have demonstrated that the right values combined with the right tools—automated or not—can be used to reclaim and redefine our urban space. Driverless technologies are not a mobility panacea, silver bullet, or magic cure-all—and you should question any mobility companies or manufacturers who claim them as such in their pitch to get you to purchase their services or products. However, AVs can be an important contributing cog in the solutions to address our contemporary transportation challenges—a useful tool that can work in tandem with other design and policy tools to do so. What they also bring is media attention, innovative minds, and large quantities of funding and investment that fuel the emerging industry. In particular, this aspect of their arrival is quite useful in their role as a technological disruptor. They offer an important moment to seize on the excitement, momentum, and available funding as a means to advance the agendas and potential outcomes outlined in this book.

And most importantly, driverless vehicles are an important tool that cannot be ignored. Just as much as they could be *an* answer, let's not forget that on the opposite extreme, AVs could also be the very *antithesis* of the answer—a poison pill used to exacerbate and reinforce all of the dependencies and negative externalities made by private automobiles on our built environments. If we ignore them, as AV-skeptics or naysayers are wont to do, then the private market will dictate their implementation and use as the technology matures.

The makers of driverless vehicles are far less concerned with the long-term spatial implications of their technological innovations, than they are with maximizing market share and satisfying the bottom lines of the stakeholders or venture capitalists fueling them. We can ignore AVs, allowing them to proliferate unregulated and potentially bereft of any consideration of the common good, or we can utilize them as tools in the fight to reclaim our cities' public realm.

Perhaps a better way to revisit Price's original question is to neither ask nor answer the question altogether. As an "answer," technology implies that in-and-of-itself, it can solve our problems for us. But if technology is not an answer but rather just a tool, then it can only assist us in solving our own problems ourselves. Misrepresenting tools as solutions, as AV-utopists are inclined to do, is dangerous especially because the language around driverless vehicles characterize the technology as "smart," "intelligent," and "autonomous," i.e. worthy of human users' trust in relinquishing control to the self-directed decisions of its hard- and soft-wares. But any technology is only as "smart" as the humans who designed it, subject just as much to human flaws, decisions, biases, preferences, and inputs as any other low-tech or high-tech tool. Again, we must always be vigilant to how language and media shape perception, integration, and use of that technology into our society and decision-making. What matters then, is how driverless technology tools are used to achieve the ends that we ourselves, as urban citizens, dictate for our own futures. Let us define what future we want and need, and then let us adapt our technological tools to further those purposes. What remains then, are tools that represent a choice and a call to action. What competing scenarios and driverless urban futures do we choose and why? This book has outlined several—with the qualification that there may yet be more that emerge—and has argued for and illustrated the compelling and transformative benefits of one of those scenarios. The choice to pursue that future is ours, both collectively as a society and individually as stakeholder citizens.

This book is a call to demand that such conversations be held amongst those leading the development and implementation of AVs in cities. It is a call for AV innovators and makers to step back for a brief moment from solving technological issues and to consider how the vehicle designs, business, and service models they are investing in will impact not only mobility systems but urban space. It is a call for urban planners, policy-makers, and public officials to put AV-regulatory guiderails and frameworks in place before they have proliferated into mass market and it becomes too late to reign in any negative externalities. It is a call for designers, architects, and engineers to shape urban infrastructure and form to facilitate human-centric, multi-modal, and publicly vibrant spaces over others, like parking. It is a call for all interdisciplinary stakeholders to work together far more than our past paradigms of development have done so, and to blur siloed disciplinary boundaries and public-private divisions in pursuit of a common future. And importantly, it is a call to you, the reader, who contributes to the evolution of our transportation systems in the daily mobility choices that you make every time you step foot outside your home—in whatever city, town, or area that may be. It is a call to each of us as citizens to consider how our own mobility behaviors shape the design of urban space and the evolution of cities.

This paradigm shift can begin from below, by citizens, for citizens. It begins with the daily mobility choices we make ourselves. Having imagined an alternative driverless city that we could aspire towards, how can we achieve this possible future? We collectively complain about the traffic congestion, parking frustrations, and exhaust pollution caused by automobile use, and yet many individuals refuse to relinquish their personal cars and expand their mobility choices. How can we use the driverless vehicle as a tool to address these woes, rather than as an exacerbator? Think about it, what can you do to instigate such a change?

More broadly, a fork in the proverbial road presents itself—a transportation crossroads of sorts. The longer we wait before acting, the higher the chance that futures filled with more automobility, car dependency, environmental degradation, mono-functional infrastructures, urban sprawl, and shrunken civic realms will manifest into reality. In fact, we may already be on the path to one of these futures because it represents the status quo paradigm of how many cities operate today. Without any intervention, this status quo will continue to be promoted, maintained, and reinforced by Motordom and those that have accepted that car dependency is the only choice that exists. This may seem like a daunting task, but it is one that has precedence as the various examples and tools provided in this book have highlighted.

Above all, how can we approach this future with an equal dose of optimism and realism? Understanding how past advances in mobility technologies have impacted cities and societies, we must not blindly ignore the pitfalls nor naively embrace the utopias of driverless and any other future technological catalysts. Instead, we must tackle both the perils and the opportunities brought on by current trajectories in technology innovations head-on, in order to shape a collective urban future fore-fronted by the public good. It is with this empowered and engaged spirit, that I hope you leave reading this book—with a desire to use any incoming technological tools to create a better future city.

Acknowledgments

When I first began research into this book's topic, I was a graduate student advancing his Master's thesis at Harvard University's Graduate School of Design (GSD). It originated as a simple curiosity in questioning the spatial impacts of a new technology, and with a seemingly simple question posed on the front cover of this book's slipcase. The answer to that question has since taken on a life of its own in the many years I spent interrogating it afterwards. Along the way, it has expanded to not only encompass research into mobility but also into transportation's relationship to all of the complex and layered urban systems, infrastructures, spaces, forms, and histories that our mobility technologies are critically implicated in. This rabbit hole of a journey, and the intertwined nature of this topic, is one that I hope the reader has also experienced as they've made their way through this book.

In the time that this book has matured into this published monograph, the project has also expanded to a scope and size previously underestimated when I began it. It has spanned multiple institutions and has mirrored my own career path into teaching and academia. As other authors know, book writing is a monumental effort, especially one that spans from research all the way to speculative design and world-building through comic illustrations. It is one that could not have been made possible without indelible support from the people and institutions acknowledged below.

One of the great privileges of being a professor is that I am inspired and challenged by the innovative and bright minds around me on a daily basis, from both my fellow faculty colleagues and our students. I have had the privilege of being able to explore, test, and germinate many of the design ideas in this book through several courses and institutions that I have taught at and been a part of. Much thanks to the students of my advanced research topical seminar course *Automated Vehicles: Imagining Urban Futures* taught at the University of Virginia in Fall 2021, and the students of the advanced research option studio course Future of Streets offered at Harvard GSD in Spring 2019. Alongside these talented undergraduate and graduate architecture and urban planning students, our weekly conversations and discussions helped test and challenge my ideas on this multi-disciplinary and multi-scalar topic. Their names are listed at the end of this section.

An indispensable thank you must go to Anthony Delaware, whose contributions and consultations as a talented graphic illustrator and comic artist were critical to the visualizations in this book. Anthony's contributions in helping me form the narrative and storytelling of the graphic novel volume of this book brought the characters imagined within to life.

Much gratitude goes to my colleagues, mentors, and friends who have provided insight and feedback as this project was being developed. Andres Sevtsuk—my thesis advisor at Harvard GSD—critically shaped my foundational approach to this topic, embedding it with an important set of values and principles through countless meetings and other collaborative projects. Additional colleagues that must be mentioned include Felipe Correa—who has always believed in the importance of this book's topic and offered me many opportunities to advance, push, and test its ideas; Rahul Mehrotra—for his intellectual guidance and support; and finally AnnaLisa Meyboom.

In particular, it was AnnaLisa's own book, *Driverless Urban Futures: A Speculative Atlas for Autonomous Vehicles*, and the conversations that we have shared about it, that have informed my own multi-scalar design speculations in this book. Along with the National Association of City Transportation Organization (NACTO)'s publication, *Blueprint for Autonomous Urbanism*, these two published works heavily influenced my research and thinking in this topic, of which credit and gratitude must go towards. They are also two works that deploy the critical tools of speculative design and drawing to envision how new mobility technologies will impact urban infrastructure and form—design work that is few and far between in this nascent field. These two works make an important contribution to a contemporary topic in which design visioning is severely lacking, and are ones in which this book importantly builds off of.

A big thank you as well to Neil Donnelly Studio—the graphic designers of this beautiful two-volume set wrapped by a slipcase—who shaped how its contents are presented and made accessible to the reader. I am always appreciative of Neil's ability to skillfully incorporate my own idiosyncratic desires for the project, while developing a larger visual language for the book.

Funding for the research and print production of this book was made respectively possible by Harvard Graduate School of Design's 2020 Irving Innovation Fellowship Grant, and the Dean's office at New York Institute of Technology's School of Architecture and Design—both funding sources of which I gratefully thank.

Lastly and most importantly, significant gratitude goes to my family—my wife Yuna, my mother Jane, my father C.J., my aunt Joan, and my sister Elisa—who have been there every step of the way. To Yuna (and our two dogs) who have seen this project evolve from start to finish, thank you for enduring the countless hours, long evenings, and innumerable weekends that I dedicated to this book. Collectively, your sacrifices, support, and love have contributed to my past and future successes in life—professionally, academically, and beyond.

Student Assistant (Diagrams):
Xinwen Cao

Autonomous Vehicles: Imagining Urban Futures
Arch 5500, Fall 2021, University of Virginia School of Architecture
Taught by Evan Shieh
Student course participants: Yanbo Chen, Cassandra Dickson, Lauryn Downing, Patrick Bay Penny, Lily Slonim, Ramya Tella, Timothy Victorio, Macey Whitt

Future of Streets
Studio 1507, Spring 2019, Harvard University Graduate School of Design
Taught by Andres Sevtsuk, Teaching Assistant: Evan Shieh
Student course participants: Weihsiang Chao, Yuebin Dong, Dora Du, Solomon Green-Eames, Sang Yoon Lee, Sunmee Lee, Yuzhou Peng, Xin Qian, Amanda Ton, Finn Vigeland, Lanchun Zeng, Bailun Zhang

Bibliography

Works Cited

"A Brief History of the Corps." The U.S. Army Corps of Engineers. 2023. https://www.usace.army.mil/About/History/Brief-History-of-the-Corps/Introduction/.

"About Intersection Safety." U.S. Dept. of Transportation Federal Highway Administration. July 21, 2023. https://highways.dot.gov/safety/intersection-safety/about/.

Adams, John S. "Residential Structure of Midwestern Cities." *Annals of the Association of American Geographers* 60, no. 1 (March 1970): 37–62. https://doi.org/10.1111/j.1467-8306.1970.tb00703.x.

"America's best (and worst) Commuter Cities." *GeoTab*, April 23, 2019. https://www.geotab.com/press-release/americas-best-commuter-cities/.

"Annual Number of Passenger Boardings of LACMTA, by Mode." *Statista*, March 23, 2023. https://www.statista.com/statistics/1297553/lacmta-network-total-annual-ridership-by-mode/.

Anthony, Sam. "Self-driving Cars still can't Mimic the most Natural Human Behavior." *Quartz*, August 29, 2017. https://qz.com/1064004/.

Artbound. "Third L.A. with Architectural Critic Christopher Hawthorne." Season 8 Episode 5, *KCET*. YouTube Video, 55:36. June 15, 2016. https://www.youtube.com/watch?v=4Ywj8c16b8o.

"Autonomous Vehicles – Get to know them better." Singapore Land Transport Authority (LTA). 2023. https://www.lta.gov.sg/content/ltagov/en/industry_innovations/technologies/autonomous_vehicles.html.

Avila, Eric. *The Folklore of the Freeway: Race and Revolt in the Modernist City.* Minneapolis, MN: University of Minnesota Press, 2014.

Banham, Reyner. *Los Angeles: The Architecture of Four Ecologies.* New York, NY: Harper & Row, 1971.

Bradford, Ben. "Report: The Average American spends the equivalent of 2.5 Work Weeks in Traffic." *Marketplace*, February 13, 2019. https://www.marketplace.org/2019/02/13/report-average-american-spends-equivalent-25-work-weeks-traffic/.

Brozen, Madeline, Chase Engelhardt, and Eli Lipmen, "Are L.A. Bus Riders Protected from Extreme Heat?" UCLA Lewis Center for Regional Policy Studies. 2023. https://www.lewis.ucla.edu/publications/do-la-bus-riders-have-shelter-from-the-elements/.

Buehler, Martin, Karl Iagnemma, and Sanjiv Singh. *The 2005 DARPA Grand Challenge: The Great Robot Race.* Berlin, DE: Springer Berlin Heidelberg, 2007.

Bushard, Brian. "Driverless Food Deliveries Grow: Uber Eats launching in California, Texas This Fall." *Forbes*, September 8, 2022. https://www.forbes.com/sites/brianbushard/2022/09/08/driverless-food-deliveries-grow-uber-eats-launching-in-calif-texas-this-fall/.

"CapMetro, City of Austin, and RATP Dev USA testing Autonomous Transit." CapMetro. 2023. https://www.capmetro.org/news/details/2020/05/11/capital-metro-city-of-austin-and-ratp-dev-usa-are-testing-autonomous-transit/.

Carpenter, Susan. "Waymo will offer Robotaxi rides in L.A. starting in October." *Spectrum News*, September 20, 2023. https://spectrumnews1.com/ca/la-west/transportation/2023/09/20/waymo-la-robotaxi-tour/.

"Certificate of Entitlement (COE)." One Motoring, Singapore Land Transport Authority (LTA). 2023. https://onemotoring.lta.gov.sg/content/onemotoring/home/buying/upfront-vehicle-costs/certificate-of-entitlement--coe-.html.

"Change Streets, Change the World." Global Designing Cities Initiative (GDCI). 2023. https://globaldesigningcities.org/.

Chester, Mikhail, Andrew Fraser, Juan Matute, Carolyn Flower, and Ram Pendyala."Parking Infrastructure: a Constraint on or Opportunity for Urban Redevelopment? A Study of Los Angeles County Parking Supply and Growth." *Journal of the American Planning Association* 81, no. 4 (2015): 268–286. https://doi.org/10.1080/01944363.2015.1092879.

Chiland, Elijah. "Measure JJJ triggers New Incentives to Encourage Affordable Housing near Transit." *Curbed Los Angeles*, March 14, 2017. https://la.curbed.com/2017/3/14/14928306/los-angeles-incentives-affordable-housing-transit-jjj/.

Clewlow, Regina R. and Gouri Shankar Mishra. *Disruptive Transportation: The Adoption, Utilization, and Impacts of Ride-Hailing in the United States.* Davis, CA: UC Davis Institute of Transportation Studies, 2017. https://escholarship.org/uc/item/82w2z91j/.

Conger, Kate. "Driver Charged in Uber's Fatal 2018 Autonomous Car Crash." *The New York Times*, September 15, 2020. https://www.nytimes.com/2020/09/15/technology/uber-autonomous-crash-driver-charged.html.

Corner, James. *Recovering Landscape: Essays in Contemporary Landscape Theory.* Princeton, NJ: Princeton Architectural Press, 1999.

Cortright, Joe. "The Myth of Pedestrian Infrastructure in a World of Cars." *Strong Towns*, September 10, 2020. https://www.strongtowns.org/journal/2020/9/9/the-myth-of-pedestrian-infrastructure-in-a-world-of-cars/.

Crute, Jeremy, William Riggs, Timothy S. Chapin, and Lindsay Stevens. *Planning for Autonomous Mobility, PAS Report 592.* Chicago, IL: American Planning Association, 2018. https://www.annapolis.gov/DocumentCenter/View/11933/PAS-Report-Autonomous/.

"Demand Responsive Parking Pricing." SFMTA. 2023. https://www.sfmta.com/demand-responsive-parking-pricing/.

Dobush, Grace. "Uber has troves of data on how people navigate cities. Urban planners have begged, pleaded, and gone to court for access. Will they ever get it?" *Medium*, September 9, 2019. https://marker.medium.com/ubers-real-advantage-is-data-e54984ff524c/.

Doll, Scooter. "Waymo to Retire Chrysler Hybrids in Transition to All-Electric Robotaxis." *Electrek*, March 30, 2023. https://electrek.co/2023/03/30/waymo-retire-chrysler-hybrids-transition-all-electric-robotaxis-zeekr-ev/.

Dutzik, Tony and Benjamin Davis. *Do Roads Pay for Themselves? Setting the Record Straight on Transportation Funding.* Baltimore, MD: Maryland PIRG Foundation, 2011. https://pirg.org/wp-content/uploads/2011/12/Do-Roads-Pay-for-Themselves_-wMD.pdf.

"Electronic Road Pricing (ERP)." One Motoring, Singapore Land Transport Authority (LTA). 2023. https://onemotoring.lta.gov.sg/content/onemotoring/home/driving/ERP/ERP.html.

Etherington, Darrell. "Lyft says nearly 250k of its Passengers ditched a Personal Car in 2017." *TechCrunch*, January 16, 2017. https://techcrunch.com/2018/01/16/lyft-says-nearly-250k-of-its-passengers-ditched-a-personal-car-in-2017/.

"Everything you need to know about Fares." NYC MTA. 2023. https://new.mta.info/fares/.

Fanelli, Mark and Daniel Savrin. "State–not Federal–Policy guides U.S. Autonomous Driving." *Automotive World*, September 19, 2022. https://www.morganlewis.com/pubs/2022/09/state-not-federal-policy-guides-us-autonomous-driving-automotive-world/.

Felix, Alison and Travis Pollack. *From App to Table: Rapid Food Deliveries in Massachusetts*. Metropolitan Area Planning Council. 2022. https://www.mapc.org/wp-content/uploads/2022/12/Rapid-Food-Deliveries-Report.pdf.

Fonseca, Ryan. "L.A. could try Congestion Pricing. Will drivers go for it?" *The Los Angeles Times*, May 31, 2023. https://www.latimes.com/california/newsletter/2023-05-30/essential-california-essential-california/.

Fraade, Jordan. "Who is Afraid of the Petextrian." *The Baffler*, January 23, 2018. https://thebaffler.com/latest/whos-afraid-petextrian-fraade/.

Gawron, James H., Gregory A. Keoleian, Robert D. De Kleine, Timothy J. Wallington, and Hyung Chul Kim. "Life Cycle Assessment of Connected and Automated Vehicles." *Journal of Environmental Health Science & Engineering* 52, no. 5 (February 2018): 3249–3256. https://doi.org/10.1021/acs.est.7b04576.

Gooden, Ben. "Woonerf: A Living Street Concept for Shared City Spaces." *Citygreen*, November 11, 2020. https://citygreen.com/woonerf-street-concept-for-shared-city-spaces/.

Griswold, Alison. "Uber CEO: Getting People to Share Rides is a 'Battle' against 'Societal Norms'." *Quartz*, April 12, 2018. https://qz.com/1250257/uber-ceo-getting-people-to-share-rides-is-a-battle-against-societal-norms/.

Hand, Ashley Z. *Urban Mobility in a Digital Age: A Transportation Technology Strategy for Los Angeles*. Los Angeles, CA: LA Department of Transportation, 2016. https://ladot.lacity.org/sites/default/files/documents/transportationtechnologystrategy_2016.pdf.

Hanson, Susan. *The Geography of Urban Transportation*. New York, NY: Guilford Press, 1995.

Hauptman, Marissa, Jonathan M. Gaffin, Carter R. Petty, William J. Sheehan, Peggy S. Lai, Brent Coull, Diane R. Gold, and Wanda Phipatanakul. "Proximity to Major Roadways and Asthma Symptoms in the School Inner-City Asthma Study." *Journal of Allergy and Clinical Immunology* 145, no. 1 (January 2020): 119–126. https://doi.org/10.1016/j.jaci.2019.08.038.

Hawkins, Andrew J. "How will driverless cars 'talk' to pedestrians? Waymo has a few ideas." *The Verge*, October 23, 2023. https://www.theverge.com/2023/10/13/23913251/waymo-roof-dome-communicate-intent-pedestrian-driver/.

Hawkins, Andrew J. "Lyft Thinks we can end Traffic Congestion and save $1 Trillion by selling our Second Cars." *The Verge*, January 10, 2018. https://www.theverge.com/2018/1/10/16870732/lyft-traffic-congestion-car-ownership-ces-2018/.

Hilberseimer, Ludwig. *The New City: Principles of Planning*. Chicago, IL: P. Theobald, 1944.

Hill, Steven. "Ridesharing versus Public Transit." *The American Prospect*, March 27, 2018. https://prospect.org/infrastructure/ridesharing-versus-public-transit/. "Home/About 626 Golden Streets." 626 Golden Streets. 2023. https://www.626goldenstreets.com/about.html.

Jacobs, Jane. *The Death and Life of Great American Cities*. New York, NY: Random House, 1961.

Johnson, Charlie and Jonathan Walker. *Peak Car Ownership: The Market Opportunity of Electric Automated Mobility Services*. Boulder, CO: Rocky Mountain Institute, 2016. https://rmi.org/wp-content/uploads/2017/03/Mobility_PeakCarOwnership_Report2017.pdf.

Kahn, Jeremy. "To Get Ready for Robot Driving, Some Want to Reprogram Pedestrians." *Bloomberg News*, August 16, 2018. https://www.bloomberg.com/news/articles/2018-08-16/to-get-ready-for-robot-driving-some-want-to-reprogram-pedestrians/.

Kenworthy, Jeffrey and Peter Newman. *Cities and Automobile Dependence: a Sourcebook*. Aldershot, UK: Gower Publishing Co., 1989.

Krieger, Alex and William S. Saunders. *Urban Design*. Minneapolis, MN: University of Minnesota Press, 2009.

Krisher, Tom, and Hilary Powell. "Tesla shouldn't call driving system Autopilot because humans are still in control, Buttigieg says." *Associated Press*, May 11, 2023. https://apnews.com/article/tesla-autopilot-buttigieg-investigation-crash-musk-1455d194b18fe17609554dd68864654c/.

Kuhn, Thomas S. *The Structure of Scientific Revolutions*. Chicago, IL: University of Chicago Press, 1962.

Le Corbusier. *Charte d'Athenes (The Athens Charter)*. New York, NY: Grossman Publishers, 1973.

Le Corbusier. *The Four Routes*. London, UK: Denis Dobson Ltd., 1947.

Liao, Yuan, Jorge Gil, Rafael H. M. Pereira, Sonia Yeh, and Vilhelm Verendel. "Disparities in Travel Times between Car and Transit: Spatiotemporal Patterns in Cities." *Scientific Reports* 10, no. 4056 (March 2020). https://doi.org/10.1038/s41598-020-61077-0.

"Low-Income Fare is Easy (LIFE)." LA Metro. 2023. https://www.metro.net/riding/fares/life/.

Lutz, Catherine and Anne Lutz Fernandez. *Carjacked: The Culture of the Automobile and Its Effect on Our Lives*. New York, NY: St. Martin's Publishing Group, 2010.

Manville, Michael and Benjamin Cummins. "Why Do Voters Support Public Transportation? Public Choices and Private Behavior." *Journal of Transportation* 42, no. 2 (March 2015): 303–332. https://doi.org/ 10.1007/s11116-014-9545-2.

Margolies, Jane. "Awash in Asphalt, Cities Rethink their Parking Needs." *The New York Times*, March 7, 2023. https://www.nytimes.com/2023/03/07/business/fewer-parking-spots.html.

Marshall, Aarian. "Can Waymo Self-Driving Cars Help Fix Phoenix's Public Transit?" *Wired*, July 31, 2018. https://www.wired.com/story/waymo-phoenix-partnership/.

Marshall, Stephen. *Streets & Patterns*. New York, NY: Taylor and Francis Group, 2005.

Mboup, Gora. *Streets as Public Spaces and Drivers of Urban Prosperity*. Nairobi, KE: UN-Habitat, 2013. https://unhabitat.org/sites/default/files/2020/08/streets_as_public_spaces_and_drivers_of_urban_prosperity.pdf.

McCarthy, Niall. "L.A. Commuters Spend the Most Time Stuck in Traffic." *Statista*, January 20. 2020. https://www.statista.com/chart/8263/la-commuters-spend-the-most-time-stuck-in-traffic/.

"Measure M." LA Metro. 2023. https://www.metro.net/about/measure-m/.

"Measure R." LA Metro. 2023. https://www.metro.net/about/measure-r/.

Melosi, Martin V. "The Automobile Shapes the City." Automobile in American Life and Society. 2005. http://www.autolife.umd.umich.edu/Environment/E_Casestudy/E_casestudy1.htm.

"Metro BikeShare." LA Metro Bike-Share. 2023. https://bikeshare.metro.net/.

"Metro Facts at a Glance." LA Metro. June 2023. https://www.metro.net/about/facts-glance/.

Millard-Ball, Adam. "Pedestrians, Autonomous Vehicles, and Cities." *Journal of Planning Education and Research* 38, no. 1 (October 2016): 6–12. https://doi.org/10.1177/0739456X16675674.

Millard-Ball, Adam. "The Width and Value of Residential Streets." *Journal of the American Planning Association* 88, no. 1 (May 2021): 30–43. https://doi.org/10.1080/01944363.2021.1903973.

Muller, Peter O. "Transportation and Urban Form: Stages of Evolution of the American Metropolis." In *The Geography of Urban Transportation*, edited by Susan Hanson, 59–85. New York, NY: Guilford Press, 1995.

Mumford, Lewis. *The City in History: its Origins, its Transformations, and its Prospects.* San Diego, CA: Harcourt, Brace & Co., 1961.

National Association of City Transportation Officials (NACTO). *Urban Street Design Guide.* Washington D.C.: Island Press, 2013.

"NextGen Bus Plan." LA Metro. 2020. https://www.metro.net/about/plans/nextgen-bus-plan/.

Norton, Peter D. *Autonorama: the Illusory Promise of High-Tech Driving.* Washington D.C.: Island Press. 2021.

Norton, Peter D. *Fighting Traffic: the Dawn of the Motor Age in the American City.* Cambridge, MA: MIT Press, 2008.

Norton, Peter D. "Street Rivals: Jaywalking and the Invention of the Motor Age Street." *Journal of Technology and Culture* 48, no. 2 (April 2007): 331-359. https://www.jstor.org/stable/40061474.

Ng, Andrew. "Self-Driving Cars Won't Work until we Change our Roads – and Attitudes." *Wired*, March 15, 2016. https://www.wired.com/2016/03/self-driving-cars-wont-work-change-roads-attitudes/.

Novini, Rana. "To Stay or to Go? What the Future Holds for NYC Outdoor Dining Sheds." *NBC New York*, May 19, 2023. https://www.nbcnewyork.com/news/local/to-stay-or-to-go-what-the-future-holds-for-nyc-outdoor-dining-sheds/4350560/.

"NYC's Temporary Open Restaurants Program." NYC Department of Transportation. August 2023. https://www.nyc.gov/html/dot/html/pedestrians/openrestaurants.shtml.

Personal Transportation Factsheet CSS01-07. Ann Arbor, MI: University of Michigan Center for Sustainable Systems, 2023. https://css.umich.edu/publications/factsheets/mobility/personal-transportation-factsheet/.

"Public Transit Statistics by Country and City." *Moovit Insights*, 2022. https://moovitapp.com/insights/en/Moovit_Insights_Public_Transit_Index-countries/.

Ratti, Carlo, Kevin S. Kung, Kael Greco, and Stanislav Sobolevsky. "Exploring Universal Patterns in Human Home-Work Commuting from Mobile Phone Data." *PLoS ONE* 9, no. 6 (June 2014). https://doi.org/10.1371/journal.pone.0096180.

Ravenscroft, Tom. "BIG and Toyota Reveal City of the Future at Base of Mount Fuji." *Dezeen*, January 7, 2020. https://www.dezeen.com/2020/01/07/big-toyota-woven-city-future-mount-fuji-japan/.

Reid, Carlton. "Biden's $1.2 Trillion Infrastructure Bill Hastens Beacons for Bicyclists and Pedestrians Enabling Detection by Connected Cars." *Forbes*, November 6, 2021. https://www.forbes.com/sites/carltonreid/2021/11/06/bidens-12-trillion-infrastructure-bill-hastens-beacon-wearing-for-bicyclists-and-pedestrians-to-enable-detection-by-connected-cars/.

"Results of our 2022 Customer Experience Survey." LA Metro. October 27, 2022. https://thesource.metro.net/2022/10/27/results-of-our-2022-customer-experience-survey/.

Richter, Felix. "Cars Still Dominate the American Commute." *Statista*, May 19, 2023. https://www.statista.com/chart/18208/means-of-transportation-used-by-us-commuters/.

Rodrigue, Jean Paul, Claude Comtois, and Brian Slack. *The Geography of Transport Systems.* 5th Ed. New York, NY: Routledge, 2020.

Rowe, Peter G. *A City and Its Stream: The Cheonggyecheon Restoration Project.* Seoul, KR: Seoul Development Institute, 2011.

Sadik-Khan, Janette, *Streetfight: Handbook for an Urban Revolution.* New York, NY: Viking Press, 2016.

Sadowski, Jathan. "Google Wants to Run Cities without Being Elected. Don't Let It." *The Guardian*, October 24, 2017. https://www.theguardian.com/commentisfree/2017/oct/24/google-alphabet-sidewalk-labs-toronto/.

Safety Study: Reducing Speeding-Related Crashes Involving Passenger Vehicles. Washington D.C.: National Transportation Safety Board, 2017. https://www.ntsb.gov/safety/safety-studies/documents/ss1701.pdf.

Sanders, James. *Renewing the Dream: The Mobility Revolution and the Future of Los Angeles.* New York, NY: Rizzoli Electa, 2023.

Schaller, Bruce. *The New Automobility: Lyft, Uber, and the Future of American Cities.* New York, NY: Schaller Consulting, 2018. http://www.schallerconsult.com/rideservices/automobility.pdf.

Scharnhorst, Eric. *Quantified Parking: Comprehensive Parking Inventories for Five U.S. Cities.* Washington D.C.: Research Institute for Housing America, 2018. https://www.mba.org/docs/default-source/research---riha-reports/18806-research-riha-parking-report.pdf.

Schneider, Todd. "Taxi and Ride-hailing Usage in New York City." Todd Schneider Monthly Data Report. September 30, 2023. https://toddwschneider.com/dashboards/nyc-taxi-ridehailing-uber-lyft-data/.

Sevtsuk, Andres. *Street Commerce: the Hidden Structure of Retail Location Patterns and Vibrant Sidewalks.* Philadelphia, PA: University of Pennsylvania Press, 2020.

"Speed & Reliability Program." Minneapolis MetroTransit. 2023. https://www.metrotransit.org/speed-reliability/.

Steinhardt, Susan. "Rideshare Users Pay More in Low-Income and Minority Neighborhoods." *George Washington University Today*, July 7, 2020.https://gwtoday.gwu.edu/rideshare-users-pay-more-low-income-and-minority-neighborhoods/.

Susaneck, Adam Paul. "American Road Deaths Show an Alarming Racial Gap." *The New York Times*, April 30, 2023. https://www.nytimes.com/interactive/2023/04/26/opinion/road-deaths-racial-gap.html.

Tafuri, Manfredo. *Architecture and Utopia: Design and Capitalist Development*. Cambridge, MA: MIT Press, 1976.

Taylor, Brian, and Martin Wachs. "Lessons of the 1st Carmageddon in L.A. by the Numbers." RAND Corporation. August 3, 2012. https://www.rand.org/pubs/commentary/2012/08/lessons-of-1st-carmageddon-in-la-by-the-numbers.html.

"Taxicab Medallion." NYC Taxicab and Limousine Commission. 2023. https://www.nyc.gov/site/tlc/businesses/medallion-owners-and-agents.page.

"The High Cost of Transportation in the United States." Institute for Transportation & Development Policy. May 23, 2019. https://www.itdp.org/2019/05/23/high-cost-transportation-united-states/.

Traffic Safety Facts Crash Statistics, Report No. 812 506. Washington D.C.: National Highway Traffic Safety Administration, 2018. https://crashstats.nhtsa.dot.gov/Api/Public/ViewPublication/813428/.

Tu, Maylin. "'It's Embarrassing': L.A. tries to address its flailing Bus Shelter Program with a New Contract." *Dot. LA*, September 20, 2022. https://dot.la/la-metro-bus-shelter-2658314436.html.

Vijay, Ritesh, Asheesh Sharma, Tapan Chakrabarti, and Rajesh Gupta. "Assessment of Honking Impact on Traffic Noise." *Journal of Environmental Health Science & Engineering* 13, no. 10 (February 2015). https://doi.org/10.1186/s40201-015-0164-4.

Wakabayashi, Daisuke. "Self-Driving Uber Car Kills Pedestrian in Arizona, Where Robots Roam." *The New York Times*, March 19, 2018. https://www.nytimes.com/2018/03/19/technology/uber-driverless-fatality.html.

Waldheim, Charles. *The Landscape Urbanism Reader*. Princeton, NJ: Princeton Architectural Press, 2006.

Warner, Sam Bass. *Streetcar Suburbs*. Cambridge, MA: Harvard University Press, 1978.

"Waymo AV Partnership gets National Spotlight." Phoenix Valley Metro. August 30, 2021. https://www.valleymetro.org/news/2021/08/waymo-av-partnership-gets-national-spotlight/.

"Welcome to Access-A-Ride Paratransit Service." NYC MTA. 2023. https://new.mta.info/accessibility/access-a-ride/.

"What is CicLAvia?" CicLAVia. 2023. https://www.ciclavia.org/.

"What is MDS Cities." LADoT. 2023. https://ladot.lacity.org/docs/what-mds-cities/.

"What is Tactical Urbanism?" Tactical Urbanist's Guide. 2023. http://tacticalurbanismguide.com/about/.

"What is the PlayStreets program?" Los Angeles PlayStreets. 2023. https://laplaystreets.com/about-the-program/.

"With Betax, you can move forward." Betax. 2023. https://www.betax.ch/.

Zheng, March. "The Environmental Impacts of Lithium and Cobalt Mining." *Earth Org*, March 31, 2023. https://earth.org/lithium-and-cobalt-mining/.

"Zone Line Maps for Buses, Trains, and the Metro." Your Public Transport, Copenhagen Metro. 2023. https://dinoffentligetransport.dk/en/plan-your-journey/zone-and-line-maps/.

Additional References

Bridges, Rutt. *Our Driverless Future: Heaven or Hell?* Independently Published, 2018.

Calthorpe, Peter. *The Next American Metropolis: Ecology, Community, and the American Dream.* Princeton, NJ: Princeton Architectural Press, 1993.

Cervero, Robert. "Mobility Niches: Jitneys to Robo-Taxis." *Journal of the American Planning Association* 83, no. 4 (August 2017): 404-412. https://doi.org/10.1080/01944363.2017.1353433.

Cervero, Robert. *The Transit Metropolis: A Global Inquiry.* Washington D.C: Island Press, 1998.

Chase, Robin. "Self-Driving Cars Will Improve Our Cities. If They Don't Ruin Them." *Wired,* August 10, 2016. https://www.wired.com/2016/08/self-driving-cars-will-improve-our-cities-if-they-dont-ruin-them/.

Chase, Robin. "The Future of Autonomous Vehicles." YouTube Video, 3:52. July 3, 2016. https://www.youtube.com/watch?v=DeUE4kHRpEk.

Chase, Robin. "Will a World of Driverless Cars Be Heaven or Hell?" *Bloomberg News,* April 3, 2014. https://www.bloomberg.com/news/articles/2014-04-03/will-a-world-of-driverless-cars-be-heaven-or-hell.

Gabbe, C.J. "The Hidden Cost of Bundled Parking." *University of California Transportation Center Access Magazine* 51 (Spring 2017). https://www.accessmagazine.org/spring-2017/the-hidden-cost-of-bundled-parking/.

Haddad, Paul. *Freewaytopia: How Freeways Shaped Los Angeles.* Santa Monica, CA: Santa Monica Press, 2021.

Hall, Peter. "Beyond the Automobile." *University of California Transportation Center Access Magazine* 30 (Spring 2007). https://www.accessmagazine.org/wp-content/uploads/sites/7/2016/07/Access-30-03-Beyond-the-Auto.pdf.

Hilburg, Jonathan. "What Role do Architects have in a Driverless Future?" *Wired,* April 12, 2018. https://www.archpaper.com/2018/04/driverless-future-architects-role/.

Howard, Ebenezer. *Garden Cities of To-Morrow: A Peaceful Path to Real Reform.* London, UK: Swan Sonnenschein & Co., 1902.

Kay, Jane Holtz. *Asphalt Nation: How the Automobile took over America and how we can take it Back.* Oakland, CA: University of California Press, 1998.

Manville, Michael. "How Parking Destroys Cities." *The Atlantic,* May 18, 2021. https://www.theatlantic.com/ideas/archive/2021/05/parking-drives-housing-prices/618910/.

Meyboom, AnnaLisa. *Driverless Urban Futures: A Speculative Atlas for Autonomous Vehicles.* New York, NY: Routledge, 2018.

National Association of City Transportation Officials (NACTO). *Blueprint for Autonomous Urbanism.* 2nd Ed. San Francisco, CA: Blurb Books, 2019.

Rose, Mark H. and Raymond A. Mohl. *Interstate: Highway Politics and Policy since 1939.* Knoxville, TN: University of Tennessee Press, 2012.

Shoup, Donald. *The High Cost of Free Parking.* New York, NY: Routledge, 2011.

Sperling, Daniel. *Three Revolutions: Steering Automated, Shared, and Electric Vehicles to a Better Future.* Washington D.C: Island Press, 2018.

Townsend, Anthony M. *Ghost Road: Beyond the Driverless Car.* New York, NY: W. W. Norton & Company, 2020.

Image Credits

Page 18, from top
Miller, H. *Power Companies Build for Your New Electric Living*. Illustrated Advertisement. *Life Magazine*. January 30, 1956.
Waymo *Self-Driving Car Prototype*. Photograph. Waymo LLC. 2015.

Page 25
Arthus-Bertrand, Yann. *A sea of Green? A working parking lot at Disney World*. Photograph. Altitude Agency. 2012.

Page 29
Ford Model T Assembly Line. Photograph. 1920.

Page 30
McConville, Patti. *There's no place like Buick*. Photographed Advertisement. Alamy Stock Photos. 1977.

Page 31, from top
Wagner, Art. *Heavy Traffic on California Freeway*. Photograph. iStock Getty Images. April 26, 2018.
Milehightraveler. *Interstate-25 traffic and exhaust fumes Denver*. Photograph. iStock Getty Images. December 6, 2013.
Weakley, Kent. *Aerial home housing development*. Photograph. iStock Getty Images. November 17, 2009.
Jilg, Karl. *How we allocate Street Space*. Illustration. Swedish Road Administration. 2014.

Page 33
Bridge, Albert. *Translink NIR Advertisement*. Photograph. Northern Ireland Railways. March 2016.

Page 34
Branger, Maurice-Louise. *Hester Street*. Photograph. Roger-Viollet Agency. 1914.

Page 35
Nation Roused Against Motor Killings. News Clipping. *The New York Times*. November 23, 1924.

Page 36, from left
New York Evening Journal. *A Traffic Problem – Jay Walking*. Illustration. National Safety Council, Library of Congress. October 1923.
Brewerton, Alfred. *Teaching the Jay Walker!* Illustration. *The Atlanta Journal*. December 16, 1915.

Page 37
Levick, Edwin. *Midtown Manhattan*. Photograph. Getty Images. 1925.

Page 38
Afshar, Vala. *Welcome to our Smart City*. Illustration. 2014.

Page 40
The Road Ahead: Reimagining Mobility. Rendering. Bloomberg Philanthropies and the National Association of City Transportation Officials (NACTO). 2017.

Page 42, from top
Wolf, Heidi, and Julio Palleiro. *Before and After: Times Square Plaza*. Photographs. NYC Department of Transportation. 2009.
Before and After: Herald Square. Photographs. NYC Department of Transportation. 2009.
Russo, Ryan. *Before and After: Pearl Street Plaza*. Photographs. NYC Department of Transportation. 2007.

Page 43, from top
Moran, Karsten. *Restaurants take over Dyckman Street*. Photograph. *The New York Times*. August 10, 2020.
Schaben, Allen J. *People dine outdoors on Main Street*. Photograph. *The Los Angeles Times*. January 27, 2021.

Page 44
Corbis Historical. *Los Angeles Track Homes*. Photograph. University of Southern California. January 1, 1957.

Page 45
Futurama. Photograph. Library of Congress. 1939.

Page 46
Shieh, Evan and Xinwen Cao. *U.S. Interstate Highway Expansion*. Drawings. Emergent Studio. 2022.

Page 47
Packwood, Dave. *Santa Monica and Harbor Freeway Interchange*. Photograph. Automobile Club of Southern California Archives. 1962.

Page 48
Before and After: Hastings Street in the Black Bottom Neighborhood. Photographs. Detroit Historical Society. 1959.

Page 51. from top
Oates, Walter. *Build Rapid Transit, No Freeways*. Photograph. Star Collection, D.C. Public Library. February 4, 1965.
Michele, Julian. *Proposed Mid-Manhattan Expressway looking East*. Illustration. MTA Bridges and Tunnels Archive. 1959.

Page 52, from top
Mondy, Russell. *Before and After: Embarcadero Freeway*. Photographs. Flickr. 1992, 2003.
Greenway Conservancy. *Before and After: Central Artery*. Photographs. Flickr. 1982, 2007.
Before and After: Inner Loop Highway. Photographs. Stantec. 2014, 2017.

Page 53
Won, Na-young. *Before and After: Cheonggyecheon*. Photographs. Seoul Metropolitan Government. 2003, 2005.

Page 59, from top
Hilberseimer, Ludwig. *Highrise City, Perspective View*. Illustration. Art Institute of Chicago Archives. 1924.
Le Corbusier. *The Radiant City, Perspective View*. Illustration. Ville Radieuse. 1924.

Page 60
Shukla, Dipti and Ritu G. Deshmukh. *Traditional City Form versus Modern City Form*. Illustration. In the *Journal of Planning, Architecture and Design* 1, no. 1 (August 12, 2021): 23–29.

Page 61, from top
Pruitt-Igoe, Aerial View. Photograph. United States Geological Survey (USGS). Unknown Date.
Brasilia, South Highway Axis. Photograph. DF. MBC. Unknown Date.

Page 69, from top
Congestion Visualization by Bicycles, Cars, and Bus. Photographs. City of Münster, Press Office. 1991.
Urban Space Efficiency. Photograph. Cycling Promotion Fund. Unknown Date.

Page 72
Radtke, Günther. *Self-Driving Cars on Superhighway*. Illustration. 1974.

Page 78, from top left
Bettmann. *Antique Travel Photograph of Hansom Cab*. Photograph. Getty Images. 1870.
Kohlstedt, Kurt. *Double-decker Horse-drawn Tram*. Photograph. Hulton Archive, Getty Images. 1894.
Heritage Images. *An Electric Motor Cab and Driver*. Photograph. Hulton Archive, Getty Images. 1900.
Camerique. *Two Conductors, Electric Streetcar*. Photograph. Getty Images. June 1, 1900.
Batuhanozdel. *Taxi on 7th Avenue at Times Square*. Photograph. iStock Getty Images. April 27, 2015.
Benedek. *Hollywood Los Angeles Metro Bus*. Photograph. Getty Images. 2023.

Page 79
Santa Monica Boulevard, Sawtelle. Photograph. Los Angeles Public Library Photo Collection. 1901.

Page 89
Silva, Marcio. *Curitiba Public Transportation System.* Photograph. iStock Getty Images. January 20, 2017.

Page 101
Chodosh, Sara and Taylor Maggiacomo. *Los Angeles High-Injury Network.* Drawing. *The New York Times.* April 26, 2023.

Page 103
Chin, Robert K. *Amazon Last Mile Delivery Package Sorting.* Photograph. Alamy Stock Photos. October 2022.

Page 108, from top
Chestnut Street west from 12th, Philadelphia. Photograph. Detroit Publishing Company. 1910.
West Side Elevated Highway, New York City. Photograph. U.S. Department of Transportation, Federal Highway Administration. 1951.

Page 112
Piazze Aperte, Before and After in Nolo. Photograph. Comune di Milano. 2020.

Page 115
All Mobility in One App. Screen Capture. Bestmap. 2020.

Page 117
Roof Dome Communication. Rendering. Waymo. 2023.

Page 119, from top
Taylor, Tony. *Woonerf Street in Dordrecht, Netherlands.* Photograph. Alamy Stock Photos. June 29, 2013.
Imagining Future Places: An AV Zone. Rendering. WSP | Parsons Brinckerhoff. 2016.
Driverless NYC. Rendering. Kohn Pedersen Fox Associates. 2018.

Page 120
Nuro's Next Generation Vehicle. Photograph. Nuro. 2023.

Page 122, from top
Hegen, Tom. *The Lithium Series.* Photograph. Tom Hegen Photography. 2021.
Dawson, Simon. *The Mutanda Mine.* Photograph. *Bloomberg News.* 2012.

Page 144, from left
Kurtz, Rory. *Drive.* Movie Advertisement. Bold Films, OddLot Entertainment, Marc Platt Productions, Motel Movies. 2011.
Cars. Movie Advertisement. Pixar Animation Studios, Walt Disney Pictures. 2006.
Fast & Furious. Movie Advertisement. Universal Pictures. 2001–2023.

Page 147, from top
1926 Pacific Electric Railway System Map. Drawing. Pacific Electric Railway Historical Society Collection. 1926.
Hayes, G.W. *Freeway System Los Angeles & Vicinity.* Drawing. Automobile Club of Southern California. 1960.

Page 148
Melpomenem. *Aerial View of a Freeway Intersection Los Angeles.* Photograph. iStock Getty Images. June 30, 2017.

Page 149, from top
Interstate 10 Under Construction. Photograph. Los Angeles Examiner Collection, University of Southern California Libraries. 1961.
Meyer, Rick. *Century Freeway, with Imperial Highway behind, Under Construction.* Photograph. Department of Special Collections, University of California Los Angeles. 1987.
Los Angeles Harbor Freeway, Under Construction. Photograph. Guy F. Atkinson Company. April 15, 1958.

Page 150
Annual Japanese Car Cruise-In. Photograph. Peterson Automotive Museum. April 28, 2022.

Page 151
Googie's Coffee Shop. Photograph. *Modern Living LA.* 1949.

Page 152
Murr, Andrew. *Dingbat on Hauser.* Photograph. Flickr. October 21, 2011.

Page 153
Cole, Steve. *Beach Parking Lot.* Photograph. iStock Getty Images. July 19, 2013.

Page 155
Los Angeles and Vicinity. Drawing. Home Owners' Loan Corporation. 1939.

Page 156
4kodiak. *Los Angeles and Subway.* Photograph. iStock Getty Images. March 12, 2018.

Page 158, from top
400-Mile Metro Rail Plan. Drawing. Los Angeles Metro. 1992.
Metro System Map. Drawing. Los Angeles Metro. 2016.

Page 159, from top
Linder, Adam. *Projected Metro System Map,* Measure M. Drawing. 2016.
Linder, Adam. *Speculative 2090 Metro System Map.* Drawing. 2016.

Page 160, from top
Laser1987. *Los Angeles Metro E Line (Expo) approaching Santa Monica Downtown.* Photograph. iStock Getty Images. December 27, 2020.
Djansezian, Kevork. *Traffic on the Interstate 405.* Photograph. Getty Images. November 23, 2011.

Page 162
Stellalevi. *Orange Metro Bus Beverly Hills.* Photograph. iStock Getty Images. January 8, 2014.

Page 165
Daemmrich, Bob. *Scooters Litter downtown Austin, Texas.* Photograph. Alamy Stock Photos. November 15, 2020.

Page 166, from top
Laser1987. *Metro Bike Shared Hub at Santa Monica Downtown.* Photograph. iStock Getty Images. December 27, 2020.
Development in Little Tokyo / Arts District. Photograph. Los Angeles Metro. 2015.
Bus Shelter Prototype. Photograph. Tranzito-Vector. 2023.

Page 169, from top
CicLAvia on 4th Street Bridge. Photograph. Los Angeles Metro. 2015.
PlayStreet Program. Photograph. Kounkuey Design Initiative. 2015.
ArroyoFest. Photograph. 626 Golden Streets. October 29, 2023.

Page 332–333, from top
McAdorey, John. *SoFi Stadium, home of the LA Rams and Chargers.* Photograph. Shutterstock. October 13, 2020.
VCG. *SoFi Stadium Inglewood.* Photograph. Getty Images. February 22, 2022.

Page 348, from top
Im, Eun-byel. *Cheonggyecheon in Jongno-gu.* Photograph. *The Korea Herald.* May 13, 2021.
Barden, Lane. *Los Angeles River in Bell Gardens.* Photograph. *The Guardian.* October 23, 2015.

Page 354
See the USA in your Chevrolet. Illustrated Advertisement. Chevrolet. 1951.

Chapters 7–10
All images (aerial and street views) courtesy of Google Earth (© 2023), unless otherwise noted above.

Diagram and Drawing Credits

All diagrams courtesy of Evan Shieh, unless otherwise noted below.

Page 13
Based on diagram in "The 6 Levels of Driving Autonomy Explained." *Synopsys*, 2023. https://www.synopsys.com/automotive/autonomous-driving-levels.html.

Page 74
Based on metrics in Clewlow, Regina R. and Gouri Shankar Mishra. *Disruptive Transportation: The Adoption, Utilization, and Impacts of Ride-Hailing in the United States.* Davis, CA: UC Davis Institute of Transportation Studies, 2017. https://steps.ucdavis.edu/wp-content/uploads/2017/10/ReginaClewlowDisuptiveTransportation.pdf.

Page 75
Based on metrics in Schaller, Bruce. *The New Automobility: Lyft, Uber, and the Future of American Cities.* New York, NY: Schaller Consulting, 2018. http://www.schallerconsult.com/rideservices/automobility.pdf.

Page 81
Based on diagram in Sevtsuk, Andres. *Street Commerce: the Hidden Structure of Retail Location Patterns and Vibrant Sidewalks.* Philadelphia, PA: University of Pennsylvania Press, 2020.

Page 82
Based on diagram in "The Environmental Impact of Today's Transport Types." *TMNT*, May 11, 2021. https://tnmt.com/infographics/carbon-emissions-by-transport-type/.

Page 83
Based on diagram by Kenworthy, Jeffrey and Peter Newman. *Cities and Automobile Dependence: a Sourcebook.* Aldershot, UK: Gower Publishing Co., 1989.

Page 84, from top
Based on metrics in National Association of City Transportation Officials (NACTO). *Transit Street Design Guide.* Washington D.C.: Island Press, 2016.
Based on diagram in "Can you think about a less efficient way to transport people through your city than cars?" Urban Cycling Institute. February 8, 2023. https://twitter.com/fietsprofessor/status/1623328144861061128/.

Page 100
Based on diagrams in "The High Cost of Transportation in the United States." Institute for Transportation & Development Policy. May 23, 2019. https://www.itdp.org/2019/05/23/high-cost-transportation-united-states/.

Page 109
Based on diagrams in Marshall, Stephen. *Streets & Patterns.* New York, NY: Taylor and Francis Group, 2005.

Page 352, from top
Based on metrics in Lopez, Juan. "Access to Public Transit." Neighborhood Data for Social Change. 2023. https://la.myneighborhooddata.org/2019/02/access-to-public-transit/.
Based on metrics in *American Community Survey.* U.S. Census. 2005.
Based on metrics in "AllTransit Gap Finder." Center for Neighborhood Technology. February 7, 2018. https://alltransit.cnt.org/gap-finder/.

Page 353, from top
Based on metrics in "Cities with the Best Public Transportation." *Hire a Helper*, May 9, 2023. https://www.hireahelper.com/lifestyle/cities-with-the-best-public-transportation/.
Based on metrics in Morris, Eric A. "Los Angeles Transportation Facts and Fiction: Driving and Delay." *Freakonomics*, March 10, 2009. https://freakonomics.com/2009/03/los-angeles-transportation-facts-and-fiction-driving-and-delay/.
Based on diagram in *Energy Technology Perspectives 2020.* International Energy Agency (IEA). 2020. https://www.iea.org/reports/energy-technology-perspectives-2020/.

Chapters 7–10
All drawings courtesy of Evan Shieh.

Published by Applied Research and Design Publishing, an imprint of ORO Editions.
Gordon Goff: Publisher

www.appliedresearchanddesign.com
info@appliedresearchanddesign.com

Author: Evan Shieh
Comic Illustrator Consultants: Anthony Delaware, George Bennett
Book Design: Neil Donnelly Studio
Production: Neil Donnelly, Cat Wentworth
Copy Editors: Patricia English, Catherine Pluimer
Project Manager: Jake Anderson

Typefaces:
Neureal by Laura Csocsán
Concorde by Günther Gerhard Lange

Funding for this project was made possible thanks to:
The 2020 Irving Innovation Fellowship Grant, Harvard University, Graduate School of Design
The Dean's Office, New York Institute of Technology, School of Architecture and Design

10 9 8 7 6 5 4 3 2 1 First Edition

ISBN: 978-1-957183-63-3

Color Separations and Printing: ORO Editions Inc.

Printed in China.

AR+D Publishing makes a continuous effort to minimize the overall carbon footprint of its publications. As part of this goal, AR+D, in association with Global ReLeaf, arranges to plant trees to replace those used in the manufacturing of the paper produced for its books. Global ReLeaf is an international campaign run by American Forests, one of the world's oldest nonprofit conservation organizations. Global ReLeaf is American Forests' education and action program that helps individuals, organizations, agencies, and corporations improve the local and global environment by planting and caring for trees.